Studies in Systems, Decision and Control

Volume 322

The series "Studies in Systems, Decision and Control" (SSDC) covers both new developments and advances, as well as the state of the art, in the various areas of broadly perceived systems, decision making and control–quickly, up to date and with a high quality. The intent is to cover the theory, applications, and perspectives on the state of the art and future developments relevant to systems, decision making, control, complex processes and related areas, as embedded in the fields of engineering, computer science, physics, economics, social and life sciences, as well as the paradigms and methodologies behind them. The series contains monographs, textbooks, lecture notes and edited volumes in systems, decision making and control spanning the areas of Cyber-Physical Systems, Autonomous Systems, Sensor Networks, Control Systems, Energy Systems, Automotive Systems, Biological Systems, Vehicular Networking and Connected Vehicles, Aerospace Systems, Automation, Manufacturing, Smart Grids, Nonlinear Systems, Power Systems, Robotics, Social Systems, Economic Systems and other. Of particular value to both the contributors and the readership are the short publication timeframe and the world-wide distribution and exposure which enable both a wide and rapid dissemination of research output.

Indexed by SCOPUS, DBLP, WTI Frankfurt eG, zbMATH, SCImago.

All books published in the series are submitted for consideration in Web of Science.

More information about this series at http://www.springer.com/series/13304

Aboul Ella Hassanien · Ashraf Darwish
Editors

Digital Transformation and Emerging Technologies for Fighting COVID-19 Pandemic: Innovative Approaches

Springer

Editors
Aboul Ella Hassanien
Faculty of Computers and Artificial Intelligence
Cairo University
Giza, Egypt

Ashraf Darwish
Faculty of Science
Helwan University
Cairo, Egypt

ISSN 2198-4182 ISSN 2198-4190 (electronic)
Studies in Systems, Decision and Control
ISBN 978-3-030-63306-6 ISBN 978-3-030-63307-3 (eBook)
https://doi.org/10.1007/978-3-030-63307-3

This Springer imprint is published by the registered company Springer Nature Switzerland AG
The registered company address is: Gewerbestrasse 11, 6330 Cham, Switzerland

This book is dedicated to all members of the scientific research group in Egypt (SRGE)

Preface

In the age of digital transformation, artificial intelligence (AI) and big data are playing a key role in processing the data to be converted into knowledge and decisions. With the rapid growth in the number of newly confirmed and suspected COVID-19 cases in 2020, COVID-19 extremely threatens public health, countries' economic, social life and the international relations around the world. In the fight against COVID-19, AI techniques and big data analytics have played a significant role in many aspects. Moreover, AI, big data, Internet of things and Blockchain-driven technologies present visualization for COVID-19 outbreak information that helps in detecting risk allocation and regional transmissions. Thus, many researchers have been developing specialized tools using AI techniques and big data analytic tools for early prediction and diagnosis of COVID-19 cases to contribute to its containment before spreading.

These models had been a great help to clinicians and decision-makers for making the appropriate recommendations for treatments. The value of integrating AI and big data tools strongly will provide clinicians and decision-makers with helpful information on COVID-19 disease, which will reduce their efforts. Thanks to its high ability to isolate and pick features, AI techniques have today been used in the medical imaging domain. Machine and deep learning technology is primarily applied to identify and distinguish pediatric viral pneumonia and bacteria.

Some recent technologies such as unmanned aerial vehicle (UAV) or drones are used to detect the infected people in different areas in the same country. In addition, robots are used in fighting the COVID-19 by protecting the workers and staff people who are dealing with infected people. On the other hand, other technology such as robotics plays an important role in saving the life of medical staff during COVID-19 pandemic and to assist patients during the treatment process.

Blockchain technology will play an important role with COVID-19 in securing many transactions and dealing in strict confidentiality, as patient's image transmission, digital and statistical data related to the patients. In addition, the adoption of Bitcoin as an official currency alternative to banknotes currencies in order to avoid the transmission and spread of infection is a great challenge for governments.

The book aims to present the AI and big data analytics solutions to help healthcare sectors and the medical staff to protect them and limit the spread of the COVID-19. Also this book presents the problems and challenges and presents to the researchers and academics some future research points from the AI point of view that can help healthcare sectors and limit the spread of COVID-19.

The content of this book is divided into three parts.

- The first part presents the effective role, importance and practical applications which include new models, systems and results of AI technology in fighting against COVID-19.
- The second part analyzes and describes the role of digital transformation and different emerging technologies for monitoring and controlling COVID-19. Some recent technologies such as big data, drones, 3D printing, robotics and the Fourth Industrial Revolution to tackle COVID-19 have been presented.
- The third part introduces the smart networks for tracking COVID-19 and Blockchain technology for securing and management of COVID-19 patient's data.

Finally, editors would like to encourage the readers and researchers to expand and explore the knowledge in the main topic of this book in order to create their implementations and obtain new results to tackle COVID-19 as a global pandemic. We acknowledge the members of the Scientific Research Group in Egypt (SRGE), for their effort and handling this book.

Cairo, Egypt
August 2020

Aboul Ella Hassanien
Ashraf Darwish

Contents

Artificial Intelligence Technology Against COVID-19

Contents

and academics some future research points from the AI point of view that can help healthcare sectors and limit the spread of COVID-19.

The content of this book is divided into three parts.

- The first part presents the effective role, importance and practical applications which include new models, systems and results of AI technology in fighting against COVID-19.
- The second part analyzes and describes the role of digital transformation and different emerging technologies for monitoring and controlling COVID-19. Some recent technologies such as big data, drones, 3D printing, robotics and the Fourth Industrial Revolution to tackle COVID-19 have been presented.
- The third part introduces the smart networks for tracking COVID-19 and Blockchain technology for securing and management of COVID-19 patient's data.

Finally, editors would like to encourage the readers and researchers to expand and explore the knowledge in the main topic of this book in order to create their implementations and obtain new results to tackle COVID-19 as a global pandemic. We acknowledge the members of the Scientific Research Group in Egypt (SRGE), for their effort and handling this book.

Cairo, Egypt
August 2020

Aboul Ella Hassanien
Ashraf Darwish

Preface

In the age of digital transformation, artificial intelligence (AI) and big data are playing a key role in processing the data to be converted into knowledge and decisions. With the rapid growth in the number of newly confirmed and suspected COVID-19 cases in 2020, COVID-19 extremely threatens public health, countries' economic, social life and the international relations around the world. In the fight against COVID-19, AI techniques and big data analytics have played a significant role in many aspects. Moreover, AI, big data, Internet of things and Blockchain-driven technologies present visualization for COVID-19 outbreak information that helps in detecting risk allocation and regional transmissions. Thus, many researchers have been developing specialized tools using AI techniques and big data analytic tools for early prediction and diagnosis of COVID-19 cases to contribute to its containment before spreading.

These models had been a great help to clinicians and decision-makers for making the appropriate recommendations for treatments. The value of integrating AI and big data tools strongly will provide clinicians and decision-makers with helpful information on COVID-19 disease, which will reduce their efforts. Thanks to its high ability to isolate and pick features, AI techniques have today been used in the medical imaging domain. Machine and deep learning technology is primarily applied to identify and distinguish pediatric viral pneumonia and bacteria.

Some recent technologies such as unmanned aerial vehicle (UAV) or drones are used to detect the infected people in different areas in the same country. In addition, robots are used in fighting the COVID-19 by protecting the workers and staff people who are dealing with infected people. On the other hand, other technology such as robotics plays an important role in saving the life of medical staff during COVID-19 pandemic and to assist patients during the treatment process.

Blockchain technology will play an important role with COVID-19 in securing many transactions and dealing in strict confidentiality, as patient's image transmission, digital and statistical data related to the patients. In addition, the adoption of Bitcoin as an official currency alternative to banknotes currencies in order to avoid the transmission and spread of infection is a great challenge for governments.

The book aims to present the AI and big data analytics solutions to help healthcare sectors and the medical staff to protect them and limit the spread of the COVID-19. Also this book presents the problems and challenges and presents to the researchers

This book is dedicated to all members of the scientific research group in Egypt (SRGE)

Editors
Aboul Ella Hassanien
Faculty of Computers and Artificial Intelligence
Cairo University
Giza, Egypt

Ashraf Darwish
Faculty of Science
Helwan University
Cairo, Egypt

ISSN 2198-4182 ISSN 2198-4190 (electronic)
Studies in Systems, Decision and Control
ISBN 978-3-030-63306-6 ISBN 978-3-030-63307-3 (eBook)
https://doi.org/10.1007/978-3-030-63307-3

This Springer imprint is published by the registered company Springer Nature Switzerland AG
The registered company address is: Gewerbestrasse 11, 6330 Cham, Switzerland

Aboul Ella Hassanien · Ashraf Darwish
Editors

Digital Transformation and Emerging Technologies for Fighting COVID-19 Pandemic: Innovative Approaches

Springer

The series "Studies in Systems, Decision and Control" (SSDC) covers both new developments and advances, as well as the state of the art, in the various areas of broadly perceived systems, decision making and control–quickly, up to date and with a high quality. The intent is to cover the theory, applications, and perspectives on the state of the art and future developments relevant to systems, decision making, control, complex processes and related areas, as embedded in the fields of engineering, computer science, physics, economics, social and life sciences, as well as the paradigms and methodologies behind them. The series contains monographs, textbooks, lecture notes and edited volumes in systems, decision making and control spanning the areas of Cyber-Physical Systems, Autonomous Systems, Sensor Networks, Control Systems, Energy Systems, Automotive Systems, Biological Systems, Vehicular Networking and Connected Vehicles, Aerospace Systems, Automation, Manufacturing, Smart Grids, Nonlinear Systems, Power Systems, Robotics, Social Systems, Economic Systems and other. Of particular value to both the contributors and the readership are the short publication timeframe and the world-wide distribution and exposure which enable both a wide and rapid dissemination of research output.

Indexed by SCOPUS, DBLP, WTI Frankfurt eG, zbMATH, SCImago.

All books published in the series are submitted for consideration in Web of Science.

More information about this series at http://www.springer.com/series/13304

Studies in Systems, Decision and Control

Volume 322

Digital Transformation and Emerging Technologies for Tackling COVID-19 Pandemic

Mohamed Elsersy, Ahmed Sherif, Ashraf Darwsih, and Aboul Ella Hassanien

Abstract Several emerging technologies were introduced to tackle the unprecedented crisis of the new COVID-19. Remarkable emerging technologies are outlined, such as machine and deep learning, Internet of things, cloud and fog computing, and blockchain technology. Those emerging technologies have been explored to support the solution proposed to ensure the integration of these technologies to fight the pandemic. Also, numerous emerging technologies used for the COVID-19 fight have been highlighted. Finally, the impact of COVID-19 is discussed, and applications showing how to mitigate this impact using the emerging technologies are outlined.

Keywords Digital transformation · Emerging technologies · Artificial intelligence · Internet of things · Big data · Blockchain · Cloud computing · The fourth industrial revolution · COVID-19 · Coronavirus

M. Elsersy (✉)
Higher Colleges of Technology, Al Ain City, Abu Dhabi, UAE
e-mail: m.elsersy@yahoo.ca
URL: https://www.egyptscience.net

A. Sherif
School of Computing Sciences and Computer Engineering, University of Southern Mississippi, Hattiesburg, USA
URL: https://www.egyptscience.net

A. Darwsih
Faculty of Science, Helwan University, Cairo, Egypt
URL: https://www.egyptscience.net

A. E. Hassanien
Faculty of Computers and Artificial Intelligence, Cairo University, Cairo, Egypt
URL: https://www.egyptscience.net

M. Elsersy · A. Sherif · A. Darwsih · A. E. Hassanien
Scientific Research Group in Egypt, Cairo, Egypt

A. E. Hassanien and A. Darwish (eds.), *Digital Transformation and Emerging Technologies for Fighting COVID-19 Pandemic: Innovative Approaches*, Studies in Systems, Decision and Control 322, https://doi.org/10.1007/978-3-030-63307-3_1

1 Introduction

Since the first warning issued on January 12, 2020, by the World Health Organization (WHO), a new Coronavirus outbreak, many technologies, and solutions have been used to combat the pandemic. The concern was raised because of the emergence of new breathing diseases that appeared in Wuhan city, China. The evidence was highly suggestive, according to the WHO, that the outbreak was associated with crowds and transfer of droplets for infected persons. Later, there was a high rate of infection among healthcare employees and a high possibility of transmission from humans to humans [1]. The COVID-19 pandemic has brought more than eleven million cases reported globally by the date of July 5, 2020, and has taken more than half a million lives worldwide [2]. At the moment, Egypt is one of the most affected nations in the vicinity, with as much as 74,500 total instances of infection and a widespread causing 3324 deaths, the highest rate within all the Arabic countries [3]. Because of the reported active cases of COVID-19, all affected nations face a situation of partial or complete lockdown wherein the range of infected cases may even surpass the capacities of all healthcare systems, including those with the most resources in the USA and Europe [4, 5]. State programs focused on interaction monitoring, isolation, and social distancing, in the absence of particular vaccines or medical drugs. On condition that the pandemic wave strength increases, the proposed solutions will adapt. Besides, the following dynamics of SARS-COV-2 transmission all around the world [6]. The effectiveness, length, and urgency of the response are all always needed in all cases.

Emerging technologies keep disrupting the manner commercial enterprise is carried out everywhere in all countries. Companies seeking to succeed in today's marketplace are rapidly recognizing the need for a complete digital transformation roadmap that incorporates numerous technology platforms through the entire framework of the organization. Researchers are seeking to implement resilient, lengthy-term, cloud-based digital infrastructures that drive their healthcare solutions' future-of-work, future-of-connection, data management, related digital experiences, and digital innovation techniques for healthcare solutions [7]. The cloud-based digital infrastructure that propels many groups into a thrilling deep dive into the different moving components and challenges of imposing a resilient, relatively aggressive future business landscape. This work is introducing the digital transformation pioneering solutions and emerging technologies across the world, along with complete analysis that will explore the latest cases of use of emerging digital technologies against COVID-19. Digital technologies have captured the imagination during the global pandemic for their potential to support people in their fight against COVID-19. Earlier than the new COVID-19, health, education, social and economic crisis, the most noticeable uses of emerging technologies, such as artificial intelligence (AI), were their various entertainment applications in growing productivity [8]. Nevertheless, now new digital technology's ability to reach beyond this has been exposed to the public at large. Emerging technologies, which include AI and IoT, assist in speeding up the vaccine's production, forecast the public health interventions will be more successful and maintaining the public up to date with accurate information [9].

A variety of technological solutions to tackle the impacts of the COVID-19 pandemic are emerging. Digital technologies including IoT, AI, blockchain technology, and big data analytics, were among those at the forefront [10]. Digital technologies will play an essential role in strengthening public health responses to the COVID-19 pandemic, according to the WHO and the Center for Disease Control and Prevention (CDC) [11]. The newly proposed technologies have also helped researchers to transfer plenty of their lives online, sustaining economic and educational structures. At the same time, the majority of people stay home, and assisting the target in the workforce remains connected. But not all regions and social communities are similarly capable of harnessing emerging technology's ability to combat the virus. According to the status of the 2019 broadband survey, while the internet usage penetration rate is 51.2%, it is still 45% in developing countries and 20% in the least developed countries [12]. Digital and information gaps have always persisted, however in a situation, wherein many human beings must stay at home, it transforms from a drawback to a debilitating weakness. A long-term study is necessary to improve exposure to emerging technology, and short-term research is needed to ensure that lack of access does not turn into an inability to maintain everyday life.

Several measures were launched against COVID-19 by UNESCO and its allies to unlock the potential of applications of new emerging technologies. Emerging technologies mentioned in the sections below clarify how these innovations are used to address the COVID-19 catastrophe. COVID-19 crisis tells us that we should promote the socially beneficial applications of emerging technology, and work on enhancing connectivity and usage in areas where they are missing. This chapter concludes that most of the patients who could not get better had a fever, cough, general exhaustion, and most possibly malaise, as depicted in Fig. 1, the symptoms and prevention measures for COVID-19 as recommended by WHO [13, 14].

The remainder of this chapter is organized as follows. Several solutions primarily based on the AI is mentioned in Sect. 2 to mitigate the impact of the COVID-19. Section 3 introduces the IoT as a promising and significant solution for the control and mitigation of the epidemic. Section 4 depicts big data analytics for the same reason. On the other hand, cloud and fog computing is an obtainable solution, as described in Sect. 5. Blockchain is a novel technology demonstrated in Sect. 6. The fourth industrial revolution is combining many technologies, as presented in Sect. 7. Section 8 sums up the contents of the chapter and discusses future challenges.

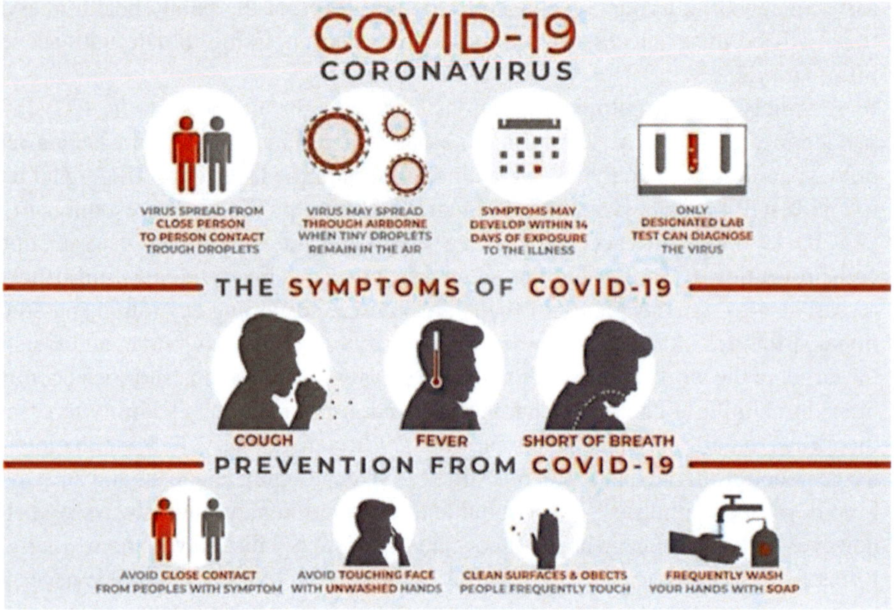

Fig. 1 Symptoms and the prevention measures from the COVID-19 pandemic

2 Artificial Intelligence

2.1 Solutions Based on Artificial Intelligence

With AI playing a leading role in solving necessary and tedious activities for enterprises, the human worker will spend more on complex jobs, which effectively provides workers with a higher degree of professional fulfillment and more effective utilization of critical thinking capacity. COVID-19 is the brand-new pandemic that is hitting worldwide. Patients across the globe face unique symptoms. The maximum number of patients with excessive signs and symptoms die, mostly the elderly and those at risk from health problems. AI has proved to be a groundbreaking technical advance and continues to be a highly successful weapon against COVID-19 when used appropriately [15].

AI can help authorities to combat COVID-19 in several ways. It used to predict risk, to develop drugs, and to bust the fake news. One of the potential ways in which AI is used against COVID-19 is the risk prediction of infection. The risk of becoming infected typically is a function of a multitude of factors. May include age, hygiene practices, the present state of health, personal medical records, pre-existing conditions, or some records of travel. Direct mathematical models do not produce fruitful results for these factors. A comprehensive analysis of these factors, however, integrated with AI techniques, can offer a more accurate and reliable forecast of individual risk profiles [16, 17].

One of the main issues with the COVID-19 is the virus' complexity, and lack of proven viral testing and treatment protocols. In this respect, Machine Learning (ML), a subset of AI, may prove to be a compelling technology to accelerate drug development [18, 19]. In past epidemics, ML has also demonstrated its efficacy in the phase of drug production. For example, Bayesian ML models were used in the course of the Ebola epidemic to speed up the process of designing molecular inhibitors against the virus [20]. With COVID-19, ML can be used in two significant ways [21]. Firstly, it can be used to quickly build and traverse existing knowledge graphs so that potential vaccines can be rapidly identified. Second, it can be used to accurately forecast the results of interactions between drug targets [16].

Furthermore, the unpredictable periods that preceded the COVID-19 epidemic have created many misconceptions and conspiracy theories. There was a lot of misinformation on social media platforms doing the rounds. To curtail the dissemination of such fake news and provide confirmed facts, technology companies such as Google, Youtube, and Facebook have used AI techniques. All these platforms endeavored to screen content for even the slightest bit of misinformation. YouTube has put in place stringent measures to take down any video that propagates false news [22]. The deep learning techniques are another subset of AI, which will be explained in the next section.

2.2 Solutions Based on Deep Learning

Deep learning (DL) as the new branch of machine learning emerged because of the vast computing power and big data sets available. DL enables multiple actors to build a standard, robust machine learning model without sharing data, thus addressing critical issues such as privacy, security, access rights for the data. Meanwhile, access to different data types in the same data sets should also be considered. DL applications range over numerous industries, counting military, telecommunications, IoT, or drug discovery. DL can be used in multiple forms of healthcare, as shown in Table 1.

Table 1 Applications of deep learning in healthcare

Basic biomedical research	Translational research	Clinical Practice
Automated experiments	Biomarker discovery	Disease diagnosis
Automated data collection	Drug-target prioritization	Interpretation of patient genomes
Gene function annotation	Drug discovery	Treatment selection
Prediction of transcription factor binding sites	Drug re-purposing	Automated surgery
Simulation of molecular	Prediction of chemical	Patient monitoring
Literature mining	Genetic variant annotation	Patient risk satisfaction of primary prevention

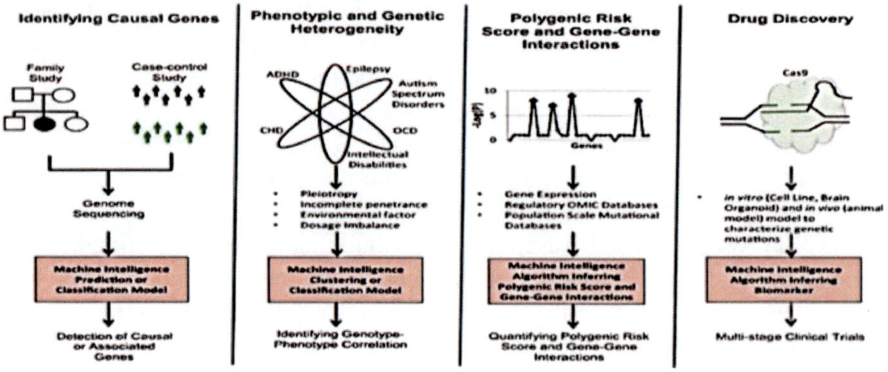

Fig. 2 Genetic interpretation as an example of AI solutions in healthcare

Several machine learning techniques are discussed to show expecting the affected person's recuperation. The support vector machine (SVM) turned into a test at the given facts with average absolute mistakes. The epidemiological information set becomes organized with the aid of researchers from many fitness reviews of actual-time instances to symbolize one of a kind attributes that make contributions as the main elements for drug discovery and re-purposing. In-intensity evaluation with other gadgets getting to know algorithms, inclusive of artificial neural networks and a regression model, changed into examined and as compared with the SVM results [23, 24].

Different classification models discussed to make the best use of the clinical data provided online to be able to predict the severity of the corona cases and classifying those into (acute, mild, and smooth) cases. SVM on many attributes of symptoms and other relevant information of patients achieved minimal classification error, which proves the feasibility of the proposed approach. Authors in [25] also demonstrated that symptoms alone could not help in deciding the severity of the cases. Moreover, data and statistics about COVID-19 are changing, and new data is added every minute. As a result, more records can be usable by the given model. As seen in Fig. 2, the genetic interpretation is an example of the use of AI in healthcare.

3 Internet of Things

3.1 Solution Based on the Internet of Things

The IoT is the new era of communication. Utilizing the IoT, physical objects such as sensors can be engaged in creating, getting, and exchanging the health data seamlessly and consistently without the intervention of the human. Existing and new IoT applications aim to increase the level of comfort, efficiency, and automation for the users [26]. With the technological advancement of the IoT, it is easy to connect various objects (or things) through the Internet. In such a provision, these objects collaboratively collect a massive amount of data from different sources and then originate a meaningful insight (or information) from the collected data. In the IoT environment, these objects are broadly classified into two distinct categories: (i) physical objects such as drones, surveillance cameras, smartphones, vehicles and smart homes, and (ii) virtual objects such as e-ticket and e-wallet [27].

The Internet of Medical Things (IoMT), also known as the healthcare IoT, is the integration between the software applications and the medical devices through the connection of the healthcare system. Nowadays, IoMT has witnessed an enormous number of its applications according to the increased demand in using mobile devices that use near field communication (NFC) readers. These readers facilitate the interaction between the tools and the healthcare system [28]. Several applications of the IoMT have been proposed, such as monitoring the patients' conditions remotely by using some wearable/implanted devices to transmit the patients' data to the connected professionals, tracking the medication records, and update the required medication according to the patient status. Due to its ability to collect, analyze, and transmit/receive health data efficiently, the healthcare providers realized the urgent need of the IoMT technologies to reduce the burden on the healthcare system [29, 30]. An example of such IoMT techniques that are used to mitigate the COVID-19 is the smart thermometer. A US health technology company called Kinsa had launched Internet-connected thermometers eight years ago to test people for severe fevers. While initially designed to monitor the flu, these thermometers nevertheless prove to be highly useful in identifying possible COVID-19 clusters across the USA. These thermometers are connected to a mobile application, which enables them to transmit their readings to the business immediately. Once received, Kinsa assimilates this data to generate daily maps showing which of the US regions are witnessing an increase in high fevers, thus enabling the US authorities to identify potential hotspots [31]. Furthermore, telemedicine/telehealth is called the method of using IoMT technology to allow remote monitoring of patients. This approach helps physicians to assess, diagnose, and treat patients without having to engage directly with them [32]. Figure 3 shows the IoT wearable technologies for monitoring healthcare.

3.2 *Solution Based on Robots*

Robots are another example of IoMT technologies. As governments and medical organizations around the world are struggling to control the COVID-19 outbreak, robots are being deployed to support patient treatment, thus alleviating the stress levels of healthcare workers. Furthermore, robot-controlled non-contact ultraviolet surface disinfection methods are also employed to limit the transmission of the disease via the contaminated surfaces. Compared to the practice of manual decontamination requiring the deployment of cleaning staff and consequently placing them at risk of contracting the virus, autonomous robots for disinfection can lead to rapid

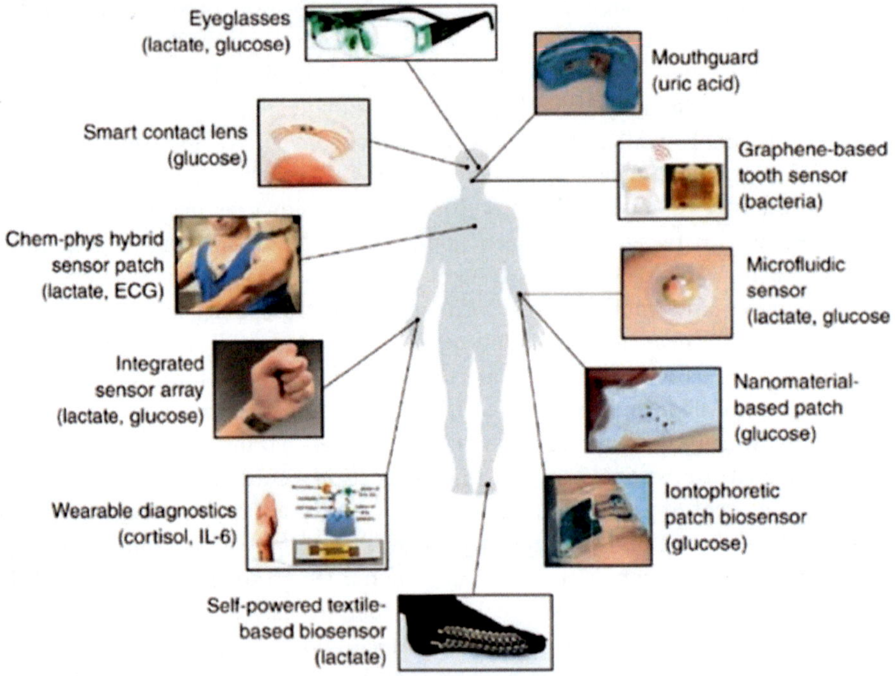

Fig. 3 IoT wearable technologies for monitoring the healthcare

and efficient disinfection [33]. Robots may likely perform activities such as disinfecting buildings, measuring people's temperatures in public spaces or at ports of entry, delivering psychological assistance to quarantined patients, gathering nasal and throat samples for research, and enabling people to attend conferences and exhibits remotely. The use of robots in each case could reduce human exposure to pathogens which will become more and more critical as epidemics escalate [34].

In addition, the use of robotics to keep IoT records could upload every other layer of safety that hackers could want to skip, which will get entry to the network. Robotics gives a robust degree of encryption that makes it, in reality, impossible to overwrite present information facts, especially during the COVID-19 pandemic. Robotics provides transparency by using allowing all of the users who are authorized to get the right of entry to the network. Robotics can offer a steady way to discover a selected source of any records manipulation or leakages and take a brief remedial movement to guide the decision-maker to assign resources according to the patients' needs. Robotics can enable speedy processing of motion and coordination amongst billions of linked gadgets.

Although offering a manner to enable with many of the stakeholders, robotics can permit IoT corporations to reduce their fees by using eliminating the processing overheads associated with IoT gateways (for e. g. traditional protocol, hardware, or communication overhead costs). Intelligent contracts, a settlement between events this is stored inside the blockchain, can, also, allow the execution of contractual preparations amongst stakeholders primarily based on specific criteria being fulfilled. For example, smart contracts can authorize bills robotically without any need for human intervention, while providing service conditions were met. Some modern blockchain-IoT gamers and their use instance the technology at the back of sensors and smart chips are evolving rapidly, making them more transportable and applicable for real-time interactions with blockchain ledgers. The combination of robotics and IoT has broad potential for the advent of a marketplace of services among devices and allows organizations to create free from gathered facts. The developing number of emerging blockchain protocols, partnerships, and IoT device carriers already indicates that there is a superb match for applications within the IoT sector. The developing number of emerging blockchain protocols, companies, and IoT device carriers already suggest that there is an excellent match for applications within the IoT sector.

3.3 Solutions Based on Unmanned Aerial Vehicles (Drones)

Unmanned Arial Vehicles (UAV) is one of the emerging technologies that can be employed to introduce one of the solutions for the COVID-19 pandemic. UAV helps to deliver the medication for remote patients. With the progress of science and technology, the drone industry is developing rapidly, and the number of drones is also increasing rapidly. This section presents the integration opportunities between UAVs, also known as drones, and solutions suggested to mitigate the impact of COVID-19. With the aid of the recently emerging technologies, UAV can control, monitor, and communicate with other drones over the Internet to overcome their limited wireless communication ranges with the ground stations, finite computation, and energy resources. Therefore, this integration of drones in providing service for healthcare will allow seamless interaction with the drones over the Internet. Besides, drones enable their control anywhere and anytime without restriction on distance. In the course of the flight, drones usually need a wireless network to control their system. The information collected during the trip from the drone needs to be transmitted back to the ground station to guide the actions taken regarding the current situation. Not only the pictures and paths observed by UAV in the area of interest can help, but also those data can decide the impact of the COVID-19.

As shown in Fig. 4, the drones can be used in the surveillance during the lockdown, carrying the medical needs to remote areas, spraying the disinfectants to protect people in the surrounding, and carrying the food supplies to the hit regions [35]. Because the quantity of interconnected robotics grows, the post-pandemic era gives a feasible technique to assist the usage of an excess of robotics in the healthcare area.

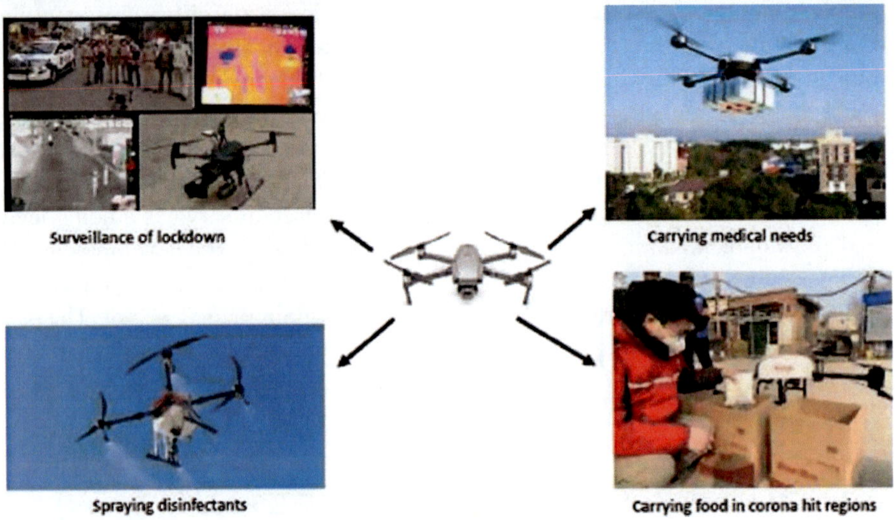

Fig. 4 Drones use for the COVID-19 pandemic

4 Big Data Analytics

The latest years have visible clinical imaging strategies like picture archiving and verbal exchange systems (percent), which emerge as a crucial part of diagnosis and treatment, in tandem with the subsequent-technology cell networks and technologies like big information analytics. It can offer more reliable data evaluation and management while requiring minimal human effort. In a consultant discipline hospital in Wuhan, Internet-enabled medical imaging platforms allowed for actual-time analysis of COVID-19 patients, and in doing so, relieved some of the burden on the medical institution's medical personnel [36].

Big data Analytics Healthcare organizations in many countries have proposed various models of healthcare information systems in order to get the best services and care for the patients. Big data in healthcare and medicine refers to these specific broad and complex data which are difficult to analyze and handle using conventional software or hardware [37, 38]. Big data analytics include heterogeneous batch processing, data quality control, analysis, modeling, interpretation, and validation [39]. Big data analytics framework offers detailed information to discover from the considerable amount of data available. Big data analytics in medicine and healthcare, in particular, allow the analysis of the large data sets of thousands of patients, the identification of clusters and the correlation between datasets, and the development of predictive models using data mining techniques [40]. Big data analytics in medicine and healthcare incorporates analysis techniques in many research fields such as bioinformatics, medical imaging, sensor analytics, medical informatics, and health informatics. The new knowledge discovered through big data analytics techniques should give patients, healthcare professionals, and health decision-makers significant advantages [41].

Data mining techniques used on electronic health records (EHRs), web and social media data enable the identification of optimal practical guidelines in hospitals, the identification of association rules in the EHRs [42], and the disclosure of disease monitoring and trends in health. Besides, data integration and analysis with different characteristics, such as social and scientific, will lead to new information and intelligence, explore new theories, and identify hidden patterns [43]. An example of how big data analytics can be used during the ongoing COVID-19 outbreak when people use social media to acquire and exchange historically and unprecedently different types of information. To respond to the epidemic, only situational information is valuable to the public and authorities. It is, therefore, important to recognize such situational details and understand how it is being propagated on social media so that effective strategies for publishing information for the COVID-19 epidemic can be aware [44]. Another example of using big data analytics to mitigate COVID-19 is using wireless cellular data to classify areas with high human density and mobility that are at risk of COVID-19 spread. Crowded regions with actively moving people are susceptible to disease spread, mainly if they include asymptomatically infected people along with healthy people [45].

5 Blockchain Technology

In this section, blockchain technology is discussed and the possible extension for this technology to mitigate the effect of COVID-19. The following applications for the IoT, starting with LoRa, industrial, heterogeneous, supply chain, and finally, the military applications. More challenges in the use of the blockchain are discussed, emphasizing the use of blockchain to secure the patients' records. People, during and after the pandemic, need more connections with each other. On top of the products and even places, people visit to track the infected and active cases. The new technology of the blockchain offered more opportunities in providing better healthcare services.

Blockchain is gaining an increasing number of prominence each day, thanks to its full programs in numerous walks of lifestyles [46]. To the uninitiated, a blockchain is a continuously enlarging document of transactions between parties [47]. Such files may be used to verify the claims of a celebration that a deal has undoubtedly happened. Seeing its software, numerous groups and governments throughout the globe have started the usage of blockchain to build apps that may assist in countering the COVID-19. Those apps aim to cope with a significant problem; that's the shortage of integration of proven information assets. In keeping with specialists, one of the foremost advantages of the use of blockchain-enabled apps is blockchain's capability of validating continuously changing records. Blockchain and IoT are each emerging technologies with outstanding capacity, but lacking extensive adoption because of technical and security worries. Numerous corporations within the marketplace are already running on use cases combining the two techniques, as collectively, they provide a manner to limit the safety and accompanying business dangers. They can gain hands-on revel in IoT and blockchain generation, provoke discussions, and share thoughts on solutions. This innovation space is designed to help corporations define their method around IoT and blockchain and encourage them to pursue it.

6 Cloud and Fog Computing

During the last months, several nations the world over have skilled numerous types of wide-ranging work and distance learning as governments had been coming to controls with COVID-19. Quarantine, lockdown, shelter in location, countries gave this isolation unique names, but this time turned into characterized through the identical want to transform many groups from retail to education. The era has been very vocal approximately the amount of digital transformation they have got supported their customers with this technology. COVID-19 changed their attitude on the way to adapt to the digital transformation and prioritized their execution. The extra trade people see, the tighter it is far to consider groups will move again to how matters have been earlier before this crisis. The mindset of no going back is not merely coming from the truth that COVID-19 will impact many in-character sports and areas from open offices, to travel and activities.

The preference to exchange is also the result of seeing the superb impact digital transformation has on an enterprise. People were transforming too throughout this time of the lockdown due to COVID-19. Humans were examined every day at the same time as they are trying to work, train, and live from home. Plenty of harder tests come from dropping jobs and a steady income. As social norms alternate, a number of the customers might spend a number of the more time customers need to learn to bake, cook, speak a new language. Others would possibly want to examine new talents to re-input the place of work after the disaster.

Very distinctive attitudes to studying: one is supporting humans with mental health, and one is helping to construct a more considerable marketable skillset. The good news is there are masses of content that has been made to be had via different agencies: webinars, digital university guides, or even online training. Consistent with the LinkedIn professional blog, human beings watched 1.7 million hours of learning content material on LinkedIn learning, accessed through the cloud and fog computing, versus. Half a million hours on common before the unfolding of the Corona. The growth in getting entry to the cloud answers equals a threefold leap in time spent getting to know. They want to accumulate new digital talents that better prepare the hard work pressure for the incoming duties of the destiny that would possibly have accelerated throughout COVID-19. Still, there are clear signs that a situation is going to become difficult or unpleasant with the expected second wave of the pandemic, or that no existence of the vaccine or medication is going to make the impact mitigation fail. Any company wants to train and retrain the generations even in college, the contemporary team of workers through cloud and fog computing solutions.

7 The Fourth Industrial Revolution

After the general data protection regulation (GDPR) is introduced in May 2019, the United States and Europe have strict medical-privacy rules. The new technology approaches should focus on the secure encryption of data, especially during its transmission between computers. By adapting the fourth industrial revolution principles, there is a possibility to cover the lack of regulation surrounding the corporate use of personal data. Breaches of healthcare systems have exposed more than 100 million people data in the USA. The fourth industrial revolution approaches for speech recognition also threaten people's privacy. Privacy groups have concerns about the deal between NHS and Amazon. It is considered a data protection breach that is going to happen soon. Google health data scandal gives the researchers a vision to start to fear the controversy over the night in gate project that will, under the scheme, undermine trust in research. National health systems give Amazon free use of health data under the Alexa advice deal. UK health service will not gain commercial benefit from future Amazon products using its data.

To sum up, the fourth industrial revolution is one of the emerging technologies with great potential, but still lacking widespread adoption due to technical and economic concerns. Several companies in the information technology market are already working on use cases combining this technology, as together they offer a way to minimize the safety and accompanying business risks. Companies should start considering the implementation of the fourth industrial revolution to address their COVID-19 pandemic issues, starting small by learning from current use cases and projects in the industry. These innovative solutions were designed to help organizations define their strategy around the fourth industrial revolution and to inspire them to pursue the new emerging technology. Table 2 introduces the primary industries hit by the COVID-19 and how these industries are managing the risk factors in these cases. These industries start from the tourism industry, which was paralyzed hard by the pandemic to the telecommunication, which was fully loaded by the pattern of employees changing to work from home style.

More advancements in chips and sensor technology provide users with the sense of affecting their surrounding environment through actuators embed into the physical elements in the real world, which gives users the path to the cyber-physical systems (CPS). In these systems, users can control physical phenomena through a cyber-interface that affects the device in the physical world—using these CPS, more additional capabilities aids in converting the insight into actions.

The IoT connects people, locations, and products, and in so doing, it gives possibilities for price creation and capture. Sophisticated chips, sensors, and actuators are embedded into bodily items, every transmitting statistic to the IoT network. The

Table 2 Industries hit by the COVID-19, their risk factors and suggested mitigation technology

Industry	Risk factors			Mitigation
Automotive	A sudden drop in sales	Automobile plants shut	A sharp drop in stock prices	Bailout plans, robotics
Aviation	High cancellations	Decreased revenues	Increased debts	Big data, travel bubbles
Tourism	A decrease in the numbers of international tourists	A drop in consumer confidence	High cancellation rates	Drones, big data
Oil and gas	A collapse in oil prices	Imbalance in real estate demand	Travel restriction	Big data
Construction	Unavailability of labor	A decline in real estate demand	Financial difficulties	Robotics, blockchain
Food	Government guidelines	Instability in food demand and supply	Potential supply chain disruption	Robotics
Telecommunication	Traffic congestion	Workforce reduction	Network reliability	AI, Drones
Healthcare	Unpredicted load	Workers at high risk of infection	Lack of resources	AI, IoT, Drones

analytics abilities of the IoT use this statistic to convert insights into movement, impacting enterprise tactics, and primary to new operating ways. However, there are still some technical and protection issues that continue to be unaddressed. Security is a fundamental problem with IoT that has hindered its big-scale deployment.

The significant scale implications of the COVID-19 pandemic on the global financial system are attributed to the response gadget followed its initial outbreak. Even though the response to the COVID-19 pandemic has been more organized than the reaction to preceding epidemics and pandemics, more problems within the contemporary pandemic reaction remain unsolved. All the underlying troubles with the contemporary response and key getting to know the factors for future public health emergency management. The lack of transparency that happened at the beginning of the pandemic is a significant fault, and solutions should establish stringent guidelines to ensure the immediate release of the information concerning any potential health emergency. Proactive monitoring for the impact of the diseases should be forced by technology to recognize the threat promptly. A higher degree of pandemic readiness is required through the use of the fourth industrial revolution. Isolation of the high-risk regions should be done immediately after the identification of the pandemic threat. Following the outbreak of the pandemic health screening measures and travel, restrictions should be introduced sooner to stop the spread of the pandemic.

8 Conclusion and Future Aspects

As a result of the COVID-19 pandemic, shortly, everything will be transformed with the digital transformation and the emerging technology into a digital entity. In this chapter, the conceptual foundation of the highlighted technologies, for fighting the crisis of the COVID-19 pandemic, is outlined. The newly faced crisis of the COVID-19 pandemic needs a complex solution that not just a fusion of several technologies such as cloud computing, big data, and IoT but also includes robotics, AI, and deep learning. These kinds of techniques aid in bringing a recent fourth industrial revolution to the current infrastructures and the used emerging technologies may transform different sectors of our daily life activities. Following the trends of the COVID-19, those emerging technologies have gained many advanced capabilities that managed it to improve the performance of the most sophisticated missions such as pandemic control and mitigation as well as applications where a rapid response is a necessity to resume action in such applications. However, COVID-19 is still considered a hot topic that all the emerging technologies should operate in a cooperative and coordinated manner to give optimized, safe, and reliable besides secure results. In this chapter, the emerging technologies associated with COVID-19 pandemic, technologies empowering the proposed solutions, real-time applications for control the pandemic, comparison with such high threats acting as a significant research area have been addressed for researchers to make secure for their implementations shortly.

References

1. https://www.who.int/csr/don/12-january-2020-novelcoronavirus-china/en/. Accessed 12 July 2020
2. World Health Organization:Novel Coronavirus (2019-ncov): Situation Report, 91'. Accessed 12 July 2020
3. Singhal, T.: A review of coronavirus disease-2019 (COVID 19). Indian J. Pediatr. 1–6 (2020)
4. Chan-Yeung, M., Xu, R.: SARS: epidemiology. Respirology **8**(Suppl S9), 14 (2003)
5. World Health Organization:Critical Preparedness, Readiness and Response Actions for COVID-19: Interim Guidance, 22 Mar 2020
6. Kissler, S.M., Tedijanto, C., Goldstein, E., et al.: Projecting the transmission dynamics of SARS-CoV-2 through the post-pandemic period. Science **368**(6493), 860–868 (2020). https://doi.org/10.1126/science.abb5793
7. International Labour Organization: The Socioeconomic Impact of COVID-19 in Fragile Settings: Peace and Social Cohesion at Risk. https://www.ilo.org/global/topics/employment-promotion/recovery-and-reconstruction/WCMS_741158/langen/index.htm. Accessed 12 July 2020
8. Guo, B., et al.: Mobile crowdsensing and computing: the review of an emerging human-powered sensing paradigm. ACM Comput. Surv. (CSUR) **48**(1), 1–31 (2015)
9. Doran, D., Severin, K., Gokhale, S., et al.: Social media enabled human sensing for smart cities. AI Commun. **29**(1), 57–75 (2015). https://doi.org/10.3233/AIC-150683
10. Ting, D.S.W., Carin, L., Dzau, V., Wong, T.Y.: Digital technology and COVID-19.Nat. Med. 1–3 (2020)

11. Digital Technology for Covid-19 Response. Apr 2020 [Online]. Available https://www.who.int/news-room/detail/03-04-2020-digitaltechnology-for-covid-19-response. Accessed 12 July 2020
12. Fong, M.W., Gao, H., Wong, J.Y., et al.:Nonpharmaceutical measures for pandemic influenza in nonhealthcare settings—social distancing measures.Emerg. Infect. Dis. (2020)
13. Toh, C.K.:Security for Smart Cities.IET Smart Cities, pp. 1–10 (2020)
14. Chamola, V., et al.: A comprehensive review of the COVID-19 Pandemic and the role of IoT, Drones, AI, Blockchain and 5G in managing its impact. IEEE Access (2020). https://doi.org/10.1109/Access2020.293,2341
15. https://www.worldometers.info/coronavirus/. Accessed 12 July 2020
16. Wittbold, K.A., Carroll, C., Iansiti, M., Zhang, H.M., Landman,A.B.:How Hospitals are using AI to Battle Covid-19[Online]. Available https://hbr.org/2020/04/how-hospitals-are-using-ai-to-battlecovid-19. Accessed 12 July 2020
17. Rees, V.: AI Technology to Screen Existing Drugs for Use Against COVID-19[Online]. Available https://www.drugtargetreview.com/news/59188/ai-technology-toscreen-existing-drugs-for-use-against-covid-19/. Accessed 12 July 2020
18. Wakefield, J.: Coronavirus: AI Steps up in Battle Against COVID-19.Apr 2020 [Online]. Available https://www.bbc.com/news/technology52120747. Accessed 12 July 2020
19. Ekins, S., Freundlich, J.S., Clark, A.M., Anantpadma, M., Davey, R.A., Madrid, P.: Machine Learning Models Identify Molecules Active Against the Ebola Virus in Vitro.F1000Research **4** (2015)
20. Schmitt, M.: How to Fight COVID-19 with Machine Learning—Towards Data Science [Online]. Accessed 12 July 2020
21. Zhang, H., Saravanan, K.M., Yang, Y., Hossain, M.T., Li, J., Ren, X., Wei, Y.: Deep Learning-Based Drug Screening for Novel Coronavirus 2019-nCov (2020)
22. Obeidat, S.: How Artificial Intelligence is Helping Fight the COVID-19 Pandemic[Online]. Available https://www.entrepreneur.com/article/348368. Accessed 12 July 2020
23. WHO Situation Report 79. WHO. Apr 2020 [Online]. Available https://www.who.int/docs/default-source/coronaviruse/situationreports/20200408-sitrep-79-covid-19.pdf?sfvrsn=4796b143_6. Accessed 12 July 2020
24. Frackowiak, M., Darlak, J., Kobylinska, A., Lund, T.: Factbox: Europe Begins Easing Coronavirus Lockdowns.Apr 2020 [Online]. Available https://www.reuters.com/article/us-health-coronaviruseurope-factbox/factbox-europe-begins-easing-coronavirus-lockdownsidUSKBN21Z1P0. Accessed 12 July 2020
25. Godin, M.: How European Countries are Lifting Coronavirus Restrictions[Online]. Available https://time.com/5822470/countries-lifting-coronavirus-restrictions-europe. Accessed 12 July 2020
26. Baldwin, R., Tomiura, E.: Thinking Ahead About the Trade Impact of COVID-19.Economics in the Time of COVID-19, p. 59 (2020)
27. WHO Situation Report 87. WHO. Apr 2020 [Online]. Available https://www.who.int/docs/default-source/coronaviruse/situationreports/20200416-sitrep-87-covid-19.pdf?sfvrsn=9523115a_2. Accessed 12 July 2020
28. Rouse, M.: What is IoMT (Internet of Medical Things) or Healthcare IoT? Definition from WhatIs.com[Online]. Available https://internetofthingsagenda.techtarget.com/definition/IoMT-Internet-of-Medical-Things
29. Medtech and the Internet of Medical Things. Deloitte Centre for Health Solutions [Online]. Available https://www2.deloitte.com/content/dam/Deloitte/global/Documents/Life-Sciences-Health-Care/gx-lshc-medtech-iomt-brochure.pdf. Accessed 12 July 2020
30. Rodrigues, J.J., Segundo, D.B.D.R., Junqueira, H.A., Sabino, M.H., Prince, R.M., Al-Muhtadi, J., De Albuquerque, V.H.C.: Enabling technologies for the internet of health things.IEEE Access **6**, 13129–13141 (2018)
31. Mcneil Jr. D.C.:Can Smart Thermometers Track the Spread of the Coronavirus? [Online]. Available https://www.nytimes.com/2020/03/18/health/coronavirus-fever-thermometers.html. Accessed 12 July 2020

32. Telemedicine Defined [Online]. Available https://www.amdtelemedicine.com/telemedicine-resources/telemedicinedefined.htm. Accessed 12 July 2020
33. Yang, G.-Z., Nelson, B.J., Murphy, R.R., Choset, H., Christensen, H., Collins, S.H., Dario, P., Goldberg, K., Ikuta, K., Jacobstein, N., et. al.: Combating COVID-19—The Role of Robotics in Managing Public Health and Infectious Diseases (2020)
34. UV Disinfection with Pulsed Xenon to Combat HAIs[Online]. Available https://www.xenex.com/. Accessed 12 July 2020
35. AI in Healthcare, 2018, website https://www.nature.com/articles/s41551-018-0305. Accessed 12 July 2020
36. https://www.cmu.edu/news/stories/archives/2020/march/coronavirus-a-wakeup-call-for-robotics-research.html. Accessed 12 July 2020
37. Eng, C.L., Tong, J.C., Tan, T.W.: Predicting host tropism of influenza a virus proteins using random forest. BMC Med. Genomics **7**(S3), S1 (2014)
38. Babayan, S.A., Orton, R.J., Streicker, D.G.: Predicting reservoir hosts and arthropod vectors from evolutionary signatures in RNA virus genomes. Science **362**(6414), 577–580 (2018)
39. Miglani, A., Kumar, N., Chamola, V., Zeadally, S.: Blockchain for internet of energy management: review, solutions, and challenges.Comput. Commun. (2020)
40. Alladi, T., Chamola, V., Sahu, N., Guizani, M.: Applications of blockchain in unmanned aerial vehicles: a review.Vehic. Commun. 100249 (2020)
41. Hassija, V., Chamola, V., Dara, N.G.K., Guizani, M.: A distributed framework for energy trading between UAVs and charging stations.IEEE Trans. Vehic. Technol. 1 (2020)
42. Hassija, V., Chamola, V., Garg, S., Dara, N.G.K., Kaddoum, G., Jayakody, D.N.K.: A blockchain-based framework for lightweight data sharing and energy trading in v2g network.IEEE Trans. Vehic. Technol. 1 (2020)
43. Hassija, V., Chamola, V., Han, G., Rodrigues, J., Guizani, M.: Dagiov: a framework for vehicle to vehicle communication using directed acyclic graph and game theory.IEEE Trans. Vehic. Technol. 1 (2020)
44. Adolph, C., Amano, K., Bang Jensen, B., et al.:Pandemic politics: timing state-level social distancing responses to COVID-19. medRxiv (2020)
45. Alladi, T., Chamola, V., Rodrigues, J.J., Kozlov, S.A.: Blockchain in smart grids: a review on different use cases. Sensors **19**(22), 4862 (2019)
46. Alladi, T., Chamola, V., Parizi, R.M., Choo, K.-K.R.: Blockchain applications for industry 4.0 and industrial IoT: a review. IEEE Access **7**, 176 935–176 951 (2019)
47. Wright, T.: Blockchain app used to track COVID-19 cases in Latin America. https://cointelegraph.com/news/blockchain-app-usedto-track-covid-19-cases-in-latin-america [Online]. 7 Apr 2020

The Role of Emerging Technologies for Combating COVID-19 Pandemic

Atrab A. Abd El-Aziz, Nour Eldeen M. Khalifa, Ashraf Darwsih, and Aboul Ella Hassanien

Abstract The outbreak of the new coronavirus disease (COVID-19) in 2019 resulted in more than 100,000 infections and thousands of deaths. The number of deaths and infections continues to rise rapidly since the virus date of appearance. COVID-19 threatens not only human health but also many aspects of life such as manufacturing, social performance, and international relations. Emerging technologies can help in the fight against COVID-19. Emerging technologies include blockchain, Internet of Things (IoT), artificial intelligence (AI), and big data technologies, and they proved its efficiency in practical fields. These fields include the fast aggregation of multi-source big data, fast visualization of epidemic information, diagnosing, remote treatment, and spatial tracking of confirmed cases. Every country in the world is still seeking realistic and cost-effective solutions to stand against COVID-19 under current epidemiological conditions. This chapter discusses the concepts of emerging technologies, applications, and contributions to combating COVID-19. Moreover, the challenges and future research directions are reviewed in detail. Also, a list of publicly available open-source COVID-19 datasets will be presented. Finally, this chapter concludes that cooperation among government, medical institutions, and the scientific community is significant and critical. Also, there is an urgent demand for improvement in the analytical algorithms and electronic devices to combat the COVID-19 pandemic.

A. A. Abd El-Aziz (✉)
Faculty of Computers and Information, KafrelSheikh University, Kafr El-Sheikh, Egypt
e-mail: Atrab_Ahmed@fci.kfs.edu.eg

N. E. M. Khalifa · A. E. Hassanien
Faculty of Computers and Artificial Intelligence, Cairo University, Cairo, Egypt

A. Darwsih
Faculty of Science, Helwan University, Helwan, Egypt

A. A. Abd El-Aziz · N. E. M. Khalifa · A. Darwsih · A. E. Hassanien
Scientific Research Group in Egypt (SRGE), Cairo, Egypt

A. E. Hassanien and A. Darwish (eds.), *Digital Transformation and Emerging Technologies for Fighting COVID-19 Pandemic: Innovative Approaches*, Studies in Systems, Decision and Control 322, https://doi.org/10.1007/978-3-030-63307-3_2

Keywords COVID-19 · Artificial intelligent · Big data · Blockchain · Internet of things

1 Introduction

In December 2019, the World Health Organization (WHO) has declared a new coronavirus disease-19 called COVID-19 as a pandemic. COVID-19 was primarily appeared in the city of Wuhan in China and has grown rapidly worldwide as shown in Fig. 1. The COVID-19 pandemic has generated an unusual economic and health crisis that required a rapid response from medical professionals, government organizations, and scientists. Due to limited resources and the urgent demand for medical supplies, this crisis has required quick and innovative solutions. The innovative use of emerging technologies has been introduced to meet growing demands. The scientific response to combat COVID-19 has been much faster and more widespread. At the end of 2019, the number of published academic research including the word "coronavirus" reached 755 articles [1, 2].

Scientific researchers have brainstormed ideas that can narrow the crisis and improve the prevention of future pandemics. Not only medical specialists but also scientists assisted with digital technologies have tackled the pandemic with novel approaches. The main digital effort was obtained from the Artificial Intelligence community in the automated detection of COVID-19 using "CT" computed tomography and X-ray scans. Many efforts have been made to analyze the emotional and social behaviors using social networks, such as the detection of COVID-19 from cough samples, obtaining academic articles for semantic analysis, and automated follow-up of contacts [3].

The main objective of this chapter is to show the effectiveness of emerging technological developments that are being executed to combat the COVID-19 pandemic. This chapter surveyed cutting-edge solutions using these emerging technologies. In

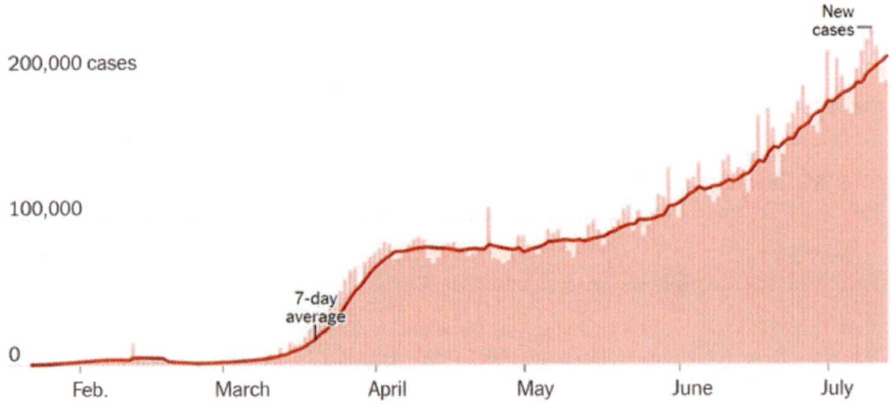

Fig. 1 Global COVID-19 trend for 11 July 2020 (*source* **NYTimes**)

addition, the challenges and problems associated with existing based approaches are highlighted. This motivates the production of a set of research directions for research communities, societies, and governments. The technologies included in this review involve Artificial Intelligence, Internet of Things, Blockchains, and Bigdata.

The chapter is organized as follows: Sect. 2 presents the background and literature review, Sect. 3 explains the open-source COVID datasets. Section 4 shows the applications of emerging technologies coronavirus. Section 5 details the challenges and limitations and Sect. 6 discusses the future research directions and finally, the chapter conclusion in Sect. 7.

2 Literature Reviews

According to the WHO, emerging digital technologies can play a vital role in promoting the public health response to the COVID-19 pandemic [4]. In the following parts, we review the background and literature contributions of the aforementioned technologies for lessening the destructive impacts of the COVID-19 pandemic.

2.1 Blockchain Technology

Blockchain technology is a distributed data structure that is replicated and distributed among the members of a network. Blockchain started with Bitcoin. Unlike traditional database systems, Blockchain technology has several comprehensive properties to guarantee transparency, precision, and immutability in data storage. These properties incorporate the service view, logical inclusion, and architectural features. The three main components of blockchain include distributed ledger (database), data block, and consensus algorithms. How could blockchain technology address the existing challenges of COVID-19 to lessen the spread of the pandemic and prepare for future pandemics?

Recently, Blockchain technology has been the subject of wide consideration among industrialists and researchers, particularly after the appearance of Blockchain 2.0 and Blockchain 3.0. Blockchain distributed accounting technologies are very useful concerning the COVID-19 health crisis. Blockchain technology allows organizations and individuals from any area to become part of a single interconnected network that promotes the secure exchange of data. The use of blockchain consensus algorithms and smart contracts leads to a decrease in the distribution of false data and fraudulent information. COVID-19 medical data such as (decisions, cases, and patient health data) could be stored as a distributed ledger that produces a single, widely accessible source of information. These medical data records can be prepared using blockchain that offers features like validation, shared database, distributed trust, and immutable records [5]. Blockchains have two categories as shown in Fig. 2 [6].

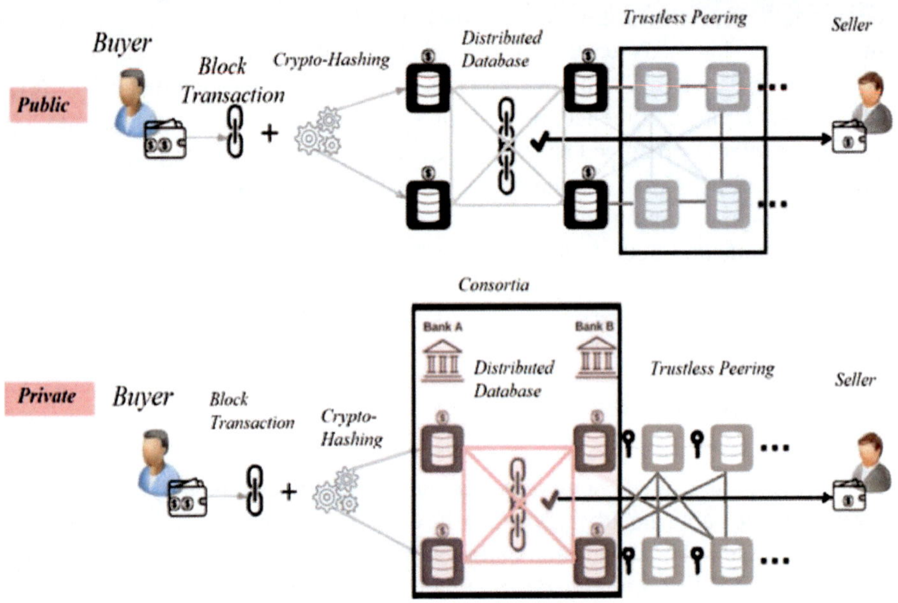

Fig. 2 Public versus private blockchain

Public blockchain: It is accessible (without permission) to everyone who can join and transact as well as participate in the agreement process. Public blockchain applications involve Ethereum and Bitcoin.

Private Blockchain: It is an invitation-only (with permission) network controlled by a central unit. A member must have permission to use a validation mechanism.

Simultaneously with the progress of blockchain, there are some innovative consensus algorithms, including the Byzantine Faulty Tolerant (BFT) and Proof of Participation (PoS). Blockchain has shown its benefit in healthcare applications. Consequently, it is likely to be implemented to solve health problems related to the coronavirus epidemic [7]. The first blockchain solution was HashLog which designed by the Georgia-based healthcare technology startup. The HashLog blockchain project was created to combat and control the spread of the coronavirus. Blockchain technology distributed ledgers ensures data logging and visualization of the coronavirus outbreak. HashLog probably offers real-time updates by tracking the movement of infected people. This allows health authorities to make decisions to fight additional infections [8]. Hyperchain is the subsequent blockchain solution for coronavirus [9]. Hyperchain has created a platform for tracking donations to help governments and healthcare institutions in China. Hyperchain can join millions of nodes where more users can reach donated and essential medical equipment. Also, this project helps to solve the lack of problems of the installation through the epidemic. VeChain is a blockchain-based platform that is designed to monitor vaccine production in China [10]. The project records and stores all activities relevant to vaccine manufacturing in the distributed books. These activities start from materials, codes to packages.

VeChain also offers a reliable method to decrease the risk of possible modifications of the vaccine information with a high-quality vaccine. Recently, Hao, et.al [11], proposed a blockchain-enabled contact tracking system called BeepTrace. The project adopts blockchain that connects the patient with authorized solvers to desensitize patient identification and location information. The proposed project records increased security and privacy with the added benefits of being battery-friendly and widely accessible.

2.2 *Artificial Intelligence*

AI is one of the emerging technologies that can quickly monitor the spread of the virus, classify high-risk patients, and is beneficial in regulating this infection in real-time. AI can further predict mortality risk by properly analyzing previous patient data. The fight against Covid19 through population detection, notification, medical help, and infection control suggestions can be done with the help of AI. This technology has the potential to enhance the competence of COVID-19 by focusing not only on patient treatment but also on disease control. AI has also been widely applied to find new molecules for COVID-19. Much research has been investigated using AI techniques to discover new medicines and drugs for the cure. Together with computer science researchers who focus on recognizing infectious patients by processing medical images such as CT and X-rays [12].

Many countries, such as Taiwan, inspired the national health insurance database with migration information and a customs database, thus confronting coronavirus patients for their symptoms and travel history. Also, at the beginning of the coronavirus outbreak, China focused on AI by using facial recognition cameras to track infected patients with travel history, drones to disinfect public places, and robots to deliver drugs and food.

Asif et al. [13], introduced a Deep Convolutional Neural Network (DCNN) model called "CoroNet" to automatically identify COVID-19 infection using chest X-ray images as shown in Fig. 3. The suggested model is based on Xception, which is a

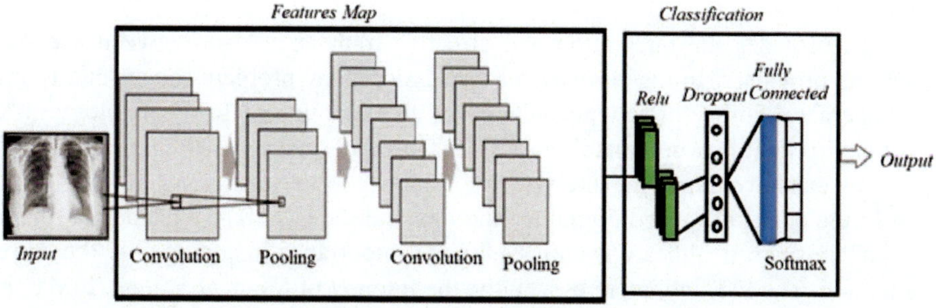

Fig. 3 DCNN structure for COVID-19

pre-trained architecture. CoroNet has achieved 98.2% accuracy for cases of 4 classes (bacterial pneumonia versus viral pneumonia versus COVID versus normal).

Pathak et al. [14], proposed a deep transfer learning classification model for patients with COVID-19. Cost-sensitive attributes with a two smoothing loss function are also employed to solve the noisy and unbalanced problems of the COVID-19 data set. The proposed technique achieves an accuracy of up to 93%. Ten popular [15] convolutional neural networks were adopted to recognize COVID-19 infection from non-COVID-19 societies. Among all networks, ResNet-101 achieved the most reliable performance. ResNet-101 achieved 99.51% accuracy.

The authors in [16] proposed an AI-based generative chemistry model to create innovative molecules that can repress COVID-19. Autoencoders and genetic algorithms were applied using molecular symbols to produce structures. Despite this, it is a hopeful approach since these AI models can exploit the great drug and automatically obtain valuable information from high-dimensional data. This method has a high potential to produce new drugs against COVID-19, as it is time-efficient and cost-effective.

The authors in [17] developed learning-based machine learning to predict the COVID-19 taxonomy. The proposed model used an unaligned method based on the decision tree approach and genomic signatures. The proposed method is a computationally economical approach that can provide a fast taxonomic classification of new pathogens by processing raw DNA sequence data.

2.3 Internet of Things Technology

Internet of Things (IoT) is a computing technology of interconnected digital and mechanical devices using the internet such as sensors, smart meters, health monitoring devices, home appliances, etc. These devices can sense, process, interact, and maintain the data transmission capacity through the defined network without any human participation as shown in Fig. 4. Each device in IoT is associated with an appropriate unique identification code. Furthermore, IoT participates in many daily applications such as smart lighting, smart home, and remote health monitoring, etc. [18].

IoT provides the integration and effective transfer of data between users and service providers. In the coronavirus epidemic, many problems occur due to ineffective accessibility to patients. Therefore, the adoption of IoT technology makes patient accessibility completely useful. IoT provides patients with remarkable care so they can have treatment safely.

In the era of advanced digital technology and the COVID-19 pandemic, the use of IoT in smart healthcare is taking on great importance in coping with the current situation. The IoT and, more especially, the Internet of Medical Things (IoMT) can offer a solution to the challenges of monitoring, detection, and control of this disease [19]. IoMT refers to a network of interconnected medical equipment and software

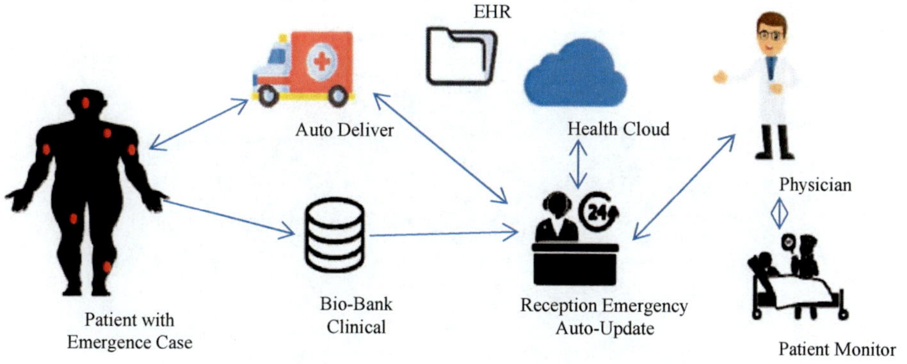

Fig. 4 IoT for healthcare

applications in the healthcare industry. One of the most emerging IoT applications is smart healthcare monitoring that is most relevant during this COVID-19 pandemic [19].

2.4 Big Data Technology

With the rapid expansion of IoT, there was a vast explosion of data generated from universal portable devices and sensors. The expansion of data volumes related to improvements in analytical techniques has led to the emergence of an era of big data. Big Data technology has been applied in a wide range of industrial application domains.

Healthcare where Automated Healthcare Records (EHR) are used using the application of smart analytics to assist medical services. Big data can be produced from various sources, such as IoT sensors, mobile devices, online social graphics, and public data in many forms, such as text or images or videos [20]. Big Data is characterized by three primary properties which called 3 V's and two additional properties (Veracity and Value) to become 5 V's as shown in Fig. 5 [21]:

- **Volume**: This feature explains the large volume of data that can vary from terabytes to exabytes.
- **Variety**: refers to the heterogeneity of the data.
- **Velocity**: refers to the data generation rate that can be estimated in the time or frequency domain.
- **Veracity**: This feature check the credibility, accuracy and quality of data sources for use.
- **Value:** This feature measures the usefulness of the data for decisions making.

Big data plays an important role in supporting the fight against infectious diseases COVID-19. Big data potentially offers promising solutions to help combat the

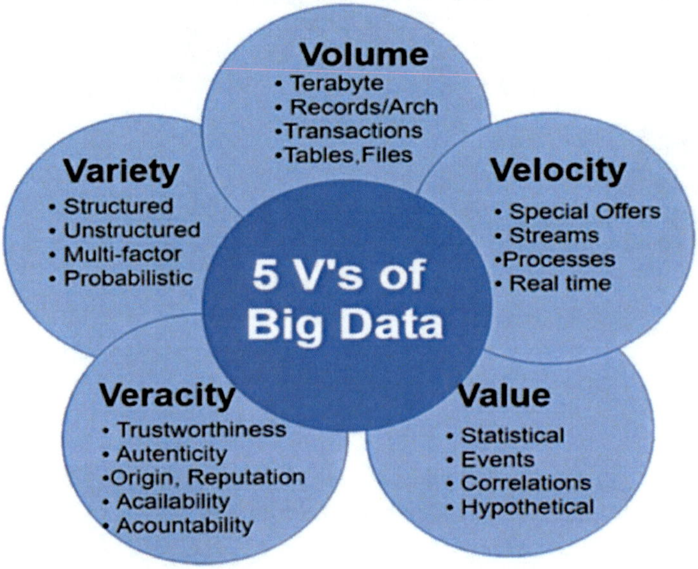

Fig. 5 Big data 5 V's characteristics

COVID-19 epidemic. Big data plays an important role in providing anti-COVID-19 strategies for drug manufacturing.

Li et al. [22], proposed a solution based on molecular coupling for drug investigations. The results proved that fifteen of the twenty-five approved medications exhibited significant inhibitory potencies and inflammatory responses and helped reposition the drug toward COVID-19. Another large data-based drug repositioning scheme was presented in [23]. The essential objective of this scheme is to implement ML to merge both the literature and the knowledge graph for the development of the COVID-19 vaccine.

The additional role of big data is tracking the spread of COVID-19. This is of utmost importance to successfully control the coronavirus pandemic [24]. Many recently emerging solutions have been introduced to assist in the process of monitoring the spread of COVID-19 using big data.

For example, the authors in [25] proposed a great data-based technique to track the spread of COVID-19. They used a large data set obtained from the China National Health Commission. Many linear models are constructed using the local population and air passengers to quantify the variance of the cases established in the cities of China. They applied a Spearman correlation analysis for daily user traffic. Another contribution on facilitate data acquisition and integrate heterogeneous data from health data resources is presented in [26]. The authors studied the fusion of large data for spatial analysis methods and Geographic Information Systems (GIS) technology.

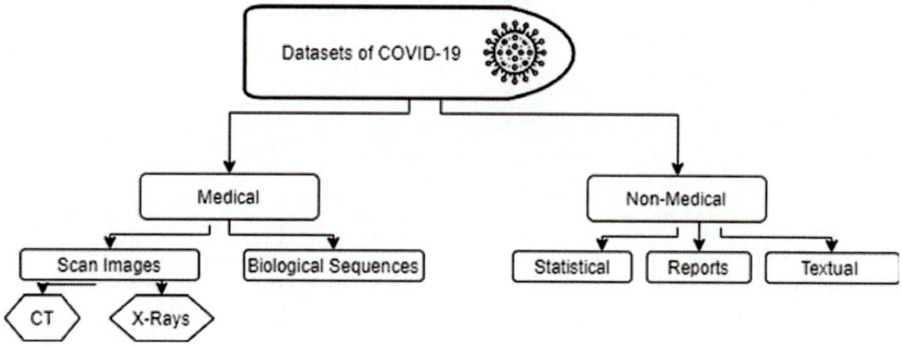

Fig. 6 COVID-19 datasets category

3 COVID-19 Datasets

This section reviews the most commonly available data sources of COVID-19. We can conclude that COVID-19 data sets are classified into two categories, medical and non-medical datasets as shown in Fig. 6.

The medical dataset includes images such as Chest (CT) or (X-ray) or biological sequences like genome. These data must preserve the privacy of patients. Non-medical data sets include daily reports and text data that include data from scientific articles and posts on social media such as Twitter. These data can be used to analyze common sentiment regarding the COVID-19 keywords and predict the COVID-19 spread rate for each region based on reported cases. Table 1 includes many sources of COVID-19 data sets used in scientific research and Fig. 7 displays samples of medical datasets.

4 Applications of Emerging Technologies in COVID-19

According to the WHO, digital technologies have many essential roles in promoting public health to combat the COVID-19 pandemic. In this section, we investigate the major applications of digital technologies to mitigate the disastrous effects of the COVID-19 pandemic as shown in Fig. 8.

4.1 Application of AI Against COVID-19

Considering its initiation, AI has confirmed to be a historic technological advance. AI becomes a very effective tool to face the COVID-19 pandemic. Below are some of the real and possible approaches where AI can help combat the COVID-19 pandemic:

Table 1 COVID-19 sources

Source	Type	Description
Medical		
[27]	X-ray and CT-scans	More than 125 images taken from various online patients
[28]	CT scans	Consisting of a total of 275 CT scans, including 183 positive and 146 negative cases
[29]	CT scans	A total of 453 confirmed cases from china hospitals
[30]	3D CT images	10 confirmed cases for scientific purposes
[31]	X-ray scans	A total of 13,800 chest radiography
[32]	X-ray scans	A total of 1341 images
[33]	SARS-CoV-2 sequences	NCBI GenBank
[34]	SARS-CoV-2 sequences	The GISAID Initiative
[35]	COVID-19 sequence database	China National GeneBank
Non-medical		
[36]	Twitter chatter	Georgia State University's Panacea Lab
[37]	Textual data	Measuring emotions
[38]	Articles	–
[39]	Statistical reported cases	–

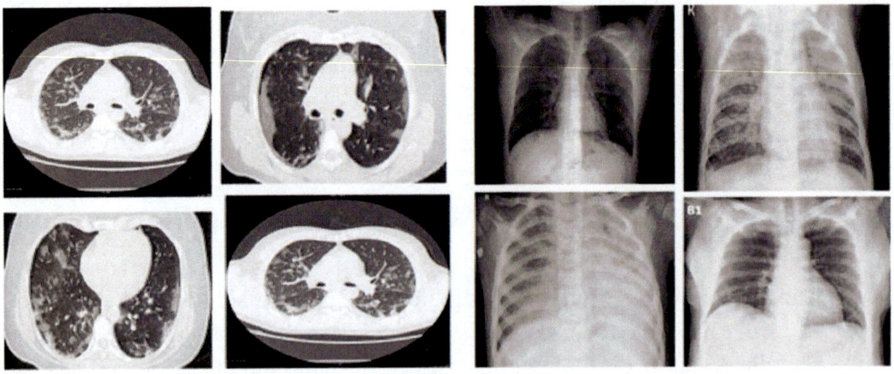

(a) (b)

Fig. 7 Samples of **a** CT [30] Dataset and **b** X-Rays [32] dataset

Medical Detection and Diagnosis: Early treatment and prediction of COVID-19 are one of the most practical solutions for combating the coronavirus pandemic. Currently, the standard method is not adequate to meet the demands of rapid detection and monitoring during the COVID-19 pandemic. Since then, these techniques have often been expensive, time-consuming, and have a low true positive rate. An

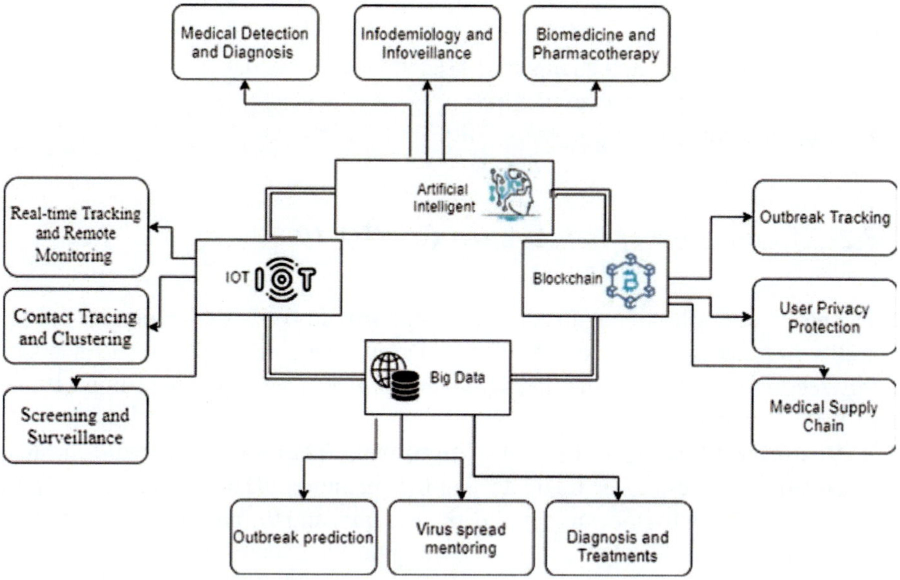

Fig. 8 Applications of emerginging technology for COVID-19

additional directive for COVID-19 detection is to employ AI methods for medical imaging processing and prediction, which has recently been developed in many investigations [40, 41].

Infodemiology and Infoveillance: The quality of information from social media platforms and the Internet is becoming very accessible, so further research can be done for the data collected and processed correctly. AI techniques have been used to produce a better understanding of the dynamics of social networks and to develop the COVID-19 situation [42]. AI applications in this area include:

- Track people's behavior using Twitter data.
- Search for health-seeking behavior of the COVID-19 explosion.
- Public attitude towards the COVID-19 revolution.

Biomedicine and Pharmacotherapy: The world has witnessed competition for efficient vaccines and treatments to combat the COVID-19 virus. This requires many efforts from health and computer science with the cooperation of AI. But, how could AI help biomedical research? The large volume of biomedical data has triggered the adoption of AI in various sectors of the biomedicine and pharmaceutical industries. Deep learning can manipulate unstructured and high-dimensional data with nonlinear relationships. These features are fitting with biomedical data like proteomics. Additionally, DL's capabilities in extracting top-level features make it a potential candidate for linking, analyzing, and understanding medical data heterogeneity like DNA microarray data. AI applications for biomedicine research include [43]:

- Genomic sequence analysis [44]
- Classification and prediction [45]

- Biomarkers [46]
- Structural biology and chemistry [47]
- Multiplatform data processing [48]
- Drug discovery and repurposing [49].

4.2 Application of IoT Against COVID-19

The health sector has recognized the potential benefits of IoT technologies due to its ability to process health data effectively. During the current COVID-19 pandemic, this section explores various IoT applications that have a potential impact on the COVID-19 pandemic.

Real-time Tracking and Remote Monitoring: The global daily real-time update in COVID-19 cases can be tracked using IoT technology. Daily updates incorporate the number of deaths, the number of patients cured, and the number of actual cases in different locations. The IoT aids doctors monitor the patient's health status remotely with medical data such as glucose level, heart rate, blood pressure level, etc. Real-time communication using IoT for medical data saves time and effort [50].

Contact Tracing and Clustering: The categorization and grouping area of the regions such as buffer zone, containment zones, red zone, green zone, orange zone, etc. it can be instantly updated based on the number of cases confirmed through IoT. The number of confirmed cases by location can be obtained in real-time when medical and healthcare systems are interconnected through IoT. The government can come up with these data and alert about health check-ups in the affected area and this can be done quickly [51].

Screening and surveillance: Facial recognition data based on thermal images can be obtained by the public and health authorities through IoT at multiple entry positions of airports, railway stations, etc. for detection and surveillance purposes. This automatic surveillance of suspected and determined cases can serve to regulate the extent of the infection [52].

4.3 Application of Big Data Against COVID-19

The escalation of big data in the healthcare sector revolutionized the medical industry by affording better patient health information. In these paragraphs, some big data applications regarding COVID-19 are provided.

Outbreak prediction: Big data has a major role in fighting COVID-19 for its significant ability in predicting the outbreak of large-scale data analysis. The outbreak prediction can be found in using the public dataset to visualize geographic areas with a possible outbreak [24].

Virus spread mentoring: Big Data is important in the process of monitoring the spread of COVID-19. That is a prevailing concern not only for healthcare organizations but also for governments in the successful management of the coronavirus pandemic. Currently, several recently emerging solutions that employ big data have been proposed to help track the spread of COVID-19 [53].

Diagnosis and Treatments: Big Data can be used to support the diagnostic and treatment processes of COVID-19. Regarding the diagnosis and treatment of COVID-19, Big Data has implemented many solutions, such as multiplex polymerase chain reactions that can diagnose SARS-CoV-2 and a method of molecular diagnostic for genomic analysis of SARS-CoV-2 strains [54].

4.4 Blockchain Applications Against COVID-19

In recent years, blockchain has got substantial attention from the healthcare industry. Blockchain has gained interest to improve transparency and authenticity of healthcare data starting from obtaining electronic health record permits to streamlining claims processing. This section discusses different applications of blockchains for the crisis.

Tracking of Outbreak: Blockchain can afford probable solutions to track the coronavirus outbreak. Certainly, blockchain can present data visualization tools for dangerous coronavirus monitoring. In the epidemic of coronavirus, blockchain is viable to preserve the thousands of victims of COVID-19 by immutably saving the signs of infection of the patients. Additionally, it can track patient bits, direct combat efforts, and store real-time data of affected regions. Tracking the spread of coronavirus is highly essential for since some patients can voluntarily report wrong their symptoms for bypassing hospital appointments. This could drive to patient quarantine failure. Blockchain helps governments and healthcare organizations monitor potential patients with high reliability, safety, and precision at every stage [55].

Protection of User privacy: In emergencies similar to coronavirus pandemics, patient privacy must be balanced with other concerns, such as protecting lives. Equalizing the value of supporting public care in the fight against central viral infections and the power to protect the privacy and individual rights of individuals is paramount to any health policy in this epidemic. Blockchain is an insurance solution to address privacy guarantee concerns during coronavirus combat. Blockchain can track data operations to uncover any illegal access or malicious use of data [56].

Medical supply chain: In this pandemic emergency, having an endless number of medicine and food stores has become a demand for the health sector. Blockchain technology can enhance supply chain partnerships by realizing a rapid flow of supply by tracing the flow from origins to the target in a committed and reliable way. Lately, China has developed a blockchain-based platform that allows people to follow the market and supply chains of medical supplies. This consists of reporting and tracking coronavirus-fighting materials, such as gloves, masks, and other emergency equipment. Therefore, while an outbreak happens, fast response and quick supply chain are the essential weapons governments have to catch the problem [57].

5 Challenges

Despite the various benefits that emerging technologies can contribute to health crises, such as the COVID-19, they are challenged by many limitations and certain challenges as shown in Fig. 9.

5.1 Shortage of Standard Datasets

Significant challenges arise from the absence of standard and large data sets. However, The authors in [58, 59] have applied the GAN network to overcome the limited size of datasets by producing more X-rays images for COVID19 detection with high detection accuracy, more datasets are required. Since then, many algorithms and technology platforms have been introduced, but they have not experimented with the same data set. To use the features and applications of emerging technologies as reliable real-time solutions to combat the COVID-19 virus, standard data sets are required [60].

Fig. 9 Main challenges for fighting COVID-19

5.2 Regulation Reflection

Government authorities play an important role in determining policies that can support the participation of various entities and overcome any barrier and restriction on how to stop the COVID-19 disease. Since the explosion of confirmed cases, many strategies have been practiced to regulate this explosion. These strategies include social distancing, large-scale testing, and confinement. Consequently, the adoption of emerging technologies, e.g. IoT, blockchain, Big-data, and AI in the fight against coronaviruses must be carefully studied with regulatory laws. While these technologies may cause advantages, they also pose legal challenges. Especially legal issues related to personal information should be considered, such as concerns with copyright defamation [61].

5.3 Security

The security of personal data is still necessary and should be reviewed. In the Covid19 pandemic, authorities can ask people to share their personal information, such as CT scans and diagnostic reports that require up-to-date policies and prompt monitoring. There is a tradeoff between security, privacy, and performance. However, Blockchain is popularly considered as a secure platform for health applications. Modern research has shown blockchain security weaknesses that are relevant to healthcare systems, such as alterations of patient data information or medical transactions, increasing privacy concerns [62].

5.4 Privacy Preservation

In coronavirus tracking apps, preserving people's privacy is extremely crucial. Governments can use mobile location to better track the spread of the virus blast. This solution should guarantee the privacy of user data. Government agencies may force privacy laws on user tracking to ensure public safety. Many healthcare organizations are now collecting patient data through electronic healthcare records. In these healthcare activities, the dispute between data collection and user privacy is necessary and must be resolved by laws and the reinforcement of related authorities [63].

6 Future Directions

Finally, current future directions and trends to investigate emerging technologies in the evolution of the health crisis are discussed in these paragraphs as shown in Fig. 10.

6.1 Covid19 Datasets

Currently, the available Covid19 data sets are saved in different formats and criteria. This could limit the extent of research contributions using emerging technologies. Therefore, the future direction is to create free benchmarking COVID-19 data sets. These benchmarks must be created according to standard protocols. Repository data sets are essential as they will help accelerate beneficial findings to incur the disease. Additionally, the available data sets are limited in size. Larger data sets are demanded to allow more useful insight and precision for deep learning algorithms in the diagnostic process. Different metrics are used for evaluation purposes. Consequently, there is a great need to generate a reference data framework. This framework will be used for evaluating and comparing existing methods. Furthermore, it should assist in universal data sets, including corresponding patient bands, data preprocessing methods, and evaluation criteria for all methods that are estimated.

6.2 AI Against Covid19

Enhance artificial intelligence algorithms for more useful analytical precision: The effectiveness of intelligent data analysis in healthcare is essentially dependent on

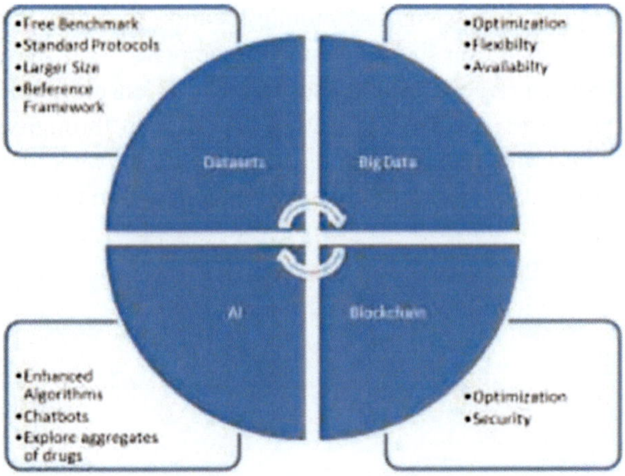

Fig. 10 Future direction of emerging technologies

artificial intelligence algorithms. Improving AI models focused on medical applications could be the solution to give the facility and the ability to manipulate multimedia healthcare data. Future research investigations are essential required in AI models. These investigations contribute to the differentiation of COVID-19 and other pneumonia types. Additionally, they could help clinicians and radiologists become aware of the virus' insight and explore upcoming coronavirus X-ray and CT images more productively. Chatbots systems can be developed using AI and NLP technologies. These chatbots can remotely allow people and patients to discuss coronavirus. Furthermore, AI can be employed on social media platforms to exterminate fake news and ensure accurate and reliable information. In the fields of medicine and biology, AI has been applied to recognize COVID-19 and explore different aggregates of drugs against the virus. AI can encourage scientists to become aware of the coronavirus quickly.

6.3 Blockchain Against COVID-19

Blockchain optimized for better implementation: Blockchain platforms' factors need to be optimized for more reliable returns. These factors include higher performance, reduced network latency, and increased security. Optimized factors can assist blockchain to become a perfect choice for healthcare emergency applications, especially the COVID-19 virus.

6.4 Big Data Against COVID-19

Since big data is the key to development, the optimization of big data algorithms for better diagnosis and treatment of COVID-19 by improving the precision and reliability of data analytics.

6.5 Merge with Other Emerging Technologies

To achieve greater efficiency in solving problems related to the epidemic, blockchain, Big Data, and AI could be incorporated with other technologies such as 5G and Drones to offer new practical solutions for fighting COVID-19 and create a complete healthcare system. Many efforts have been investigated in the way of improving solutions to provide more precise COVID-19 [64] prediction. Considering the COVID-19 outbreak, many tech Solutions have been offered to alleviate its impact. Nonetheless, to obtain the transformative potential of these technologies, there is a great need for a cellular network. This network can overcome the latency, bandwidth, and flexibility issues essential to today's network technology. Finally, some possible future directions have been pointed out. This chapter will shed valuable light on researching digital technologies to fight COVID-19. Also, it will motivate researchers to make more efforts to combat future coronavirus-like epidemics.

7 Conclusion

Emerging technologies provide comprehensive functions and efforts for healthcare to compete and alleviate the impact of the COVID-19 pandemic. In health crises, these technologies are useful for maintaining quality surveillance with real-time information. Technologies adopted in this study in terms of diagnosis, monitoring, tracking, prediction, and treatment of COVID-19. Modern developments have shown that collaboration between medical researchers, government authorities, and engineers is essential to the evolution of rapid and less expensive approaches to tackling the pandemic. We started with a comprehensive review of the characteristics of emerging technologies, methodologies, and the contributions of researchers for fighting against the COVID-19. Following this, a concise discussion of the challenges and future directions of emerging technologies to eliminate the disastrous impact of COVID-19 are presented. Research communities and healthcare specialists must continue to collaborate and share experience to provide quick solutions in crisis. Recognized treatments using emerging technologies must also support standard clinical trials.

References

1. Coronavirus Disease 2019: WHO Health Emergency. Available online https://www.who.int/emergencies/diseases/novel-coronavirus-2019. Accessed July 2020
2. Chen, N., Min, Z., Xuan, D., Qu, J., Gong, F., Han, Y., Qiu, Y., et al.:Epidemiological and clinical characteristics of 99 cases of 2019 novel coronavirus pneumonia in Wuhan, China: a descriptive study. Lancet **395**(10223), 507–513 (2020)
3. Randhawa, G.S., Soltysiak, M.P.M., El Roz, H., de Souza, C.P.E., Hill, K.A., Kari, L.:Machine learning using intrinsic genomic signatures for rapid classification of novel pathogens: COVID-19 case study. Plos One **15**(4) (2020)
4. Rhodes, T., Lancaster, K., Rosengarten, M.:A Model Society: Maths, Models and Expertise in Viral Outbreaks, pp. 253–256 (2020)
5. Christidis, K., Devetsikiotis, M.: Blockchains and smart contracts for the internet of things. IEEE Access **4**, 2292–2303 (2016)
6. Nguyen, D.C., Pathirana, P.N., Ding, M., Seneviratne, A.: Blockchain for 5G and beyond networks: a state of the art survey. J. Netw. Comput. Appl. 102693 (2020)
7. Nguyen, D., Ding, M., Pathirana, P.N., Seneviratne, A.: Blockchain and AI-Based Solutions to Combat Coronavirus (COVID-19)-Like Epidemics: A Survey. techrxiv (2020)
8. Acoer Coronavirus Tracker, Powered by Hedera Hashgraph, Now Freely Available to General Public with Added Clinical Trial Data [Online]. Available https://www.hedera.com/blog/acoer-coronavirustracker-powered-by-hedera-hashgraph-now-freely-available-to-generalpublic-with-added-clinical-trial-data. Accessed July 2020 Hyper Chain-Blockchain Charity

Platform to Fight Against the Coronavirus [Online]. Available https://www.hyperchain.cn/en. Accessed July 2020
9. China [Online]. Available https://www.nasdaq.com/articles/vechainannounces-blockchain-vaccine-tracing-solution-china-2018-08-16. Accessed July 2020
10. Xu, H., Zhang, L., Onireti, O., Fang, Y., Buchanan, W.B., Imran, M.A.: BeepTrace: Blockchain-Enabled Privacy-Preserving Contact Tracing for COVID-19 Pandemic and Beyond. arXiv:2005.10103 (2020)
11. Mei, X., Lee, H.-C., Diao, K., Huang, M., Lin, B., Liu, C., et al.: Artificial intelligence-enabled rapid diagnosis of patients with COVID-19. Nat. Med. (2020). https://doi.org/10.1038/s41591-020-0931-3
12. Asif Iqbal, K., Shah, J.L., Bhat, M.M.:Coronet: a deep neural network for detection and diagnosis of COVID-19 from chest X-ray images. Comput. Methods Progr. Biomed. 105581 (2020)
13. Pathak, Y., Shukla, P.K., Tiwari, A., Stalin, S., Singh, S., Shukla, P.K.: Deep transfer learning based classification model for COVID-19 disease. IRBM (2020)
14. Ardakani, A.A., Kanafi, A.R., Rajendra Acharya Nazanin Khadem, U., Mohammadi, A.: Application of deep learning technique to manage COVID-19 in routine clinical practice using CT images: results of 10 convolutional neural networks. Comput. Biol. Med. 103795 (2020)
15. Zhavoronkov, A., Aladinskiy, V., Zhebrak, A., Zagribelnyy, B., Terentiev, V., Bezrukov, D.S., Yan, Y.: Potential COVID-2019 3C-Like Protease Inhibitors Designed using Generative Deep Learning Approaches. https://doi.org/10.26434/chemrxiv.11829102.v2,2020
16. Randhawa, G.S., Soltysiak, M.P., El Roz, H., de Souza, C.P., Hill, K.A., Kari, L.: Machine Learning Using Intrinsic Genomic Signatures for Rapid Classification of Novel Pathogens: COVID-19 Case Study. https://doi.org/10.1101/2020.02.03.932350
17. Yang, T., Gentile, M., Shen, C.F., Cheng, C.M.: Combining point-of-care diagnostics and internet of medical things (IoMT) to combat the COVID-19 pandemic. Diagnostics (Basel) **10**(4)
18. Bai, L., Yang, D., Wang, X., Tong, L.: Chinese experts' consensus on the internet of things-aided diagnosis and treatment of coronavirus disease 2019 (COVID-19). Clin. eHealth **3**(1), 7–15 (2020)
19. Manogaran, G., Lopez, D., Thota, C., Abbas, K.M., Pyne, S., Sundarasekar, R.: Big data analytics in healthcare internet of things. In: Innovative Healthcare Systems for the 21st Century. Springer, pp. 263–284 (2017)
20. Nyarko, R.:Security of Big Data: Focus on Data Leakage Prevention (DLP)(2018)
21. Li, Z., Li, X., Huang, Y.-Y., Wu, Y., Zhou, L., Liu, R., Wu, D., Zhang, L., Liu, H., Xu, X., et al.: FEP-Based Screening Prompts Drug Repositioning Against COVID-19. bioRxiv 2020
22. Ge, Y., Tian, T., Huang, S., Wan, F., Li, J., Li, S., Yang, H., Hong, L., Wu, N., Yuan, E., et al.: A Data-Driven Drug Repositioning Framework Discovered a Potential Therapeutic Agent Targeting COVID-19. bioRxiv 2020
23. Is Big Data Effective in Response to Coronavirus Outbreak? 2020 [Online]. Available https://www.analyticsinsight.net/big-data-effective-response-coronavirus-outbreak/
24. Zhao, X., Liu, X., Li, X.: Tracking the Spread of Novel Coronavirus (2019-ncov) Based on Big Data. medRxiv 2020
25. Zhou, C., Su, F., Pei, T., Zhang, A., Du, Y., Luo, B., Cao, Z., Wang, J., Yuan, W., Zhu, Y., et al.: COVID-19: challenges to GIS with big data. Geogr. Sustain. (2020)
26. Cohen, J.P., Morrison, P., Dao, L.: Covid-19 Image Data Collection. arXiv:2003.11597 (2020)
27. Zhao, J., Zhang, Y., He, X., Xie, P.: Covid-CT-Dataset: Act Scan Dataset About Covid-19. arXiv:2003.13865 (2020)
28. Wang, S., Kang, B., Ma, J., Zeng, X., Xiao, M., Guo, J., Cai, M., Yang, J., Li, Y., Meng, X., et al.: A Deep Learning Algorithm Using CT Images to Screen for Corona Virus Disease (covid-19). medRxiv 2020
29. https://coronacases.org. Accessed July 2020
30. https://github.com/lindawangg/COVID-Net. Accessed July 2020
31. https://www.kaggle.com/tawsifurrahman/covid19-radiography-database. Accessed July 2020

32. https://www.ncbi.nlm.nih.gov/genbank/sars-cov-2-seqs/. Accessed July 2020
33. https://www.gisaid.org/. Accessed July 2020
34. https://db.cngb.org/datamart/disease/DATAdis19/. Accessed July 2020
35. https://www.panacealab.org/covid19/. Accessed July 2020
36. https://github.com/ben-aaron188/covid19worry. Accessed July 2020
37. https://www.semanticscholar.org/cord19/download. Accessed July 2020
38. https://www.kaggle.com/lachmann12/correcting-under-reported-covid19-case-numbers. Accessed July 2020
39. Corman, V.M., Landt, O., Kaiser, M., Molenkamp, R., Meijer, A., Chu, D.K., Bleicker, T., Brünink, S., Schneider, J., Schmidt, M.L., et al.: Detection of 2019 novel coronavirus (2019-nCoV) by real-time RTPCR. Eurosurveillance **25**(3) (2020)
40. Fomsgaard, A.S., Rosenstierne, M.W.: An Alternative Workflow for Molecular Detection of SARS-CoV-2-escape From the NA extraction Kit-Shortage. medRxiv 2020
41. Ganasegeran, K., Abdulrahman, S.A.: Artificial Intelligence Applications in Tracking Health Behaviors During Disease Epidemics, pp. 141–155. Springer International Publishing, Cham (2020)
42. Hu, F., Jiang, J., Yin, P.: Prediction of Potential Commercially Inhibitors Against SARS-CoV-2 by Multi-Task Deep Model. arXiv:2003.00728 (2020)
43. Rothan, H.A., Byrareddy, S.N.: The epidemiology and pathogenesis of coronavirus disease (COVID-19) outbreak. J. Autoimmunity 102433 (2020)
44. Salman, F.M., Abu-Naser, S.S., Alajrami, E., Abu Nasser, B.S., Alashqar, B.A.M.: Covid-19 Detection Using Artificial Intelligence(2020)
45. Moein, S.T., Hashemian, S.M.R., Mansourafshar, B., Khorram-Tousi, A., Tabarsi, P., Doty, R.L.: Smell dysfunction: a biomarker for COVID-19. Int. Forum Allergy Rhinol. 2020
46. Lu, J., Liu, X.: Structural biology aids the research of new anti-COVID-19 drugs. Curr. Org. Synth. **17**(3), 162–163 (2020)
47. Ferrara, E.: #Covid-19 on Twitter: Bots, Conspiracies, and Social Media Activism. arXiv:2004.09531 (2020)
48. Elmezayen, A.D., Al-Obaidi, A., Şahin, A.T., Yelekçi, K.: Drug repurposing for coronavirus (COVID-19): in Silico screening of known drugs against coronavirus 3CL hydrolase and protease enzymes. J. Biomol. Struct. Dyn. 1–13 (2020)
49. Yang, T., Gentile, M., Shen, C.F., Cheng, C.M.: Combining point-of-care diagnostics and internet of medical things (IoMT) to combat the COVID-19 pandemic. Diagnostics (Basel) **10**(4) (2020)
50. Shi, F., Wang, J., Shi, J., Wu, Z., Wang, Q., Tang, Z., He, K., Shi, Y.S.D.: Review of artificial intelligence techniques in imaging data acquisition, segmentation and diagnosis for COVID-19. IEEE Rev. Biomed. Eng. **2**(1), 220–235 (2020)
51. Srinivasa Rao, A.S.R., Vazquez,J.A.:Identification of COVID-19 can be quicker through artificial intelligence framework using a mobile phone-based survey when cities and towns are under quarantine. Infect Control Hosp. Epidemiol. 1–5 (2020)
52. Giordano, G., Blanchini, F., Bruno, R., Colaneri, P., Di Filippo, A., Di Matteo, A., Colaneri, M., et al.: A SIDARTHE Model of COVID-19 Epidemic in Italy. arXiv:2003.09861 (2020)
53. Li, C., Debruyne, D.N., Spencer, J., Kapoor, V., Liu, L.Y., Zhou, B., Lee, L., Feigelman, R., Burdon, G., Liu, J., et al.: High Sensitivity Detection of Coronavirus SARS-CoV-2 Using Multiplex PCR and a Multiplex-PCR-Based Metagenomic Method. bioRxiv 2020
54. Fernández-Caramés, T.M., Fraga-Lamas, P.: A review on the use of blockchain for the internet of things. IEEE Access **6**, 32979–33001 (2018)
55. Coronavirus Tracking and Surveillance May Have Consequences for the Protection of Personal Data Privacy [Online]. Available https://blockchain.news/news/coronavirus-tracking-surveillanceconsequences-protection-personal-data-privacy
56. Gonczol, P., Katsikouli, P., Herskind, L., Dragoni, N.: Blockchain implementations and use cases for supply chains-a survey. IEEE Access **8**, 11856–11871 (2020)
57. Khalifa, N.E.M., Taha, M.H.N., Hassanien, A.E., Elghamrawy, S.: Detection of Coronavirus (COVID-19) Associated Pneumonia Based on Generative Adversarial Networks and a Fine-Tuned Deep Transfer Learning Model Using Chest X-ray Dataset. arXiv:2004.01184.2020

58. Loey, M., Smarandache, F., Khalifa, N.E.M.: Within the lack of chest COVID-19 X-ray dataset: a novel detection model based on GAN and deep transfer learning. Symmetry **12**(4), 651 (2020)
59. How DAMO Academy's AI System Detects Coronavirus Cases 2020 [Online]. Available https://www.alizila.com/how-damo-academys-ai-system-detects-coronavirus-cases/. Accessed July 2020
60. How Next-Generation Information Technologies Tackled COVID-19 in China 2020 [Online]. Available www.weforum.org/agenda/2020/04/how-next-generation-information-technologies-tackled-covid-19-in-china/. Accessed July 2020
61. Coronavirus Accelerates China's Big Data Collection but Privacy Concerns Remain [Online]. Available https://www.scmp.com/tech/appssocial/article/3052232/coronavirus-accelerates-chinas-big-datacollection-privacy. Accessed July 2020
62. Agbo, C.C.: Mahmoud, Q.H., Eklund, J.M.: Blockchain technology in healthcare: a systematic review. Healthcare **7**(2), 56 (2019)
63. Sujath, R., Chatterjee, J.M., Hassanien, A.E.:A machine learning forecasting model for COVID-19 pandemic in India. Stoch. Environ. Res. Risk Assess. 1 (2020)
64. Ewis, A., Dagnew, G., Reda, A., Elmarhomy, G., Elhosseini, M.A., Hassanien, A.E., Gad, I.: ARIMA Models for Predicting the End of COVID-19 Pandemic and the Risk of a Second Rebound (2020)

An Optimized Classification Model for COVID-19 Pandemic Based on Convolutional Neural Networks and Particle Swarm Optimization Algorithm

Walid Hamdy, Ismail Elansary, Ashraf Darwish, and Aboul Ella Hassanien

Abstract With the daily rapid growth in the number of newly confirmed and suspected COVID-19 cases, COVID-19 extremely threatens public health, countries' economic, social life, and international relations around the world. There are different medical methods to detect and diagnose this disease such as viral nucleic acid screening by using specimens of the lower respiratory tract. However, the availability of sufficient laboratory screening in the infested counties represents a critical challenge especially with the fast-spreading of COVID-19. Therefore, alternative diagnostic procedures that depend on Artificial Intelligence (AI) techniques are required in the meantime to fight against this epidemic. This paper focuses on using chest CT for diagnosis of COVID-19, as an alternative or assistive method to the reverse-transcription polymerase chain reaction (RT-PCR) tests. Motivated by this, this paper introduces a new model based on deep learning for detecting patients infected with COVID-19 using chest CT. In this paper, a new proposed model for diagnosis of COVID-19 based on using Convolutional Neural Networks (CNN) and Particle Swarm Optimization (PSO) algorithm to classify the CT chest images of patients into infected or not infected. In this paper, the network hyper-parameters in the CNN are optimized by using the PSO algorithm to eliminate the requirement of manual search and enhance the network performance. The used chest radiography dataset in this paper is described which leveraged to train COVID-Net and includes include more 16,500 chest radiography images across more 13,500 patient cases

W. Hamdy (✉) · I. Elansary
Faculty of Science, Port Said University, Cairo, Egypt
e-mail: walid.hamdy@ckes.cu.edu.eg

A. Darwish
Faculty of Science, Helwan University, Cairo, Egypt

A. E. Hassanien
Faculty of Computers and Artificial Intelligence, Cairo University, Giza, Egypt

W. Hamdy · I. Elansary · A. Darwish · A. E. Hassanien
Scientific Research Group in Egypt (SRGE), Cairo, Egypt

A. E. Hassanien and A. Darwish (eds.), *Digital Transformation and Emerging Technologies for Fighting COVID-19 Pandemic: Innovative Approaches*, Studies in Systems, Decision and Control 322, https://doi.org/10.1007/978-3-030-63307-3_3

from two open access data repositories. The experimental results of this work exhibited that the suggested system accuracy ratio of 98.04% is competitive to the other models.

Keywords COVID-19 · Coronavirus · COVID-19 classification · Convolutional neural networks · Particle swarm optimization algorithm

1 Introduction

At the beginning of April 2020, more than 1000000 confirmed cases of COVID 19 and more than 50000 deaths are documented globally as humanity is currently witnessing the COVID 19 pandemic. Isolation and social distance are an annoying and unpractical solution in fighting COVID-19. Unfortunately, if it works, the COVID-19 vaccine will not be available before at least 18 months. Besides, COVID-19 epidemics will mostly mutate to a more violent type [1].

It has become very hard and critical to handle the huge amount of data resulting from the rapid rising in the COVID-19 confirmed and suspected cases. To the date (Mar 30th, 2020), there have been 769,826 confirmed cases and 36,934 deaths all around the world [2] and the rate is rapidly increasing. International collaboration based on the international data in the form of images about COVID-19 cases can help the countries of the world which have not yet experienced outbreaks to diagnose this virus by analyzing the CT images of patients. A broad range of means of COVID-19 medical diagnosis can be conducted to detect the confirmed cases of COVID-19.

Machine learning and deep learning play an important role in processing the data and convert it into knowledge and decisions [3, 4]. The outbreak of the latest COVID-19, detected in late 2019, needs specific attention due to its potential universal risk. Different Machine Learning Algorithms and deep learning-based AI methods are used for data analysis and decision-making processes in addition to the clinical practice and medical therapies. That means AI-based systems help to detect COVID-19 outbreaks and predict the existence of its global spread as well. The basic theory of AI-based devices is that they need sufficient training data (for all possible cases) [5].

In recent years, deep learning technologies have been used in various aspects of applications such as the analysis of images. The commonest architecture and mostly used in deep learning are the CNN as indicated in [6]. CNN is trained in a supervised way with learning discriminatory features. CNN exploits a few artificial neurons number compared to other traditional neural networks which made it suitable to utilize with the recognition and processing of images. On the other side, CNN requires a large sample number for the training stage. Also, CNN has hyper-parameters and a huge diversity of specific architectures which is counted as expensive and challenging to recognize these optimal hyper-parameters manually [7]. They are sensitive to the setting of these hyper-parameters, which have a significant impact on the performance of CNN. Besides, the hyper-parameters need to be adjusted for each dataset because the hyper-parameters vary from one dataset to another. Selecting the appropriate values for the hyper-parameters of a certain dataset depends on trial and error because there is no mathematical formula which means that the hyper-parameters are tuned manually. The selection of hyper-parameters values requires specialized data, which forces the non-specialists to utilize a grid search or a random search to find the best values of the hyperparameters that obtain better performance of the CNN.

This paper presents a new optimized model to classify COVID-19 CT images based on utilizing CNN architecture with the PSO algorithm to overcome the hyper-parameters troubles in CNN and to optimize its parameters. The optimization methods in this optimized model depend on eliminating the requirement of a manual search for optimal hyper-parameters for the images classification of COVID-19 into four different categories.

The rest of the paper is organized as follow, Sect. 2 presents the recent related work on COVID-19 based machine and deep learning techniques. Section 3 describes the COVID-19 epidemic, CNN, and PSO in detail. The new proposed in this paper is presented in Sect. 4. Section 5 concludes this paper and presents future work.

2 Related Work

Chest CT may be counted as an essential tool for the current COVID-19 diagnosis in epidemic areas due to it has a high sensitivity to detect COVID-19. U-Net ++ neural network model has been developed for processing over 6000 CT scans to classify the cases into COVID-19 infected or not infected [8]. The model produced 93.55% specificity and 100% sensitivity. Besides, the model presented negative-positive value (NPV) of 100%, a positive value (PPV) of 84.62%, and total accuracy of 95.24%. The model makes it much easier for radiologists by reducing the reading time of CT scans by 65%. On the other hand, the 3-category classification model is designed to distinguish the COVID-19 cases where both Song et al. (2020) and Xu et al. (2020) have applied feature extraction using Feature Pyramid Network and several fully-connected layers for cases classifications [9, 10]. The feature extraction using the Feature Pyramid Network model achieves an acceptable total accuracy of 86.7% using CT scans. The fully-connected layers model can accurately detect the COVID-19 cases with AUC of 0.99 and a sensitivity of 0.93.

A U-Net [11] architecture and a Resnet-50 [12] have been integrated to reveal lung abnormalities. This hybrid model combines and develops 2D and 3D deep learning, algorithms them with a clinical study to classify the cases into COVID-19 vs non-COVID-19 cases per chest CT scan. Based on 157 cases (China and U.S), the results were 0.996 AUC, 98.2%, and 92.2% sensitivity and specificity respectively. The system can quantitively identify measures and visualize the tiny opacities regarding their volume and diameter [13]. Using V-net architecture [14], a framework has been adapted to segment new CT scans so that the time needed to analyze the new images decreased from 30 min to 5 min [15]. This system decreases the segmentation time to four minutes using only 3 iterations and the similarity coefficient was 91.6%. The ongoing trend is correlating the clinical studies with the radiological findings. The accuracy of AI-based models improves decision-making processing [16]. Therefore,

AI-based image analysis attains high detection of COVID-19 as regard accuracy and time.

A brief report on the epidemiology and imagery of COVID-19 syndromes was provided in [17]. The authors described the main characteristics and varieties in chest radiography (Chest CT) for SARS-infected and MERS COVID-19 (which was first reported in Saudi Arabia in 2012), as well as for COVID-19 (a recent study in The Lancet). Overall, the radiological findings are highly nonspecific and may overlap with the symptoms of other infections such as influenza H1N1, pneumonitis cytomegalovirus, or atypical pneumonitis.

Based on the claim that there is a huge similarity between a novel COVID-19 and pulmonary syndromes of the COVID-19 family the authors in [18] reviewed the radiological images obtained during the period of the acute respiratory syndrome (SARS) and the Middle East respiratory syndrome (MERS) outbreaks. This review aims to understand the role of the radiologist in the outbreak of COVID-19 as well as helping to predict the imaging studies and manifestations of COVID-19. Although SARS and MERS have high similarities in clinical pictures, imaging studies, and virulence factors, they have several serious variations. The authors illustrate the similarities and differences between them in imaging features. Also, the authors study the Long-Term Follow-Up Imaging results in both SARS and MERS.

In [19], the authors focus on the major role of chest imaging whether in diagnosing the stage of the disease or in following up the patients. Similar to other COVID-19 chest imaging, COVID-19 infected patient shows widespread asymmetrical lung opacities. For 40 of 41 COVID-19 patients, the first report described that on the initial chest CT there is bilateral lung involvement. Also, the report described for a patient who is in the ICU shows a consolidative pattern. While for patients who were not in the ICU there is a predominantly ground-glass appearance [18].

3 Basics and Background

1. ***COVID-19 Pandemic overview***

Infected cases data of COVID-19 can be detected using several medical means. These means of COVID-19 medical diagnosis include clinical characteristics and radiologist's diagnosis. Clinical characteristics involve human temperature monitoring and reverse transcription-polymerase chain reaction (RT-PCR) [20]. Based on the human body temperature, patients' body temperatures range within ranges of 36·5–38·8°C [21]. Accordingly, available thermal sensors or thermal images can be collected as instantaneous data to investigate the potential COVID-19 [22]. Reverse transcription-polymerase chain reaction (RT-PCR) is a laboratory method that is used for testing COVID-19 using a blood sample of the case. Using real-time RT-PCR test confirms the actual number of confirmed cases globally [17, 23]. Moreover, point-of-care testing, Immunoglobulin M (IgM), and detection of antibodies can be applied to detecting infected cases and other related COVID-19 works are presented in [24–27].

The main test to detect the COVID-19 confirmed cases is RT-PCR. However, this approach needs analysis time and specimen collection. Therefore, medical imaging is important for screening COVID-19 cases [28]. Two days after the onset of flu-like symptoms from COVID-19, about 50% of COVID-19 confirmed cases have normal CT. Based on chest CT scans, pneumonia patients due to COVID-19 have lung abnormalities. However, COVID-19 RT-PCR may be initially negative due to the low sensitivity of COVID-19 RT-PCR. About 50–75% of COVID-19 patients have lung abnormalities such as multifocal ground-glass opacities or peripheral focal based on the early COVID-19 infection. During the COVID-19 progresses, significant consolidation and paving dominate the finding of the CT scans and reach its peak within 9–13 days.

Radiologist's diagnosis involves computed tomography (CT) scans, chest X-ray (CXR) radiographs [29]. COVID-19 symptoms can be effectively detected using CT or X-ray images. However, CT scans and X-ray images are time-consuming and exhaustive even for expert radiologists. Computer vision and deep learning algorithms like CNN can effectively help radiologists for detecting COVID-19 confirmed cases. Based on the chest CT scans during recovery, radiologists can detect the (COVID-19) pneumonia and the stage of patient recovery or deterioration. Automated AI models can accurately provide early detection for the diagnosis of the cases of COVID-19 by detecting the early lung damage signs in the images. The inception of artificial neural networks (ANNs) has been applied for binary classification for infected with COVID-19 or health persons using 1,119 CT images [30]. The modified Inception transfer-learning model produced an accuracy of 89.5% with 0.88 and 0.87 for specificity and sensitivity, respectively using the internal validation dataset. On the other hand, the model produced an accuracy of 79.3% with 0.83 and 0.67for specificity and sensitivity, respectively using an external validation dataset. Therefore, the findings demonstrate the reliability of deep learning to find COVID-19 radiological features based on terms of time and accuracy.

2. ***Convolutional Neural Networks***

Lately, CNN has significant development in different fields of applications to solve problems associated with image categorization [31]. This is due to that CNN is one of the best methods for extracting features from images of significant tasks. CNN structure, which is built from scratch, includes two main categories: convolutional and classifier.

3.1 Convolutional Part

In the Convolutional part, images features extraction is done utilizing a group of layers that are listed below.

(i) The convolutional layer is an essential component in the architecture of CNN. It contains a set of filters used to detect the presence of certain patterns or features useful for image characterization, such as textures and edges called feature maps.

(ii) Activation function **layers** are used to apply a non-linear transformation to the output x of the previous layer. There are different types of activation functions as the rectified linear unit (ReLU) which is one of the most commonly utilized activation methods as it does not suffer from exploding or vanishing problems and it provides rapid computations. ReLU is mathematically represented as:

$$\mathrm{F(x)} = \max(0, \mathrm{x}). \tag{1}$$

Gradient w.r.t. the input

$$\frac{\partial F(x)}{\partial x} = \begin{cases} 0\ if\ x \leq 0 \\ 1\ if\ x > 0 \end{cases} \tag{2}$$

(iii) Batch normalization layers (BN) aim to speed up the network training by normalizing every scalar feature x_i in a mini-batch $B = \{x_{1...m}\}$ separately as illustrated in Eq. (3) [18].

$$\widehat{x}_i = \frac{x_i - \mu_B}{\sqrt{\sigma_B^2 + \upsilon}} \tag{3}$$

where υ is a small positive constant which prevents division by zero, μ_B represents the mean of the mini-batch and σ_B^2 is representing the diversity of the mini-batch which are given by

$$\mu_B = \frac{1}{m}\sum_{i=1}^{m} x_i \tag{4}$$

and

$$\sigma_B^2 = \frac{1}{m}\sum_{i=1}^{m} (x_i - \mu_B)^2 \tag{5}$$

Then, for every training mini-batch, the normalized value is scaled and shifted using the scale and shift parameters γ and β as illustrated in Eq. (6)

$$y_i = \gamma \widehat{x}_i + \beta \tag{6}$$

where γ and β are parameters trained along with the original system parameters.

(iv) Pooling layers: these layers aim to down-sample the dimensionality of the feature maps to focus only on the most important features and reduce the number of parameters and computational time in the network.

3.2 *Classifier Part*

In this part, the images are classified depending on the extracted features from the convolutional part using a set of layers that are mentioned below.

(i) Fully connected layers (FC) also known as dense layers, contain many neurons. After the feature extracted from the convolutional part, they are mapped by one or more of the connected layers to the outputs of the network. Each FC layer is followed by an activation function layer, which usually is followed by a BN layer. The last connection has the same number of nodes as the class number.
(ii) Last layer activation function. The function of activation applied to the last layer is different from the other layers, and the sigmoid or softmax function is commonly used. Multi-class classification problems use softmax function to allocate decimal probabilities to every class and the objective class will have a large probability. Moreover, those decimal probabilities must add up to 1.0. Moreover, this additional constraint helps training to converge quickly than it otherwise would. Softmax of the ith output unit is calculated mathematically using the next formula:

$$\hat{y}_i = \frac{e^{x_i}}{\sum_i^N e^{x_i}} \quad \text{for } i = 1 \ldots N \tag{7}$$

where:

x_i = the ith dimension output
N = the number of dimensions that are often equal to the class number.
$\hat{y}_i$ is the probability associated with the ith class. When making a prediction, a sample is assigned to the class that has the largest probability as follows:

$$\hat{y}_i = \max_{\mathrm{i} \in [1.N]} \hat{y}_i \tag{8}$$

Binary-class classification problems use the sigmoid function. Moreover, the sigmoid function takes any range real number and returns the output value which falls between 0, 1. Also, it is mathematically calculated by the following formula:

$$S(x) = \frac{1}{1 + e^{-x}} \tag{9}$$

(iii) Dropout layers are a regulation technique, applied only in the network training to prevent it from overfitting by dropping an input neurons subset as well as all their connections temporarily from the previous FC layer.

3.3 Training a CNN Network

The training process is the process of identifying the right weights of the convolutional and classifier parts to minimize the differences between the predictions of output and the actual labels of the training set.

(i) A loss function also indicated as a cost method. A loss function like cross-entropy (CE) is used to measure the contrast between the true label distribution y and the predicted label distribution $\hat{y}$ from the network. In binary-class classification problems, the CE is mathematically calculated as Eq. (10), but in multi-class classification problems, it is calculated as Eq. (11).

$$L_{\text{cross - entropy}}\left(\hat{y}.y\right) = -y\left(\log\left(\hat{y}\right) + (1 - y)\log\left(1 - \hat{y}\right)\right) \tag{10}$$

$$L_{\text{cross - entropy}}\left(\hat{y}.y\right) = -\sum_{i=0}^{N} y_i \log\left(\hat{y}_i\right) \quad \text{for } i = 1 \ldots N \tag{11}$$

where: N = Number of classes

(ii) The optimization function is used to minimize the loss function, such as in [31]. Although the optimization function is the optimal component of deep learning success, its efficiency depends predominately on the initial learning rate choice and the learning rate schedule (LRS). LRS is a serious problem in the training of deep neural networks. Moreover, LRS is a method that minimizes over time the learning rate which is causing an efficient training operation. Therefore, smaller training updates are utilized with the weights in the training process later. Furthermore, various schedules were presented like EDLR. EDLR could be calculated mathematically as Eq. (12).

$$Lr = Lr_0 * \left(r_{decay}\right)^{\frac{s_{golbal}}{s_{decay}}} \tag{12}$$

where Lr refers to the current rate of learning, Lr_0 refers to the initial rate of learning, r_{decay} is learning decay rate, $r_{decay} \in (0.1)$. s_{golbal} refers to the current learning steps and s_{decay} represents the number of steps per round.

3. **Particle Swarm Optimization Algorithm**

PSO algorithm is one of evolutional algorithm which was firstly introduced in 1995. PSO is widely utilizing in diverse domains.

PSO algorithm consists of various sequential steps. First, initialization which selects at random the particles as searching agents (x) as well as the velocities (v). Secondly, the particles then integrated into the cost method to detect local bests (p_{best}) and global best (g_{best}). Local best is known as the place on which the expense is the smallest one for every particle. Meanwhile, the global best is the place on which the expense is the smallest one between the local bests. Finally, the particles are updated using Eqs. (13, 14).

$$v_{n+1} = v_n + c_1 r_1 (p_{best} - x_n) + c_2 r_2 (g_{best} - x_n) \tag{13}$$

$$x_{n+1} = x_n + v_{n+1} \tag{14}$$

where c_1 and c_2 are constants, r_1 and r_2 are random numbers, and n is iteration.

4 The Proposed COVID-19 Classification Optimized Model

The suggested model is an extremely efficient method to develop the CNN network's performance by integrating two pre-trained CNN networks to do the terminal prediction. Moreover, the hyper-parameters of every model are optimized utilizing the PSO approach to have better performance for every model. The suggested algorithm steps are summarized in four main stages as shown in Fig. 1: (a) data prepare stage, (b) hyper-parameters optimizing stage, (c) learning stage, (d) evaluation stage. Extra specifics for every stage are illustrated in the next sections.

1. **Dataset description**

The dataset which is used to train and evaluate the suggested approach is selected from a total of 16,756 CT images across 13,645 patient cases. To generate the training dataset, we integrated and adjusted two diverse available datasets: (1) COVID-19 image data collection [32] and (2) RSNA Pneumonia Detection Challenge dataset [33]. Samples of the used dataset are given in Fig. 2.

This subset is selected as it contains the most enormous variety of cases with a reasonable image number in every class. Images number in each class is shown in Table 1.

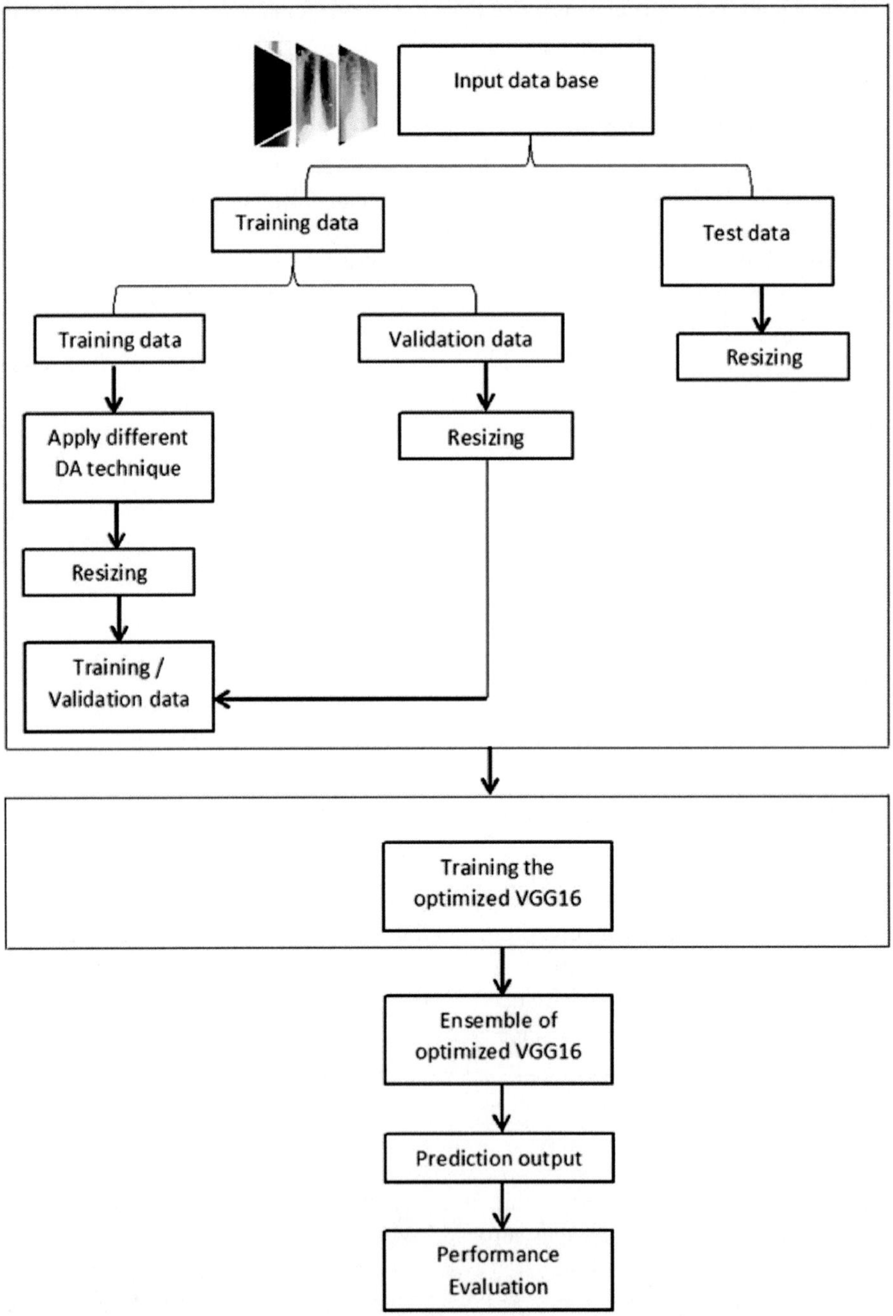

Fig. 1 The proposed approach block diagram, DA = Data augmentation, and HPO Phase = Hyper-parameters optimization phase

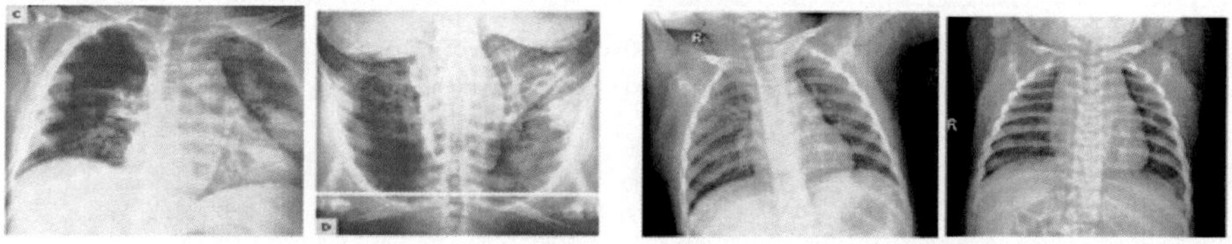

(a) (b)

Fig. 2 Samples of representative images for each of the class of the dataset **a** COVID-19, **b** NORMAL

Table 1 The number of samples in each class

Classes	Number
NORMAL	1500
COVID-19	1050

4.1 Data Preprocessing Phase

A validation group was generated by setting 20% from the set of the training model for monitoring the model's accuracy during the training stage. Table 2 exhibits samples numbers in each class and the total sample number in the sets of validation and training, after splitting the group of training to a validation and training group. Moreover, each class in the set of training has various samples range, indicating that the set of training is Unbalanced and the total performance of the trained network may be affected. Thus, RUS and ROS are used to handle the imbalance of the training set. At the RUS process, some set was deleted at random of the majority classes (COVID-19, healthy) until the sample number meets that of the minority category. ROS is the invert of RUS, where some model was copied at random. The replicated samples were cropped at random before they were gathered to the minor category perfectly method to not lose significant features.

Once the ROS and RUS techniques were executed in the set of training, it became balanced as each class includes a similar range of 500 images. Several data augmentation methods were applied for the models of training like width shift, rotation, shear, zoom, height shift, and horizontal flip to overcome the overfitting and get better generalization. Eventually, before utilizing the training, validation, and test

Table 2 The samples number in each class and the all sets number in the training model and validation model

Category	Number of images in the training sample	Number of images in the validation sample
NORMAL	200	100
COVID-19	200	100
Total	400	200

sets in the next stages, each image was adjusted to 256 × 256 resolution and rescaled by dividing the value of the color of every pixel by 255 for getting values in the range [0, 1].

2. **The flowchart of the proposed algorithm**

The suggested algorithm as shown in Fig. 3 is received the input parameters relative to a certain problem, like the training data to be applied and parameters concerning the CNN structures to be generated, like the maximum layers number particles during initialization. Algorithm 1 shows the general framework for the suggested CNN-PSO. In this algorithm, the global best particle is based on the best blocks found in the swarm by following the PSO algorithm. Therefore, there is no necessity for manually optimizing the parameters of every block. Moreover, the optimization of the parameter does not start over, also good blocks are passing over to the next generation in the form of the global best particle. Currently, only the particle evaluation requires restarting at each iteration, but the proposed algorithm guarantee to keep the good blocks. Figure 3 depicts the flowchart of the suggested algorithm using CNN-PSO to classify the given dataset.

3. **Hyper-parameters optimizing phase**

It has to be mentioned that the implementation of any pre-trained CNN architectures have limitations. Noticeable constraints are that cannot alter most of the CNN hyper-parameters and that there are a few hyper-parameters required to adjust, like the mini-batch, unit's number in every dense layer, size of the learning rate and the rate of dropout. In this paper, the PSO approach was applied to optimize the rate of the dropout layer and the size of the minibatch in the VGG16 function. The particles were encoded to a set of two coordination corresponding to the hyper-parameters that should be improved, as illustrated in Table 3. Moreover, the foremost swarm particles were initialized at random: p1, p2. p1 was initialized with integer values in the range [20, 128], and p2 was initialized with decimal values in the range [0.1, 0.9].

Table 4 is used to set the parameters of the PSO approach for every pre-trained model. The swarm size is set for being 5 particles. The maximum iteration number is set for being 6.

After any iteration for the PSO, the efficiency of the particle's solutions was estimated based on the loss rate acquired of the particles at the validation class after the K number of CNN training epochs. Subsequent doing a lot of experiments, it was clear out that CNN's convergence ability is depressed at the training epochs number is less than three. Moreover, the PSO training stage executed in a long time when the number of training epochs is greater than six. Therefore, the value of K has been set to six, for each PSO iteration, a six of CNN training epochs was proceeded to calculate particle's solutions fitness. After the PSO algorithm finished the suggested hyper-parameters were given in the Gbest (Niter) and listed in Table 5.

After the PSO optimizing the chosen hyper-parameters of VGG16. These pre-trained systems had trained at an excess training sample utilizing the same framework,

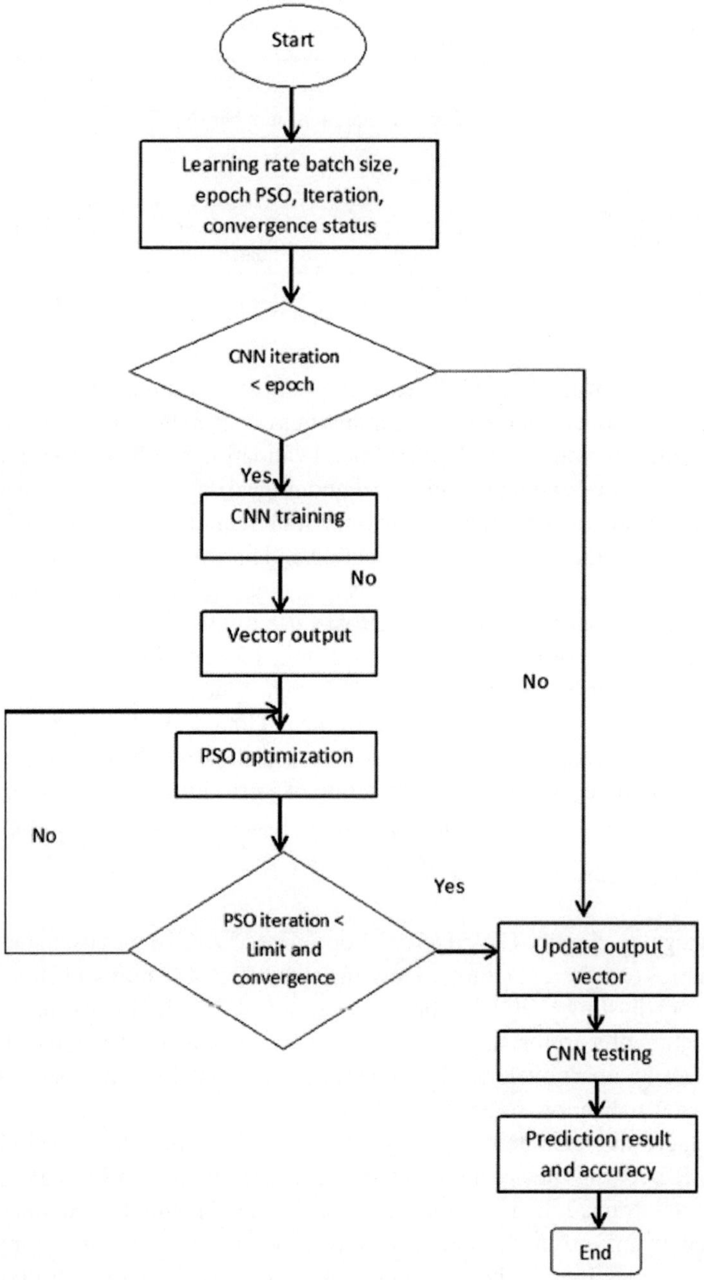

Fig. 3 The flowchart of the suggested algorithm using CNN + PSO

Table 3 Hyper-parameters optimization stage in the proposed method

Particle coordinate	Hyper-parameter
p_1	mini-batch size
p_2	The rate of dropout

Table 4 The parameters values of the PSO algorithm

Parameter	Value
Swarm size	5
Maximum iterations number (N_{iter})	6

Table 5 The values of the hyperparameters that used by the PSO model

The pre-trained sets	The size of the small batch	Dropout number
VGG16	23	0.25

where each one of them predestined at the sample of validation during the training phase to the M epochs number. After numerous experiments, it was achieved that the proposed method was reached their bestead validation results in the 9th recurrences during the features extraction stage and about the 10th recurrences during the fine-tuning stage and it was not their improvement thereafter. So, the value of M has been set to 10 in the stage of the feature extraction and has been set to 15 in the fine-tuning stage. All of the considered pre-trained stage, compiled with the CE in conservation by Eq. (7). Also, an addition optimizer was used with the constant rate of learning = 1e-4 in the stage of the extraction of features and with EDLR in the stage of fine-tuning. The EDLR has been put up by Eq. (8) when the rate of the initial learning was a sample to (1e-4), and the proportions of the decline of the learning proportions the sample to 0.5 and S_{decay} was a sample to five steps. During the stage of training of all pre-trained samples, we used some function of Keras library to economize the better weight that accomplished performance in the loss rate of the validation models.

4. **Learning phase**

At this stage, to classify COVID-19 the optimized VGG16 was used. Depending on the scenarios of transfer learning. Thus, the extraction of features and fine-tuning was applied for adjusting VGG16 to the novel dataset which used in this approach. During the extraction of features, the convolutional part was glaciated, but, the classifier part was replaced with a novel one. The novel classifier contains a dropout stage, a flatten stage, a normalization stage, a batch, and two dense stages also indicated to as joined stages as illustrated in Fig. 4. The first fully joined layer includes neurons set with RELU and the second one contains four units with a softmax method. The moment that trained the novel classifier for the epochs number, the fine-tuning was happened by unfreezing the latter two convolutional part stages and re-training those by the novel classifier as shown in Fig. 3. After pre-trained models training finished, they integrated for doing eventual prediction by utilizing the AE techniques where averaging their softmax category.

5. **Evaluation phase**

During this stage, the classification issues usually utilize five evaluation indicators, matrix confusion, accuracy, precision, F1-score, and recall was applied to assess

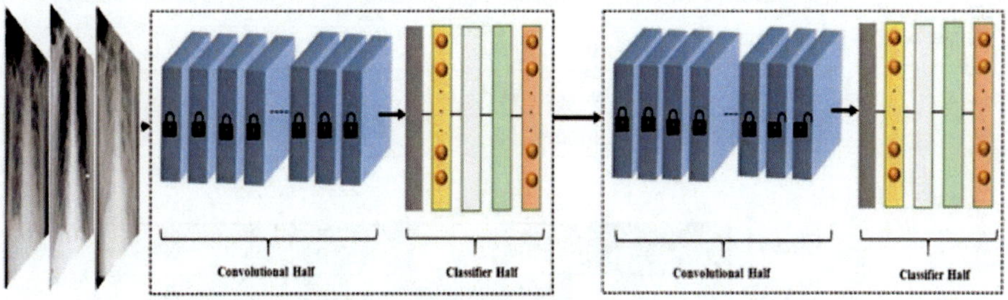

Fig. 4 Illustration of the transfer learning methods which is used for every VGG16

the prediction power of the suggested algorithm. Accuracy is the ratio of correct predictions number to all predictions done, which is given as a proportion and it is measured as Eq. (15). Precision calculates the model's capacity for predicting values correctly for a specific class and is measured as Eq. (16). Recall calculates favorable patterns part is classified correctly and is calculated as Eq. (17). The F1-score is the weighted precision average and recall. For measuring the comprehensive performance of whole evaluation techniques utilized excluding the matrix confusion, the micro and macro average was applied. The matrix confusion is a table that is used to illustrate the classification model performance at the test group for which the valid values are recognized by the chart the predicted result through the actual result.

$$\text{Accuracy} = \frac{\text{Correct prediction number}}{\text{Total prediction number}} \tag{15}$$

$$\text{Precision} = \frac{\text{particular class predicted correctly}}{\text{all classes predictions}} \tag{16}$$

$$\text{Recall} = \frac{\text{Correctly Predicted class}}{\text{All Real classes}} \tag{17}$$

6. Experiment result, discussion, and performance evaluation

In this section, we present the experimental result on CNN- PSO algorithm and also compare the performance with the proposed method by CNN algorithm. In this work, the Keras library was used for building the structure of the suggested algorithm. Moreover, Keras library is a high-level API that is coding using Python for building and developing the architectures of deep learning and able to run on top of the libraries of the numerical computational like TensorFlow. We apply some methods of Keras library which is used to implement the methods for resizing and rescaling the samples.

Every pre-trained stage and its ensemble was assessed at the test group using the five pointers. From Table 6, it is found that the accuracy attained using the optimized VGG16 is 97.17% and by their ensemble is 98.04%. It could be noticed the AE stage produced better accuracy than the individual pre-trained stage.

Table 6 Pre-trained model accuracy

Model	Accuracy (%)
Optimized VGG16	97.17
AE model	98.04

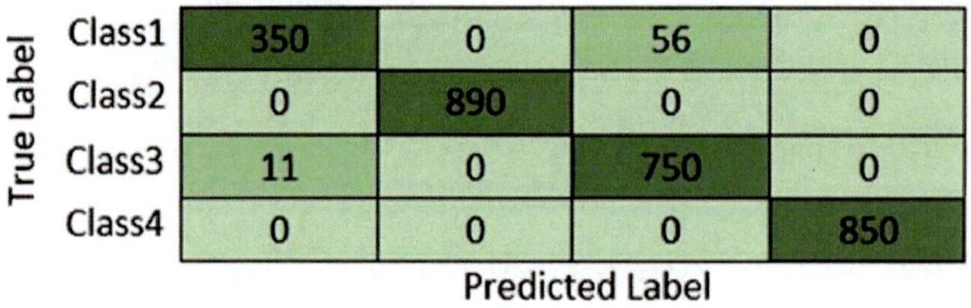

Fig. 5 The proposed model confusion matrix

Figure 5 shows the confusion matrix of the proposed model. From that Figure, we can deduce the accurately classified image is demonstrated through the major diagonal elements of the confusion matrix. From the matrix confusion from the optimized, this is shown that the model accurately classified 900 objects from the test group but it misclassified 12 objects. Our model has achieved better results than other individual models.

Table 7 reports the performance of our approaches against other models. We can see that CNN + PSO improves the performance over models CNN, which demonstrates the effectiveness of the CNN + PSO model. Compared with other models, CNN + PSO gains the best performance in 5 tasks and show very competitive results on the others.

In Fig. 6, we illustrate the curves of the best accuracy model found by CNN + PSO on the Convex dataset for 200 epochs comparative by CNN model. We can see the accuracy of our model raised over the CNN model which means that the proposed CNN + PSO is truly adequate for finding good models for a given image classification dataset

Table 7 The results obtained from the comparative models between CNN model and CNN + PSO model

Epoch number	CNN + PSO	CNN
Epoch 1	98.04	94.59
Epoch 2	97.6	91.89
Epoch 3	97.6	92.89
Epoch 4	97.17	94.59
Epoch 5	98.01	93.66

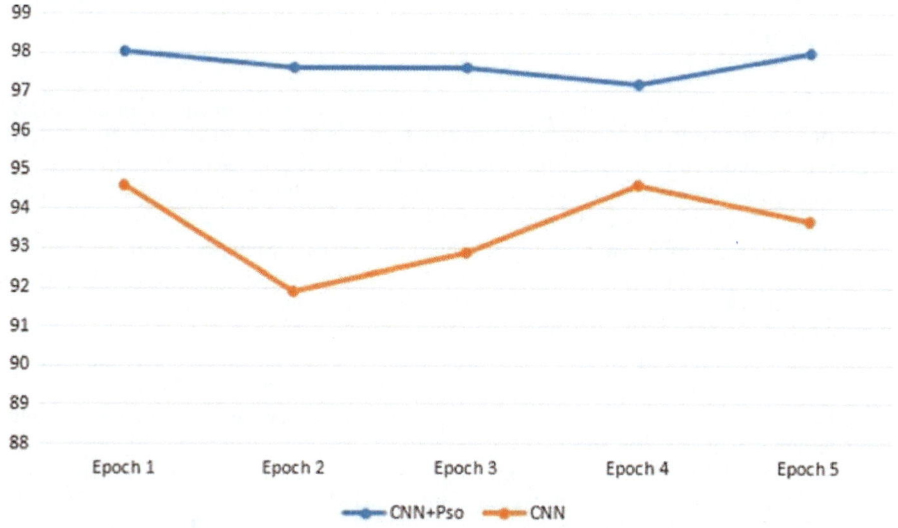

Fig. 6 accuracy curves for comparative between the CNN model and CNN + PSO model

5 Conclusion and Future Work

The outbreak of the 2019 COVID-19 has caused around 1 million confirmed cases and thousands of deaths according to WHO statistics. Therefore, researchers have been developing different models based on machine and deep learning techniques to classify the medical CT images of patients to be infected or not infected. This is an early prediction and diagnosis of COVID 19 cases can contribute to its containment before spreading. CNN is considered as one of the most important deep learning methods that can deal with medial images. Therefore, in this paper, CNN has been used. However, CNN suffers from the hyperparameters problem. Moreover, the PSO algorithm is used to optimize the CNN parameters to enhance the performance of the proposed model. The used dataset in this paper used to train and evaluate the proposed model is selected from a total of 16,756 CT images across 13,645 patient cases. For the training dataset, an integrated and adjusted two diverse available datasets have been used for the implementation; COVID-19 image data collection and RSNA Pneumonia Detection Challenge dataset The proposed model in this paper can classify the CT images of patients with a high level of accuracy 98.04% which makes this model is competitive to the other proposed models based on some accuracy, precision and recall measures.

To extend and improve the proposed approach in this paper in the future, this work in the future, much training of the used dataset have to be gathered from various sources and geographical areas. Also, new optimization algorithms can be integrated with CNN to improve the overall performance of the proposed model.

References

1. Elmousalami, H.H., Hassanien, A.E.: Day level forecasting for coronavirus disease (COVID-19) spread: analysis, modeling, and recommendations. arXiv preprint arXiv:2003.07778 (2020)
2. WHO: Coronavirus disease 2019 (COVID-19) Situation Report–66 (2020). https://www.who.int/docs/default-source/coronaviruse/situation-reports/20200330-sitrep-70-covid-19.pdf?sfvrsn=7e0fe3f8_2. Accessed March 31th 2020
3. Hutter, L.K., Vanschoren, J.: Automated Machine Learning. Springer (2019)
4. Khalifa, N.M., Taha, M.H., Ella Hassanien, A., Elghamrawy, S.: Detection of coronavirus (COVID-19) associated pneumonia based on generative adversarial networks and a fine-tuned deep transfer learning model using chest X-ray dataset. arXiv preprint arXiv:2004.01184
5. Vaishy, R., Javaid, M., Khan, I.H., Haleemb, A.: Artificial Intelligence (AI) applications for COVID-19 pandemic. Diab. Metab. Syndr. Clin. Res. Rev. **14**(4), 337–339 (2020)
6. LeCun, Y., Bottou, L., Bengio, Y., Haffner, P.: Gradient-based learning applied to document recognition. Proc. IEEE **86**(11), 2278–2324 (1998)
7. Srivastava, Hinton, G.E., Krizhevsky, A., Sutskever, I., Salakhutdinov, R.: Dropout: a simple way to prevent neural networks from overfitting. J. Mach. Learn. Res. **15**(1), 1929–1958 (2014)
8. Allam, Z., Jones, D.S.: On the Coronavirus (COVID-19) outbreak and the smart city network: universal data sharing standards coupled with Artificial Intelligence (AI) to benefit urban health monitoring and management. Healthcare **8**(1) (2020). Multidisciplinary Digital Publishing Institute
9. Chen, J., Wu, L., Zhang, J., Zhang, L., Gong, D., Zhao, Y., Hu, S., Wang, Y., Hu, X., Zheng, B. et al.: Deep learning-based model for detecting 2019 novel coronavirus pneumonia on high-resolution computed tomography: a prospective study. medRxiv preprint medRxiv:2020.02.25.20021568 (2020b)
10. Xu, X., Jiang, X., Ma, C., Du, P., Li, X., Lv, S., Yu, L., Chen, Y., Su, J., Lang, G. et al.: Deep learning system to screen coronavirus disease 2019 pneumonia. arXiv preprint arXiv:2002.09334 (2020)
11. Song, Y., Zheng, S., Li, L., Zhang, X., Zhang, X., Huang, Z., Chen, J., Zhao, H., Jie, Y., Wang, R., et al.: Deep learning enables accurate diagnosis of novel coronavirus (COVID-19) with CT images. medRxiv preprint medRxiv:2020.02.23.20026930 (2020)
12. Ronneberger, O., Fischer, P., Brox, T.: U-net: convolutional networks for biomedical image segmentation. In: International Conference on Medical Image Computing and Computer-Assisted Intervention, pp. 234–241. Springer (2015)
13. He, K., Zhang, X., Ren, S., Sun, J.: Deep residual learning for image recognition. In: Proceedings of the IEEE Conference on Computer Vision and Pattern Recognition, pp. 770–778 (2016)
14. Gozes, O., Frid-Adar, M., Greenspan, H., Browning, P.D., Zhang, H., Ji, W., Bernheim, A., Siegel, E.: Rapid AI development cycle for the coronavirus (COVID-19) pandemic: initial results for automated detection and patient monitoring using deep learning CT image analysis. arXiv preprint arXiv:2003.05037 (2020)
15. Milletari, F., Navab, N., Ahmadi, S.-A.: V-net: fully convolutional neural networks for volumetric medical image segmentation. In: 2016 Fourth International Conference on 3D Vision (3DV), pp. 565–571. IEEE (2016)
16. Shan, F., Gao, Y., Wang, J., Shi, W., Shi, N., Han, M., Xue, Z., Shi, Y.: Lung infection quantification of COVID-19 in CT images with deep learning. arXiv preprint arXiv:2003.04655 (2020)
17. Chen, H. et al.: Clinical characteristics and intrauterine vertical transmission potential of COVID-19 infection in nine pregnant women: a retrospective review of medical records. The Lancet **395**(10226), 809–815 (2020)
18. Hosseiny, M. et al.: Radiology rerspective of Coronavirus disease 2019 (COVID-19): lessons from severe acute respiratory syndrome and middle east respiratory syndrome. AJR Am. J. Roentgenol. 1–5 (2020)

19. Kooraki, S. et al.: Coronavirus (COVID-19) outbreak: what the department of radiology should know. J. Am. Coll. Radiol. JACR (2020)
20. Wang, D., Hu, B., Hu, C. et al.: Clinical characteristics of 138 hospitalized patients with 2019 novel coronavirus-infected pneumonia in Wuhan, China. JAMA 2020 Feb 7 [Epub ahead of print] [Google Scholar]
21. Bao, C., Liu, X., Zhang, H., Li, Y., Liu, J.: COVID-19 computed tomography findings: a systematic review and meta-analysis. J. Am. Coll. Radiol. (2020). https://doi.org/10.1016/j.jacr.2020.03.006
22. Novel, Coronavirus Pneumonia Emergency Response Epidemiology: The epidemiological characteristics of an outbreak of 2019 novel coronavirus diseases (COVID-19) in China. Zhonghua liu xing bing xue za zhi = Zhonghua liuxingbingxue zazhi **41**(2), 145 (2020)
23. Lim, J., et al.: Case of the index patient who caused tertiary transmission of COVID-19 infection in Korea: the application of lopinavir/ritonavir for the treatment of COVID-19 infected pneumonia monitored by quantitative RT-PCR. J. Korean Med. Sci. **35**(6) (2020)
24. Fang, Y., et al.: Sensitivity of chest CT for COVID-19: comparison to RT-PCR. Radiology 200432 (2020)
25. Lan, L., et al.: Positive RT-PCR test results in patients recovered from COVID-19. Jama (2020)
26. Li, Z., et al.: Development and clinical application of a rapid IgM-IgG combined antibody test for SARS-CoV-2 infection diagnosis. J. Med. Virol. (2020)
27. Lange, J.M., et al.: Diagnostic value of specific IgM antibodies in primary HIV infection. AIDS (London, England) **2**(1), 31–35 (1988)
28. Kanne, J.P., Little, B.P., Chung, J.H., Elicker, B.M., Ketai, L.H.: Essentials for radiologists on COVID-19: an updated radiology scientific expert panel. Radiology 200527 (2020)
29. Pan, F., et al.: Time course of lung changes on chest CT during recovery from 2019 novel coronavirus (COVID-19) pneumonia. Radiology 200370 (2020)
30. Wang, S., Kang, B., Ma, J., Zeng, X., Xiao, M., Guo, J., Cai, M., Yang, J., Li, Y., Meng, X., et al.: A deep learning algorithm using CT images to screen for coronavirus disease (COVID-19). medRxiv preprint medRxiv:2020.02.14.20023028 (2020a)
31. Darwish, A., Ezzat, D., Ella Hassanien, A.: An optimized model based on convolutional neural networks and orthogonal learning particle swarm optimization algorithm for plant disease diagnosis. J. Swarm Evol. Comput. **52**, 100616 (2020). https://doi.org/10.1016/j.swevo.2019.100616
32. Cohen, J.P., Morrison, P., Dao, L.: COVID-19 image data collection. arXiv:2003.11597 (2020). https://github.com/ieee8023/covid-chestxray-dataset
33. Radiological Society of North America: RSNA pneumonia detection challenge, vol. 2(4). https://www.kaggle.com/c/rsnapneumonia-detection-challenge/data (2019)

Automatic Scoring and Grading of COVID-19 Lung Infection Approach

Kamel. K. Mohammed, Heba M. Afify, Ashraf Darwish, and Aboul Ella Hassanien

Abstract Although the successful detection of COVID-19 from lung computed tomography (CT) image mainly depends on radiologist's experience, specialists occasionally disagree with their judgments. The performance of COVID-19 detection models needs to be improved. According to COVID-19 symptoms and human immune approach response, there are four types of its contagion such as asymptomatic, mild, severe, and recovered. In this chapter, an automatic scoring of COVID-19 lung infection grading approach is presented. The proposed approach is based on a combination of image segmentation techniques and the Particle Swarm Optimization (PSO) algorithm to access accurate evaluation for infection rate. Fuzzy c-means, K-means and thresholding-based segmentation algorithms are used for isolating the chest lung from the CT images. Then, PSO is used with the three segmentation algorithms for clustering the region of interest (ROI) that consists of COVID-19 infected regions in lung CT. Then, scoring the infection rate for each case. Finally, four infection classes related to the obtained infection COVID-19 is determined and classified.

Kamel. K. Mohammed (✉)
Center for Virus Research and Studies, Al-Azhar University, Cairo, Egypt
e-mail: tawfickamel@gmail.com

H. M. Afify
System and Biomedical Engineering Department, Higher Institute of Engineering in El-Shorouk City, Cairo, Egypt
e-mail: hebaaffify@yahoo.com

A. Darwish
Faculty of Science, Helwan Uinversity, Helwan, Egypt

A. E. Hassanien
Faculty of Computers and Artificial Intelligence, Cairo University, Giza, Egypt
e-mail: aboitcairo@gmail.com

Kamel. K. Mohammed · H. M. Afify · A. Darwish · A. E. Hassanien
Scientific Research Group in Egypt (SRGE), Cairo, Egypt

A. E. Hassanien and A. Darwish (eds.), *Digital Transformation and Emerging Technologies for Fighting COVID-19 Pandemic: Innovative Approaches*, Studies in Systems, Decision and Control 322, https://doi.org/10.1007/978-3-030-63307-3_4

Keywords Infection segmentation approach · COVID-19 detection · Lung CT · Particle swarm optimization (PSO) algorithm · Infection rate

1 Introduction

Currently, the novel coronavirus (COVID-19) is considered one of the deadly diseases that spread between humans worldwide. These COVID-19 virus-related deaths are higher than the other types of pneumonia virus: SARS-CoV and MERS-CoV [1]. The infected people gradually increased and they classified to COVID-19 asymptomatic carrier, susceptible, and recovered [2]. Some risk criteria may impact on COVID-19 infection such as age, diabetes, heart disease, lung disease, hypertension, smoking, blood type, obesity, and genetic factors. Moreover, the problem is how to measure the infection/spreading rate for infected people which included COVID asymptomatic carrier and recovered people in case of re-infection probability. Therefore, the information of COVID-19 statistics is still unknown and controversial due to lake of tests and hospitals' incapacity to realize large patients [3].

The premature diagnostic methods of COVID-19 are urgent to guarantee the recovery rate of the infected people [4]. Due to the current epidemic evolution, the real-time polymerase chain reaction (RT-PCR) is not an effective test for some cases [5]. Therefore, the computed tomography (CT) image is more accurate to rapidly screen the infected lung by COVID-19 [6]. Also, the pathological data from the chest CT image is more obvious than that from the chest X-ray image. Hence, research works supported a chest CT image for a good screening of COVID-19 but taking into account radiology experts and a long amount of time. Some researchers investigated the linkage between CT scan and disease severity rate for improving the treatment program [7]. In some cases, the radiologists have reported the decision according to the wrong interpretation of CT images caused by a mutation in the COVID-19 virus [8]. Therefore, the automated computational models for COVID-19 images play an essential role to measure the infection probability according to infected areas and lung segments.

The progress of image medical processing and machine learning technology exploits in COVID-19 disease to solve the challenge for early prediction.

The main target for handling the COVID-19 epidemic is utilized computer-aided diagnosis to reach maximum accuracy with less time for reducing the number of deaths and improving the economy. In COVID-19 beginning, the research works are mostly divided into COVID-19 classification and calculation of lung total severity score (LTSS) [8]. Image classification between COVID-19 against non-COVID-19 focused on execration some features from CT images using the learning merit from previous diagnosis [9].

Recently, deep learning algorithms are developed for the prediction of COVID-19 on CT images with limited accuracy of 86.7% [10]. Narin et al. [11] realized a prediction approach of COVID-19 on CT images with high accuracy of 98%. On the other hand, the calculation of LTSS for patients with COVID-19 is critical due to different levels of infection, providing a boost for subsequent medical treatment. The quantification of the COVID-19 infection rate is applied by an artificial intelligence approach on CT images based on pixel information [12]. The CT image segmentation is developed to separate the infection locations in the lung using deep learning algorithms [13]. The lung CT severity index (CTSI) is designed on CT-COVID-19 patients with a sensitivity of 82.6% [14] and developed to provide a sensitivity of 98.0% [15].

Wang et al. [16] used CT images of COVID-19 to estimate the virus succession according to incubation days and confirmed that 6–11 infection days are achieved higher infection rate compared to the first day in infection. Shi et al. [17] applied the clinical and laboratory information from COVID-19 to determine the severity score during 1–3 weeks. According to viral activity, the dynamic demand for COVID detection is based on the creation of the follow-up process in the progressive stage and early-stage for this disease by calculation of infection rate. The recognition of two traditional epidemiologic models is employed as key for discovering the daily level of new infections and studying the prevention. According to the type of infected people, the SIR model [18] consists of susceptible (S), infected (I) and recovered (R) while SEIR model [19, 20] consists of susceptible (S), exposed (E), infected (I) and recovered (R). However, these traditional models are not appropriate for COVID-19 patients due to social communication and short lead-time.

In this chapter, an automatic scoring of COVID-19 lung infection grading approach based on combination of image segmentation techniques and the Particle Swarm Optimization (PSO) algorithm. Fuzzy c-means, K-means and thresholding-based segmentation algorithms are used for isolating the chest lung from the CT images. Then, optimized these three segmentation techniques using PSO for clustering the region of interest (ROI) that consists of COVID-19 infected regions in lung CT. Scoring the infection rate for each case, and classify the infection into four classes based on the obtained infection COVID-19 rate.

2 Materials and Methods

2.1 CT COVID-19 Image Segmentation Methods

Generally, the function of image segmentation is to avoid poor image contrast, conflicting intensities, and noise disorder [21]. There are some kinds of image segmentation such as clustering (e.g. Fuzzy C-means (FCMs), K-means), and thresholding. The clustering algorithms are difficult to choose the first cluster centers that lead to low segmentation performance [22]. Therefore, the optimization algorithm (e.g. PSO) is used to improve the accuracy of the segmented image by selecting an adequate solution with low computational time.

2.1.1 Fuzzy C-Means Algorithm

This algorithm is commonly used for image iterative segmentation to create the new clusters from image pixels that near to the existing classes. The target of FCM is to discover the ROI by clustering variable numbers without noise [23]. This FCM algorithm is based on three parameters including the objective function, fuzzy membership function, and partition matrix. The reduction of the fuzzy objective function (J_q) is produced by fuzzy partitions and prototypes for numerical data as shown in the following Eqs. (1–3):

$$J_q = \sum_{i=1}^{N} \sum_{j=1}^{C} u_{ii}^{a} p_i - c_j^2 \quad \text{where } 1 \leq q < \infty \tag{1}$$

$$u_{ij} = \frac{1}{\sum_{k=1}^{c} \left(\frac{\| p_i - c_j \|}{\| p_i - c_k \|} \right)^{\frac{2}{q-1}}} \tag{2}$$

$$c_j = \frac{\sum_{i=1}^{N} u_{ij}^{a} p_i}{\sum_{i=1}^{N} u_{ij}^{q}} \tag{3}$$

where q is any real number greater than 1, u_{ij} is the degree of membership of p_i in cluster j, pi is the ith element of the d-dimensional measured data, c_j is the center of the cluster, and normal function is related to the similarity between any measured data and the center c_j. k is the number of iteration steps. The output of FCM consists of various measures of the right cluster [23].

2.1.2 K-Means Algorithm

The K-means segmentation is applied for portioning images into the K cluster [24]. This algorithm is used to locate image pixels to a unique cluster without overlapping by reducing the distance between the pixel (X_j) and cluster center (C_i) in target function (J) as in Eqs. (4) and (5). The disadvantages of this algorithm are invalid for the nonlinear dataset and noisy data.

$$J = \sum_{i-1}^{k} \sum_{X_j \varepsilon C_I} X_j - C_i^2 \tag{4}$$

$$C_i = \frac{1}{n_i} \sum_{j=1}^{n_i} X_j \tag{5}$$

where i = 1, 2, …, K, n_i is the number of pixels in cluster i.

The K-means algorithm is applied by $K = 4$ on the COVID-19 infected lung CT image.

2.1.3 Thresholding Algorithm

This algorithm is a simple technique founded on the separation of image background from the lung by a unique intensity value from image histogram [25]. In this chapter, the multilevel thresholds are ranged from 50 to 230 for improving the image segmentation. The image background created from pixel values that less than or equal to the threshold level while the image foreground created from pixel values that larger than the threshold level.

2.2 Particle Swarm Optimization (PSO) Algorithm

The output of COVID-19 image segmentation that obtained from three methods incorporated with the PSO algorithm. The PSO algorithm is applicable to increase the difference in intensity values between different pixel groups in the segmented image [26]. The proposed solution is namely particle and the relationship between particles is developed by neighbor particles. It is based on particle velocity (v) and particle position (x). This algorithm searches for an excellent position according to all particles and the excellent positions according to each particle in the study problem. This algorithm is described by the following Eqs. (7) and (8):

$$v_{ij}(t+1) = w \cdot v_{ij}(t) + c_1 \cdot rand1_{ij} \cdot (pbest_{ij} - x_{ij}) \\ + c_2 \cdot rand2_{ij} \cdot (gbest_{ij} - x_{ij}) \quad (7)$$

$$x_{ij}(t+1) = x_{ij}(t) + v_{ij}(t+1) \quad (8)$$

where j is called particle index that consists of 1, 2… N, N is the population size, x_i is the position of the ith particle, and v_{ij} represents the jth element of the velocity vector of the ith particle. While *pbest* is called an excellent previous position yielding the best fitness value for the ith particle, and *gbest* is the global excellent particle according to all particles. $rand1_{ij}$ and $rand2_{ij}$ are two random numbers that formed by the range of [0, 1]. The inertia parameter is w while c_1 and c_2 are two learning parameters for organizing the impact of the social and cognitive parts. According to our computation, the PSO algorithm takes $c_1 = 1$, and $c_2 = 2$.

The contribution of the PSO algorithm is to optimize a segmentation problem by attempting iteratively to enhance a proposed solution with respect to a given quality measure. The PSO algorithm is used the image intensity distributions for accurately distinguishing between two clusters including infected region versus non-infected region in COVID-19 images. The PSO algorithm is estimated the particle success during some iterations [27]. The traditional image segmentation algorithms only are not able for recognizing the correct separation between infected regions and non-infected regions, especially for noisy images. A significant benefit of utilizing PSO over conventional clustering algorithms is that multiple proposed solutions can be retained, recombined, and compared concurrently. Consequently, the combined algorithm is presented by the traditional image segmentation algorithms and PSO algorithm for providing the segmentation efficiency on COVID-19 images. In this chapter, we also used the output of PSO for computing the infection rate in segmented COVD-19 images.

3 The Proposed Scoring of COVID-19 Lung Infection Grading Approach

The proposed approach consists of four phases namely COVID-19 segmentation, PSO-based clustering to determine the region of interested (i.e. the infected region), Scoring the infection rate for each case, and determine the level infection. Figure 1 shows the 4 phases of the approach.

3.1 PSO–Based FCM Algorithm

The traditional FCM algorithm could optimize to a local minimum of the objective function and leads to non-optimized segmentation. To avoid this problem, the PSO–FCM algorithm created a significant improvement in image segmentation [28]. Therefore, the PSO–FCM algorithm is presented to optimize the global minimum of the objective function. By minimizing the objective function in Eq. 1, the PSO algorithm is developed for updating the cluster centroids. In this chapter, the PSO–FCM algorithm is used to determine two clusters in COVID-19 images by both FCM and PSO algorithms.

3.2 PSO–Based K-Means Algorithm

The PSO–K-means algorithm [29] is utilized for computing the inertia parameter w in Eq. (7) by the following Eq. (9). In this chapter, the maximum inertia and minimum inertia parameters are 0.9, and 0.4 respectively.

$$w = w_{max} - iter \frac{w_{max} - w_{min}}{iter_{max}} \tag{9}$$

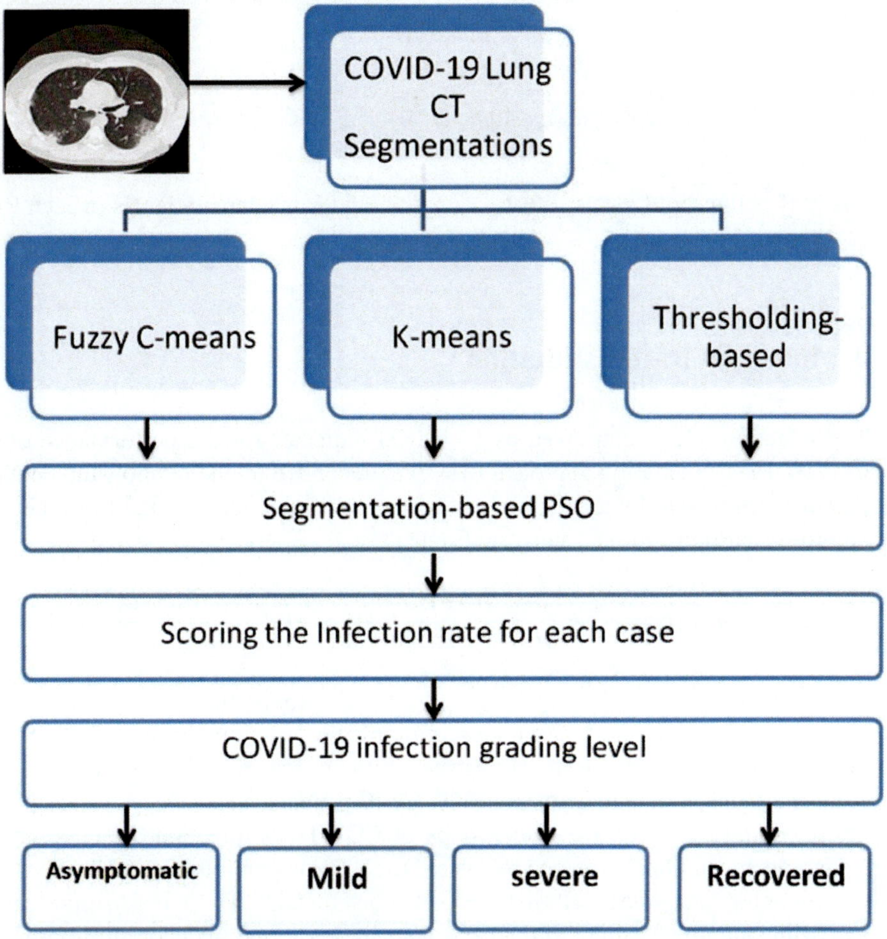

Fig. 1 The proposed approach and its phases

where $iter_{\max}$, and $iter$ are maximum, and current iteration time respectively.

3.3 PSO–Based Thresholding Algorithm

The problem of image thresholding is solved by a combined algorithm using both thresholding and PSO algorithms that based on searching for the thresholds t_j^c that maximized the objective function [30]. In the preliminary step, the particle velocity is adjusted to null and its position is randomly adjusted inside the boundaries of the search space. The search space relied on the number of intensity levels. If the image consists of 8-bits, the particles distributed between 0 and 255. The objective function is based on variance σ_B^2 of the image intensity distributions as in Eq. (10).

$$\begin{aligned} \varphi^c &= \max \sigma_B^2 \left(t_j^c\right) \\ 1 &< t_1^c < \ldots < t_{n-1}^c < L \\ C &= \{R, G, B\} \end{aligned} \tag{10}$$

where t_j^c is threshold levels; j = 1; ...; n − 1, and L is intensity levels in each RGB component.

4 Infection Rate Estimation

The infection rate is employed as a medical indicator for the performance of the COVID-19 segmentation approach [31]. It is depended on the relationship between pixels of segmented lung and pixels of an extracted area that created from the PSO algorithm as shown in the following Eq. (11):

$$Infection\, rate = \frac{Pixel\, density\, of\, the\, infected\, area\, in\, PSO}{pixel\, density\, of\, segmented\, lung} * 100 \tag{11}$$

The infection rate is calculated to help the radiologist for assigning the type of infected patients and the type of supporting medical equipment (e.g. ventilator) or drug dose that essential for managing the treatment schedule. Further, it is used for confirming the clinical diagnosis in COVID-19 patients.

According to WHO, the four groups of COVID-19 infection are classified as asymptomatic, mild, severe, and recovered [32]. The asymptomatic, mild, and severe infections have positive results in the RT-PCR test for COVID-19. It also reported that increasing numbers from COVID-19 patients have asymptomatic or mild symptoms and the asymptomatic patient is very difficult for screening [33]. Therefore, the WHO recommended that any person who has contacted suffered COVID-19 patients is called asymptomatic patients.

In this chapter, we assumed the four infection classes related to the obtained infection percentage of the COVID-19 segmented image as shown in Table 1. Also, the clinical characteristics, treatment, RT-PCR test, and infection rate for each class are explained in Table 1. The asymptomatic patient is a COVID-19 disease carrier without symptoms and achieved a small percentage of infection rate. The mild infection may be achieved infection rate with less than 50% while the severe infection achieved infection rate with higher than 50%. The recovered patient achieved a zero percentage of infection rates. During the treatment, the infection rate may be changed according to a patient reaction towards clinical care. This infection rate is recorded through different infection days to measure the advancement of COVID-19 patients. Also, Lai et al. [34] confirmed that the measuring of disease severity level is useful in solving problems regarding the low capabilities of the hospital. It means that the treatment of mild infection will be outside a hospital environment under medical

Table 1 Classification of infection types according to clinical characteristics, treatment, RT-PCR test, and infection rate

Infection types	Clinical characteristics	Treatment	RT-PCR test	Infection rate %
Asymptomatic	Carrier for COVID-19 virus No clinical symptoms Transmit the virus to others	Without medical assistance Outside the hospital environment	Positive	1–10
Mild	Some common symptoms like Influenza e.g. fever, cough, anorexia, muscle pain, sore throat, dyspnea, nasal congestion, headache Mild pneumonia in lung CT scan Transmit the virus to others	Medical assistance Outside hospital environment	Positive	11–50
Severe	Serious symptoms e.g. respiratory failure Lesions progressed > 50% in lung CT scan Transmit the virus to others	ICU monitoring treatment	Positive	51–100
Recovered	No clinical symptoms	Outside the hospital environment	Negative	0

supervision. The treatment of early-stage in COVID-19 infection supported good results for providing urgent recovery [35].

5 Results and Discussion

5.1 Data Description

The CT image dataset for COVID-19 [36] is available on open access website (www. kaggle.com/plameneduardo/sarscov2-ctscan-dataset). These data collected from different patients infected by COVID-19 in hospitals from Sao Paulo, Brazil. The used samples of COVID CT images are presented in Fig. 2.

The proposed approach performed some steps to extract ROI in the segmented lung for the COVID-19 CT image under the MATLAB 2019a program. The first step is to obtain the segmented lung that extracted from the COVID-19 CT image and followed by the PSO algorithm to appear ROI. The PSO algorithm is performed on the

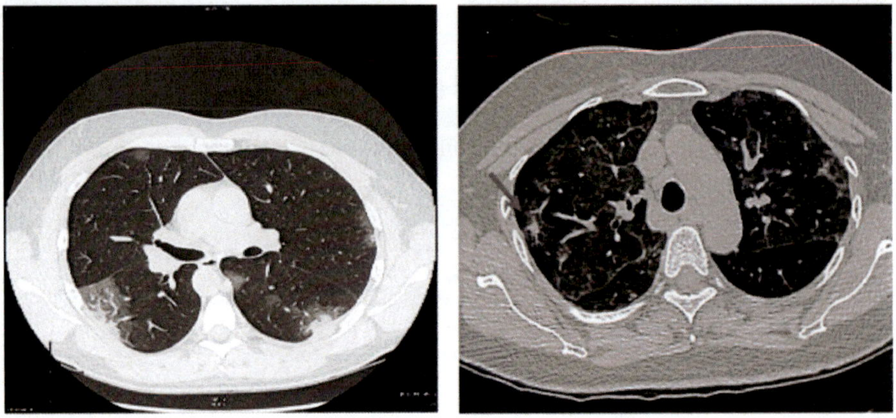

Fig. 2 Samples of COVID-19 CT images

segmented image; therefore it is not dependent on spatial information in the original image. The second step is a pixel-based level evaluation approach by calculating the infection rate.

5.2 *Experimental Results for Segmentation Algorithms*

The segmentation results of three different types of segmentation algorithms after PSO algorithm on COVID-19 CT image are shown in Figs. 3, 4 and 5.

Figure 3 illustrated the applying PSO algorithm on COVID-19 segmented images based FCM algorithm. Figure 4 illustrated the applying PSO algorithm on COVID-19 segmented images based K-means algorithm by $K = 4$. Figure 5 illustrated the applying PSO algorithm on COVID-19 segmented images based thresholding. Figure 6 illustrated the applying lung thresholding segmentation on CT images of COVID-19.

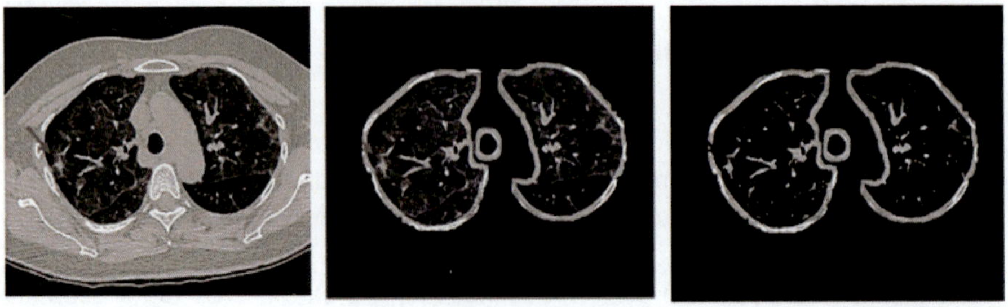

(a) **(b)** **(c)**

Fig. 3 PSO–FCM algorithm on COVID-19 image **a** original image, **b** segmented image based FCM algorithm, and **c** segmented image after PSO algorithm

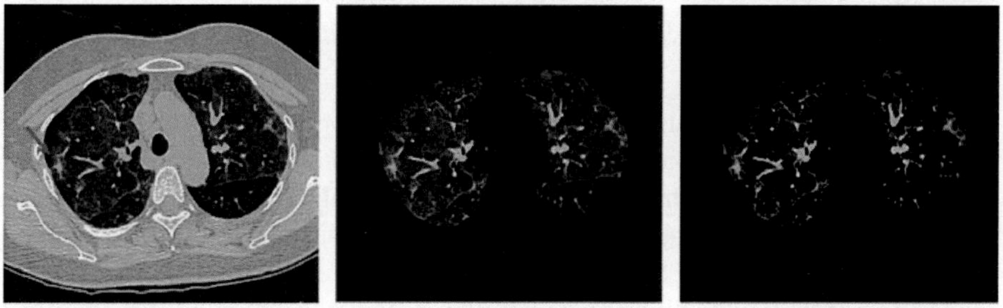

(a) (b) (c)

Fig. 4 PSO–K-means algorithm on COVID-19 image **a** original image, **b** segmented image based k-means, and **c** segmented image after PSO algorithm

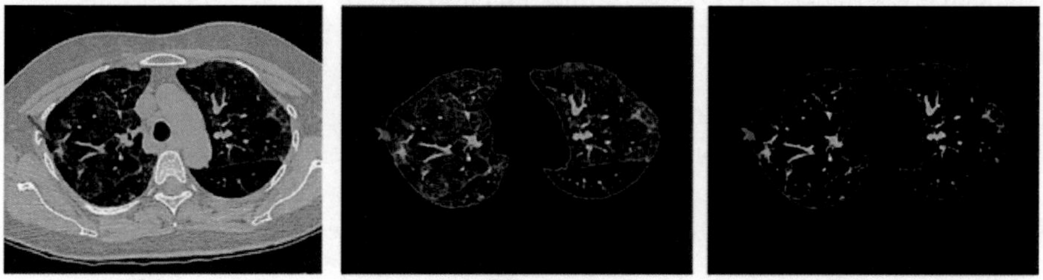

(a) (b) (c)

Fig. 5 PSO–thresholding algorithm on COVID-19 image **a** original image, **b** segmented image based thresholding, **c** segmented image after PSO algorithm

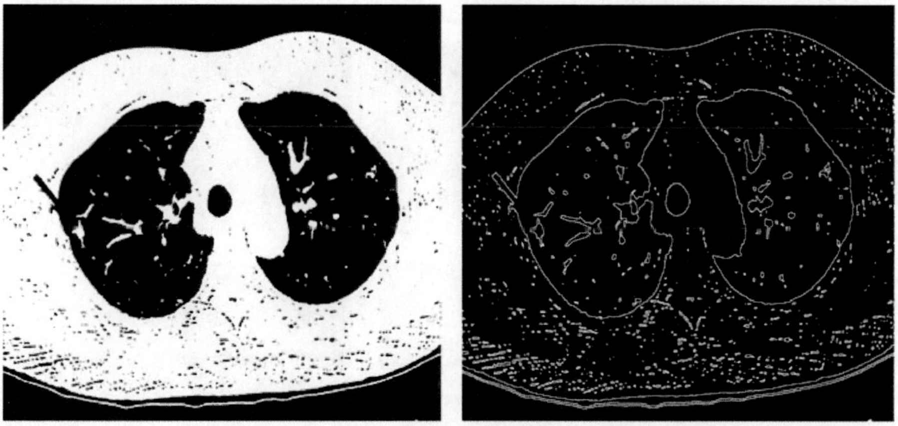

(a) (b)

Fig. 6 Thresholding technique on lung CT image of COVID-19 **a** threshold-based segmentation, **b** morphological operation

5.3 Experimental Results for Infection Rate

The empirical results for the infection segmentation approach on COVID-19 patients for severe and mild cases are displayed in Table 2. From the segmentation approach, the proposed results confirmed that the thresholding algorithm achieved the highest rate of accuracy rather than FCM and k-means algorithms. The integration of multi-level thresholding with PSO is efficient for extracting several ROI in COVID-19 detection. From the visible comparison, the FCM algorithm is not feasible for the segmentation of COVID-19 revealed by lung CT image because of its high thickness boundary in the PSO image. The k-means algorithm is not feasible for the segmentation of COVID-19 revealed by lung CT image due to unclear infected areas. For a selected CT image of COVID-19, the infection rate for FCM, k-means, and thresholding is 87.47%, 30.06%, and 34.42% respectively. The computed infection rate is related to the ratio of pixels between the segmented lung and infection areas. Therefore, the proposed approach is used this infection rate for rapid identification of the infection type among COVID-19 patients. Moreover, the proposed approach explained the medical care priority according to infection rate and image clustering efficiency of lung CT suffered from COVID-19 disease.

Table 2 Comparison between FCM, k-means, and thresholding after PSO algorithm for image infection segmentation approach in COVID-19 patients to assign infection types (severe, mild)

	Original CT-COVID-19	Segmented lung	PSO	Infection rate %	Infection types
Fuzzy C-means	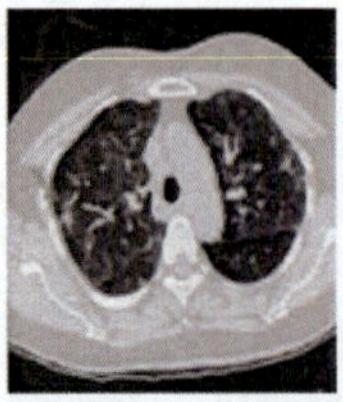	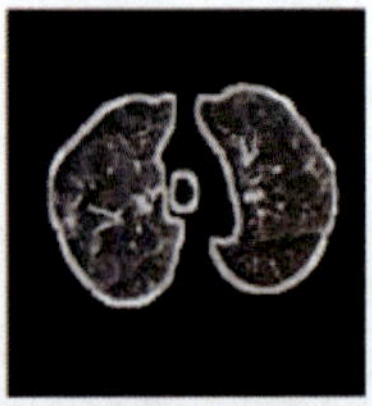	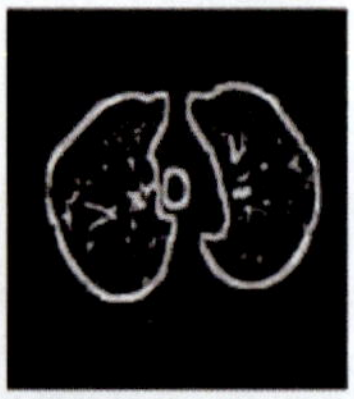	87.47	Server
k-means	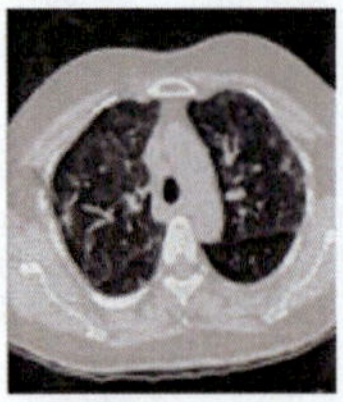	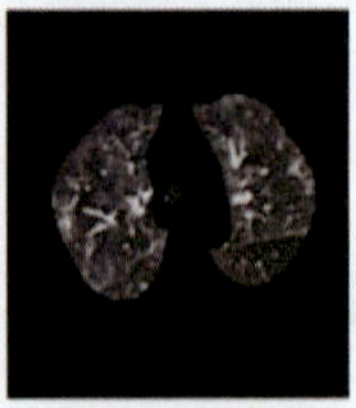	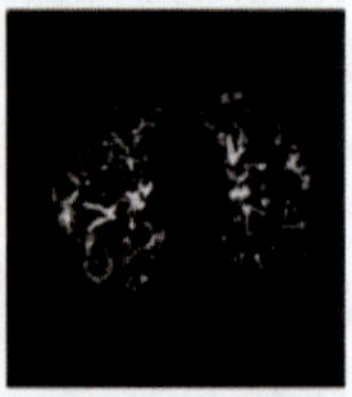	30.06	Mild
Thresholding	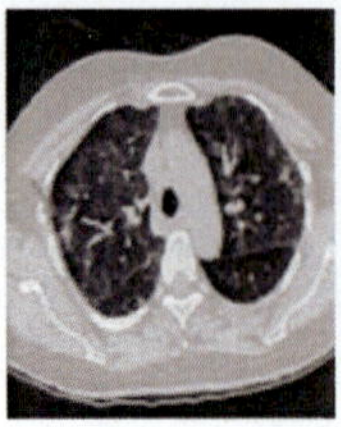	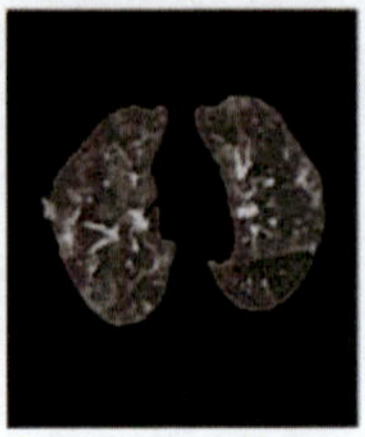	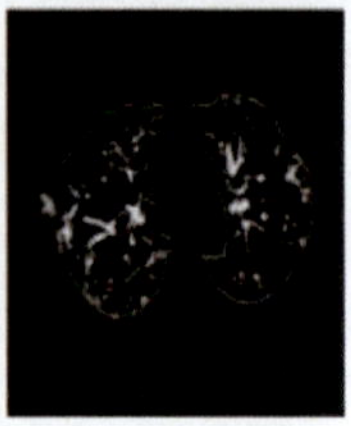	34.42	Mild

Table 3 clarified another example of an image infection segmentation approach in COVID-19 patients for asymptomatic and mild cases. The thresholding algorithm has more effective for refining the COVID-19 segmented image than the other algorithms. Figure 7 showed the difference between the histogram of original CT-image COVID-19 and lung segmented by thresholding.

By applying K-means and PSO algorithm on original CT-image COVID-19, the infection ratio is calculated by $\frac{46,493}{360,887} * 100 = 12.88\%$ as shown in Fig. 8. This infection ratio is referred to a mild infection in the lung CT image.

6 Conclusions

The prediction models of the COVID-19 outbreak suffered from inaccuracy due to a shortage of data and limited research works. In this chapter, the performance of the infection segmentation approach for the lung CT image of COVID-19 is based on integrating between image segmentation and PSO algorithm. This is exploited by calculating the infection rate that referred to better detection accuracy of COVID-19 infected. Therefore, the infection rate is considered a classification tool for infected patients with COVID-19. Finally, this proposed approach can be extended to distinguish normal and abnormal cases as well as different types of lung diseases by computing the infection rate. The relationship between the extracted infection rate from a segmented image and drug dosage will need to seriously study for preventing the risk of COVID-19 disease.

Table 3 Another example of image infection segmentation approach in COVID-19 patients to assign infection types (asymptomatic, mild)

	Original CT-COVID-19	Segmented lung	PSO	Infection rate %	Infection types
Fuzzy C-means	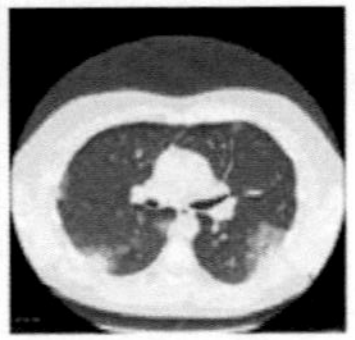	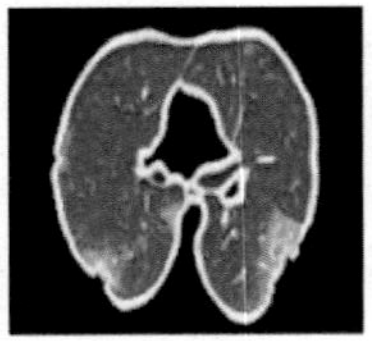	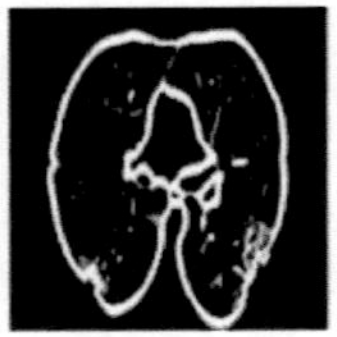	9.07	Asymptomatic
k-means	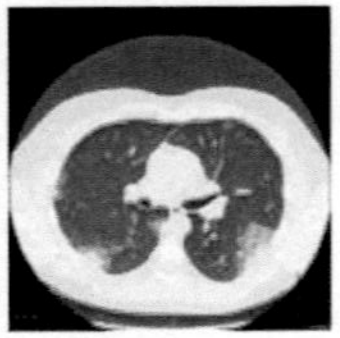	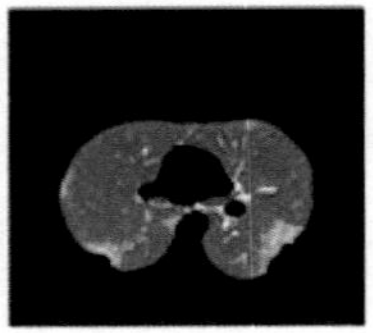		12.88	Mild
Thresholding	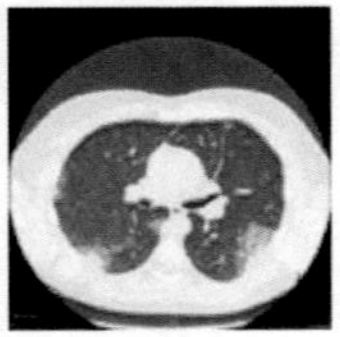	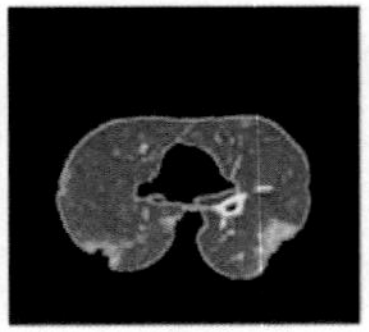	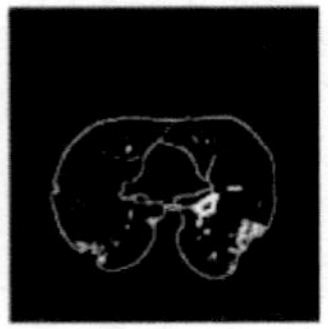	11.98	Mild

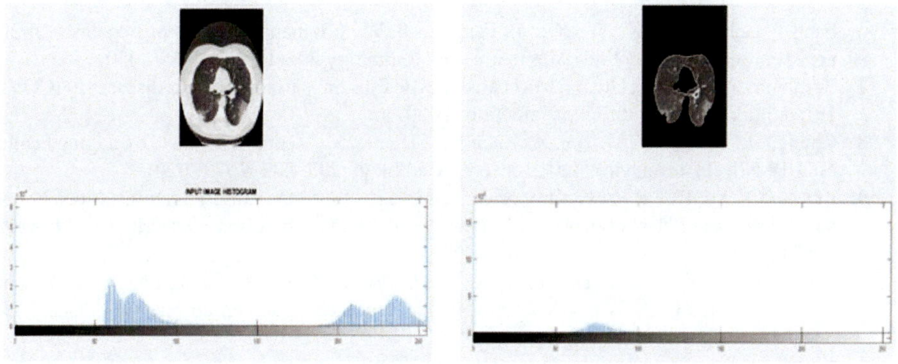

(a) (b)

Fig. 7 Results obtained by thresholding segmentation: **a** histogram of the original image, and **b** histogram of lung segmented

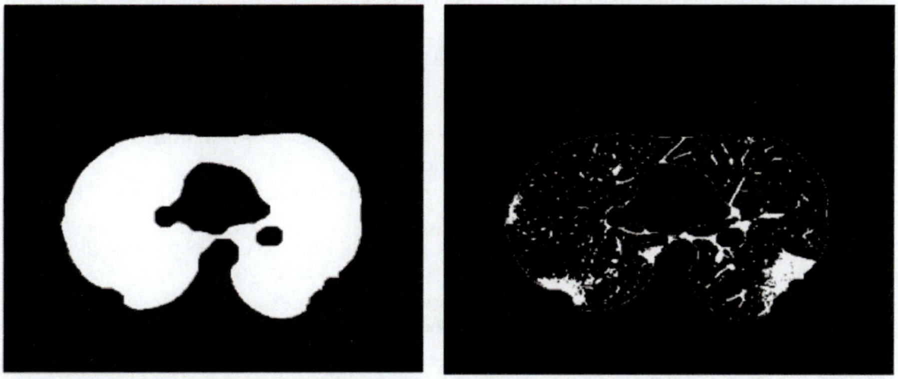

(a) (b)

Fig. 8 Segmentation approach using K-means and PSO algorithm: **a** lung segmented by K-means with a pixel density of 360,887 and **b** infected area by PSO with a pixel density of 46,493

References

1. Perlman, S., Netland, J.: Coronaviruses post-SARS: update on replication and pathogenesis. Nat. Rev. Microbiol. **7**(6), 439–450 (2009)
2. Yin Leung, K., Trapman, P., Britton, T.: Who is the infector? Epidemic models with symptomatic and asymptomatic cases. Math. Biosci. **301**, 190–198 (2018)
3. Li, R., et al.: Substantial undocumented infection facilitates the rapid dissemination of novel coronavirus (SARS-CoV2). Science (2020)
4. Singh, D., Kumar, V., Vaishali, Kaur, M.: Classification of COVID-19 patients from chest CT images using multi-objective differential evolution-based convolutional neural networks. Eur. J. Clin. Microbiol. Infect. Dis. (2020)
5. Ali, T., et al.: Correlation of chest CT and RT-PCR testing in coronavirus disease 2019 (COVID-19) in China: a report of 1014 cases. Radiology (2020)

6. Fang, Y., Zhang, H., Xu, Y., Xie, J., Pang, P., Ji, W.: CT manifestations of two cases of 2019 novel coronavirus (2019-nCoV) pneumonia. Radiology **295**(1), 208–209 (2020)
7. Bernheim, A., Mei, X., Huang, M., et al.: Chest CT findings in coronavirus disease-19 (COVID-19): relationship to duration of infection (2020)
8. Chung, M., Bernheim, A., Mei, X., Zhang, N., Huang, M., Zeng, X., et al.: CT Imaging Features of 2019 Novel Coronavirus (2019-nCoV). Radiology **295**, 202–207 (2020)
9. Liu, K.-C., Xu, P., Lv, W.-F., Qiu, X.-H., Yao, J.-L., Jin-Feng, G.: CT manifestations of coronavirus disease-2019: a retrospective analysis of 73 cases by disease severity. Eur. J. Radiol. 108941 (2020)
10. Xu, X., Jiang, X., Ma, C., Du, .P, Li, X., Lv, S., Yu, L., Chen, Y., Su, J., Lange, Li, Y., Zhao, H., Xu, K., Ruan, L., Wu, W.: Deep learning approach to screen coronavirus disease 2019 pneumonia, pp. 1–29. arXiv preprint arXiv:2002.09334 (2020)
11. Narin, A., Kaya, C., Pamuk: Automatic detection of coronavirus disease (COVID-19) using X-ray images and deep convolutional neural network. arXiv preprint arXiv:2003.10849 (2020)
12. Gozes, O. et al.: Rapid AI development cycle for the coronavirus (COVID-19) pandemic: initial results for automated detection and patient monitoring using deep learning CT image analysis. arXiv preprint arXiv:2003.05037 (2020)
13. Shan, F., Gao, Y., Wang, J., Shi, W., Shi, N., Hana, Xue, Z., Shi, Y.: Lung infection quantification of COVID-19 in CT images with deep learning, pp. 1–19. arXiv preprint arXiv:2003.04655 (2020)
14. Li, K., Fang, Y., Li, W., Pan, C., Qin, P., Zhong, Y., Liu, X., Huang, M., Liao, Y., Li. S.: CT image visual quantitative evaluation and clinical classification of coronavirus disease (COVID-19). Eur. Radiol. (2020)
15. Fang, Y., Zhang, H., Xie, J., Lin, M., Ying, L., Pang, P., Ji, W.: Sensitivity of chest ct for COVID-19: comparison to RT-PCR. Radiology (2020)
16. Wang, Y., et al.: Temporal Changes of CT Findings in 90 Patients with COVID-19 Pneumonia: a longitudinal study. Thorac. Imag. (2020)
17. Shi, H., et al.: Radiological findings from 81 patients with COVID-19 pneumonia in Wuhan, China: a descriptive study. Lancet Infect Dis. **20**(4), 425–434 (2020)
18. Koike, F., Morimoto, N.: Supervised forecasting of the range expansion of novel non-indigenous organisms: alien pest organisms and the 2009 H1N1 flu pandemic. Global Ecol. Biogeogr. **27**, 991–1000 (2018). https://doi.org/10.1111/geb.12754
19. Zha, W.T., Pang, F.R., Zhou, N., Wu, B., Liu, Y., Du, Y.B., Hong, X.Q., Lv, Y.: Research about the optimal strategies for prevention and control of varicella outbreak in a school in a central city of China: based on an SEIR dynamic model. Epidemiol. Infect. (2020)
20. Wang, D., Tan, D., Liu, L.: Particle swarm optimization algorithm: an overview. Soft. Comput. **22**, 387–408 (2018)
21. Pham, D., Xu, C., et al.: A survey of current methods in medical image segmentation. Annu. Rev. Biomed. Eng. **2**, 315–337 (2000)
22. Klir, J.G., Yuan, B.: Fuzzy Sets and Fuzzy Logic, Theory, and Applications. Prentice-Hall Co. (2003)
23. Icer, S.: Automatic segmentation of corpus callosum using Gaussian mixture modeling and fuzzy C means methods. Comput. Methods Programs Biomed. **112**, 38–46 (2013)
24. Siddiqui, F.U., Mat Isa, N.A.: Enhanced moving K-means (EMKM) algorithm for image segmentation. IEEE Trans. Consum. Electron. **57**(2), 833–841 (2011)
25. Liao, P.S., Chen, T.S., Chung, P.C.: A fast algorithm for multilevel thresholding. J. Inform. Sci. Eng. **17**(5), 713–727 (2001)
26. Fernandes-Junior, F.E., Yen, G.G.: Particle swarm optimization of deep neural network architectures for image classification. Swarm Evol. Comput. **49**, 62–74 (2019)
27. Gautam, K., Singhai, R.: Color image segmentation using particle swarm optimization in lab color space. IJEDR **6**(1), 373–377 (2018)
28. Pang, L., Xiao, K., Liang, A., Guan, H.: A improved clustering analysis method based on fuzzy C-means algorithm by adding PSO algorithm. In: International Conference on Hybrid Artificial Intelligence Approachs (HAIS), pp 231–242 (2012)

29. Xiaoqiong, W., Zhang, Y.E.: Image segmentation algorithm based on dynamic particle swarm optimization and K-means clustering. Int. J. Comput. Appl. (2018)
30. Ghamisi, P., Couceiro, M.S., Benediktsson, J.A., Ferreira, N.M.F.: An efficient method for segmentation of images based on fractional calculus and natural selection. Exp. Approach Appl. **39**, 12407–12417 (2012)
31. Rajinikanth, V., Dey, N., Raj, A.N.J., Ella Hassanien, A., Santosh, K.C., Sri Madhava Raja, N.: Harmony-search and Otsu based approach for coronavirus disease (COVID-19) detection using lung CT scan images (2020)
32. Gao, Z., Xu, Y., Sun, C., Wang, X., Guo, Y., Qiu, S., Ma, K.:, A approach atic review of asymptomatic infections with COVID-19. J. Microbiol. Immunol. Infect. (2020)
33. Meng, H., Xiong, R., He, R., Lin, W., Hao, B., Zhang, L., Lu, Z., Shen, X., Fan, T., Jiang, W., Yang, W., Li, T., Chen, J., Genga, Q.: CT imaging and clinical course of asymptomatic cases with COVID-19 pneumonia at admission in Wuhan, China, J. Infect. **81**(1), e33–e39 (2020)
34. Lai, X., Wang, M., Qin, C., Tan, L., Ran, L., Chen, D., Zhang, H., Shang, K., Xia, C., Wang, S., Xu, S., Wang, W.: COVID-19 infection among health careworkers in a tertiary hospital in Wuhan, China. JAMA Netw. Open. **3**(5), e209666 (2020)
35. Li, Q., Guan, X., Wu, P., et al.: Early transmission dynamics in Wuhan, China, of novel coronavirus-infected pneumonia. N. Engl. J. Med. **382**(13), 1199–1207 (2020)
36. Zhao, J., Zhang, Y., He, X., Xie, P.: COVID-CT-dataset: act scan dataset about COVID-19. arXiv preprint arXiv:2003.13865 (2020)

Artificial Intelligence Strategy in the Age of Covid-19: Opportunities and Challenges

Walid Hamdy, Ashraf Darwish, and Aboul Ella Hassanien

Abstract With the frequent speedily rise in the number of recently reported and suspected cases of COVID-19, COVID-19 is a significant threat to public health, cultural, social and foreign relations around the world. Accurate diagnosis has to turn into a critical issue affecting the containment of this disease, especially at the countries which outbreak the virus. In the fight against COVID-19, Artificial Intelligence (AI) techniques have played a significant role in many aspects. In this chapter, a systematics review of the recent work related to COVID-19 containment using AI and big data techniques is introduced, showing their main findings and limitations to make it easy for researchers to investigate new techniques that will help the healthcare sector worker and reduce the spread of COVID-19 pandemic. The chapter also presents the problems and challenges and present to the researchers and academics some future research points from the AI point of view that can help healthcare sectors and curbing the COVID-19 spread.

Keywords Artificial intelligence · Machine learning · Deep learning · Coronavirus COVID-19 · CT scan images

---

W. Hamdy (✉)
Faculty of Science, Port Said University, Cairo, Egypt
e-mail: walid.hamdy@ckes.cu.edu.eg
URL: http://www.egyptscience.net

A. Darwish
Faculty of Science, Helwan University, Cairo, Egypt
e-mail: ashraf.darwish.eg@ieee.org

A. E. Hassanien
Faculty of Computers and Artificial Intelligence, Cairo University, Giza, Egypt
e-mail: aboitcairo@cu.edu.eg

W. Hamdy · A. Darwish · A. E. Hassanien
Scientific Research Group in Egypt (SRGE), Giza, Egypt

A. E. Hassanien and A. Darwish (eds.), *Digital Transformation and Emerging Technologies for Fighting COVID-19 Pandemic: Innovative Approaches*, Studies in Systems, Decision and Control 322, https://doi.org/10.1007/978-3-030-63307-3_5

1 Introduction

Currently, AI is causing a new revolution in the healthcare sector and therefore can add value to the application of the current COVID-19 outbreak. In addition, AI can deliver effective insights and provide effective solutions at a timely to help stop the current COVID-19. AI can be applied with COVID-19 in treatment and diagnostics of this virus, epidemiological tracking and predicting of the spread of the COVID-19, medical imaging processing for the virus, and the broader human impact of COVID-19 and pandemics on governments, economies, and human behaviour [1]. AI can contribute to controlling the COVID-19 pandemic AI tools may fight COVID-19 in several ways, for example, it can predict the spread of coronavirus, map its genetic development as it travels from one human to another, accelerate diagnosis, and at developing potential treatments, while helping policymakers deal with related issues, such as food supplies, affecting transport, and travel [2]. Besides, mathematical models based on AI play an important part in helping policymakers understand how COVID-19 is spread and when the infection rate is peaked. Deep learning is perfect at predicting general behaviour, however, it's not perfect at extrapolating to a crisis while nearly everything is new, if people interact in a new way, thus AI cannot predict that. Until you can see this, you cannot learn from it. But in all of these cases, AI is only effective if it has enough examples to learn from. Since COVID-19 has taken the world to an uninhabited region, "deep learning" systems, which computers use to gain new capabilities, do not necessarily have the data they need to produce useful outputs.

The general procedure for AI and applications that helps general physicians for the identification of the symptoms of COVID-19 show in Fig. 1. The flow diagram illustrates the flow of AI-based treatment. The physician isn't only focused on the treatment of the patient but also to control of disease by the application of AI. The main symptoms and test analyses are done with the support of AI by the highest accuracy. As shown it reduces the overall number of steps that are taken in the entire process, making them more accessible for procurable by their nature [3].

In a world pandemic such as COVID-19, technology such as AI and data science have become critical to helping societies cope effectively with an outbreak. Nowadays, AI techniques have been used in the medical imaging domain due to its high capability of feature extraction and selection. Mainly, deep learning technology is applied to detect and differentiate viral pneumonia and bacterial in paediatric chest radiographs. In the following we highlight the main areas where AI can be used to deal with the COVID-19.

The rest of this chapter is organized as follows. Section 2 highlights the related work and the AI services against COVID-19. Section 3 explores an application about using AI to fight against COVID-19. Section 4 presents the problems, challenges and future trends. Section 5 concludes this chapter.

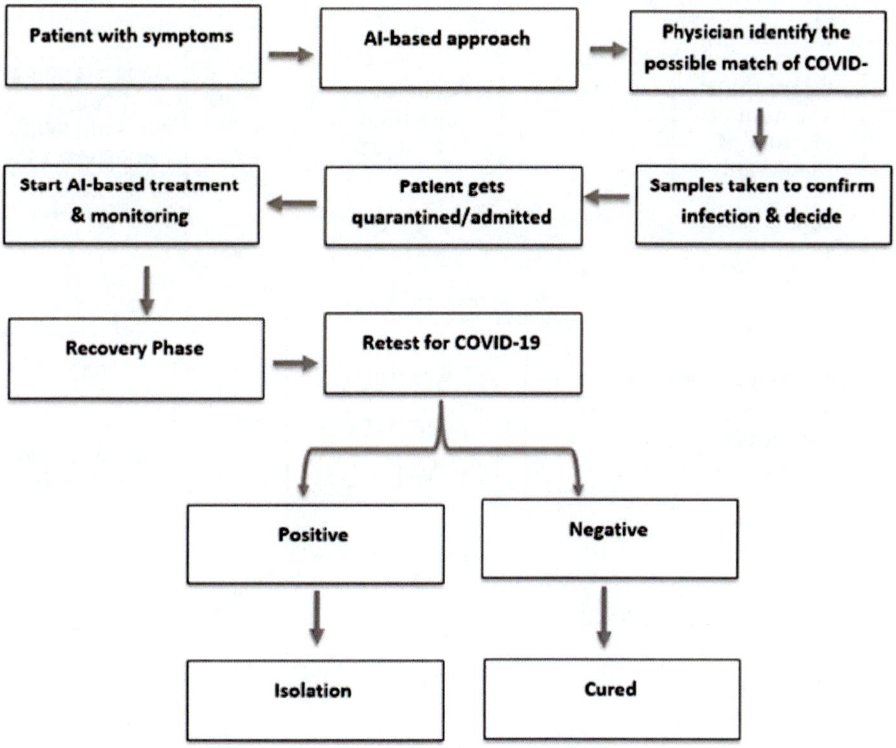

Fig. 1 AI procedures to help healthcare workers s to identify the symptoms of COVID-19

2 Artificial Intelligence Strategy to Understand, Track and Monitor COVID-19

AI requires several key factors for a successful response of Public Health to avoid a new infection. Those factors as shown in Fig. 2 include establishing the natural history of infection, an understanding of transmissibility and risk populations; identifying and characterizing the causative organism; in some instances, epidemiological modelling to suggest effective prevention and control measures; and, including incubation period and mortality rate. This information can be gathered from those who work at the sites of interaction associated with effective prevention with the World Health Organization. The work of this strategy in favor of SARS, and again a major source of information for COVID-19. Using of AI techniques to understand, analyse, and predict how and where to spread the virus or slowing down. Figure 2 describes the AI strategy to fight against COVID-19. In this chapter, we offer a review of early and fast for the stampede of AI and discuss the actual contribution and potential of AI in the fight against COVID-19, so as to the current restrictions on these contributions. The aim of this chapter is to draw quick takeaways from the discussion of the fast-growing group of diverse work in order to work as an entry point to rapid responses at policy, research, and analysis medical.

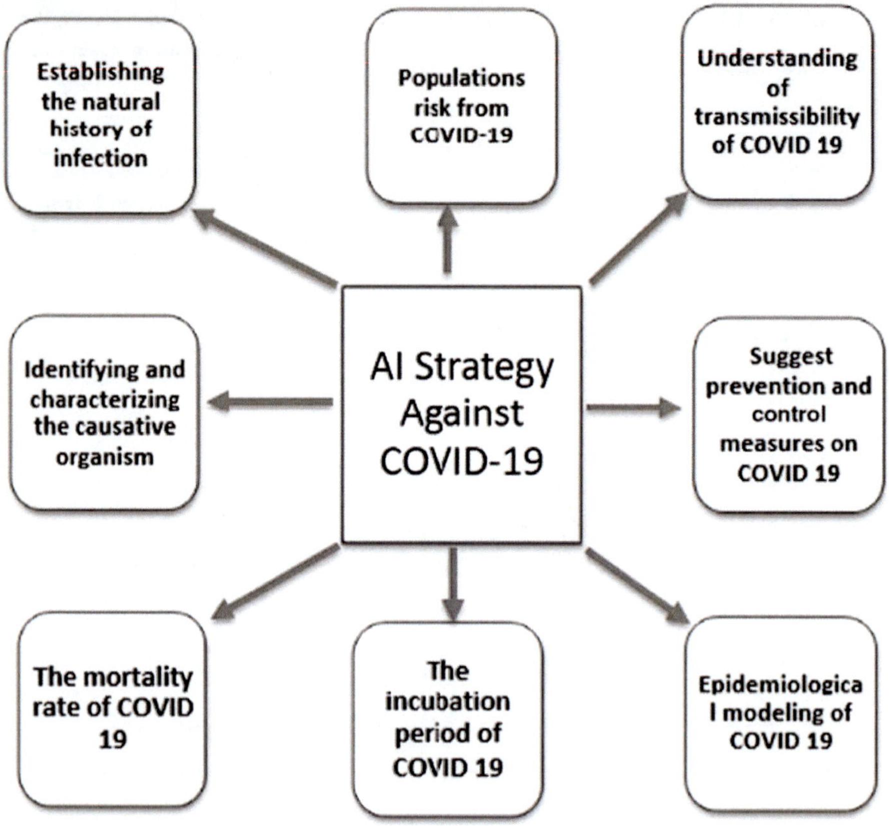

Fig. 2 AI Strategy against COVID-19

Automated contact tracing, for example, to create detailed graphs for social interaction through the analysis of the countless number of citizens' data such as locations of cell phones and identify the public facial and undo the movement of persons to determine the source of the virus likely. It can then notify the individuals who encountered the source can then be notified, tested, or quarantined. Other AI applications that fall within this area include epidemic forecasting and monitoring the development of herd immunity.

In [4] infected most patients of COVID-19 with pneumonia are named novel coronavirus pneumonia (NCP) and move rapidly to respiratory failure. However, the rapid diagnosis and identification of high-risk patients for early involvement is a challenge. Development a system of AI that can diagnose NCP and to distinguish it from pneumonia is common and normal controls. The AI system can help radiologists and physicians in making a diagnosis quick and able to identify clinical signs of significant wear characteristics of the lesion NCP. The clinical diagnosis accurate, which can help doctors to excel in the clinical management of the early and appropriate allocation of resources appropriately. The provision of a system of AI that globally to assist doctors on the anti-COVID-19.

In [5], Deep Learning used a data set of x-ray images for patients who suffer from bacterial pneumonia and confirmed Covid-19 disease, and Regular accidents were used to automatically detect Coronavirus disease. To evaluate the performance of the structure of the neural network sets the background of the evolving proposed over recent years to classify medical images. Specifically, a procedure was adopted labelled transfer learning. By learning about the way of transport, is the discovery of imbalances in different data sets of medical images of the small target can be achieved, often leads to remarkable results.

In the United States [6], the transition of society is the pattern of adult transmission of the virus. Therefore it has become necessary to assess the effectiveness of the primary forms for social communication used by local and national governments. Now available sufficient data at the provincial level to provide a fair assessment of the effectiveness of the guiding principles of the presidential elections on March 16, 2020, the establishment of a form of 'social distancing.' by limiting the gatherings to 10 people or less. Data are also sufficient to assess the effectiveness of the orders of the SIP at the level of the county compared to counties that did not issue any orders to the SIP after the instructions, 16 March. Machine Learning model building a random model for automatic control of forests to predict the growth rate of the composite after application SIP. Found jungle random to the population, longitude, and population per square mile was the most important attribute when predicting the impact of the SIP.

In [7], The machine learning approaches used to analyze the genome sequence of the 257 clinical isolates available for SARS-CoV-2: At least 5 different groups of SARS-CoV-2, there is a great deal of genetic diversity emerging between clinical isolates. Each continent appears to have multiple introductions to different viral strains (no geographical grouping other than previous epidemics). ORF1ab, spike, ORF3a, and E proteins are most susceptible to mutations. The RBD of Sunbulah protein is a mutation hotspot (a key driver of diversity). Greater transmission/death in the Lombardy region (Italy), compared to other European countries, the African continent, or China, cannot be linked to any specific molecular difference pattern.

2.1 Machine and Deep Learning Techniques

This section is designed to explore the role of machine learning to fight COVID-19, according to WHO report No. 96 [8], Coronavirus disease 2019 (COVID-19) is an infectious disease that affects 2,719,897 confirmed people worldwide and has caused 187,705 deaths as of Apr 25 in 2020. AI based Deep learning [9, 10], play an important role in health care, Deep learning is a modern AI machine-learning technique, and has developed a great deal of interest in its medical applications over the past few years. It is designed to mimic the layers of human brain neurons to process and extract information, enabling computers to learn without explicit programming. Three different CNN-based models (ResNet50, InceptionV3 and Inception-ResNetV2) were proposed for the identification of patients infected with coronavirus pneumonia using X-ray radiographs in the chest. These three models include ROC, and predict s and uncertainty matrices and are analysed using 5-fold cross validation [11]. Results show that the ResNet50 pre-trained model provided the highest precision of 98 per cent among the three models. The output of state-of-the-art CNN architectures introduced for the classification of medical images over the last few years. The method called Transfer Learning was expressly adopted. The identification of different anomalies in small medical image datasets is an achievable target with transfer learning, which often produces impressive results [5]. These architectures were compared based on their results on the collection of X-ray images from patients with severe bacterial pneumonia, reported Covid-19 disease and usual events was used to diagnose Coronavirus disease automatically. While the best accuracy, sensitivity and specificity obtained is 96.78%, 98.66% and 96.46% respectively. A review of recent studies was presented in [12], How AI provides secure, reliable and effective imaging solutions for COVID-19 implementation is discussed in this work. Intelligent imaging systems, clinical diagnosis, and pioneering work are analysed in-depth, and the full pipeline of AI applications in COVID-19 is covered. Two methods of imaging, i.e. x-rays and CT scans, are used to illustrate this efficacy from AI medical imaging of COVID-19.

Another study using chest CT images of 219 COVID-19 patients, 224 influenza patients and 175 healthy individuals was presented in [13]. Which a V-Net-based deep learning model is first used for segmenting the regions of candidate infections. Infection regions corrections are sent to Resnet-18 network along with relative infection distance features from the edge, and the output is one of these three groups. The results of the benchmark data set experiments showed that, from the viewpoint of CT cases as a whole, the overall accuracy is 86.7%. In [14], CT images of a total of 2685 participants were obtained retrospectively. Within this study, 1658 confirmed cases of COVID-19 were identified with positive nucleic acid conformation tests by the national CDC. The remaining 1027 cases were CAP patients. VB-Net [15] is used to segmentation of the image into the left/right lung, 5 lobes of the lung, and 18 lung segments. A number of handmade features are identified and used to train a random forest model. Experimental tests indicate a sensitivity of 90.7%, a precision of 83.3% and an accuracy of 87.9%. This approach has been incorporated into the u cloud network as an online service and is currently accessible to more than 20 clinical facilities in China. In the meantime, our research portals, including computer hardware, such as GPUs and software, have been deployed and used in more than 50 hospitals in Wuhan, China, where more than 20 are based.

2.2 *Artificial Intelligence Solutions to Fight Against COVID-19*

This subsection describes the main solutions and services that AI can provide to fight COVID-19. Figure 3 summarizes the main functions of AI to fight against COVID-19.

1. Identifying, tracking and forecasting outbreaks. AI can learn to detect an outbreak by analyzing social media platforms, news reports and governmental documents. Tracking the risks of infectious diseases with AI is precisely the service that some countries started to apply.
2. Virus diagnosis. Some AI companies have launched an AI solution for coronavirus that helps front-line healthcare workers monitor and detect coronavirus.

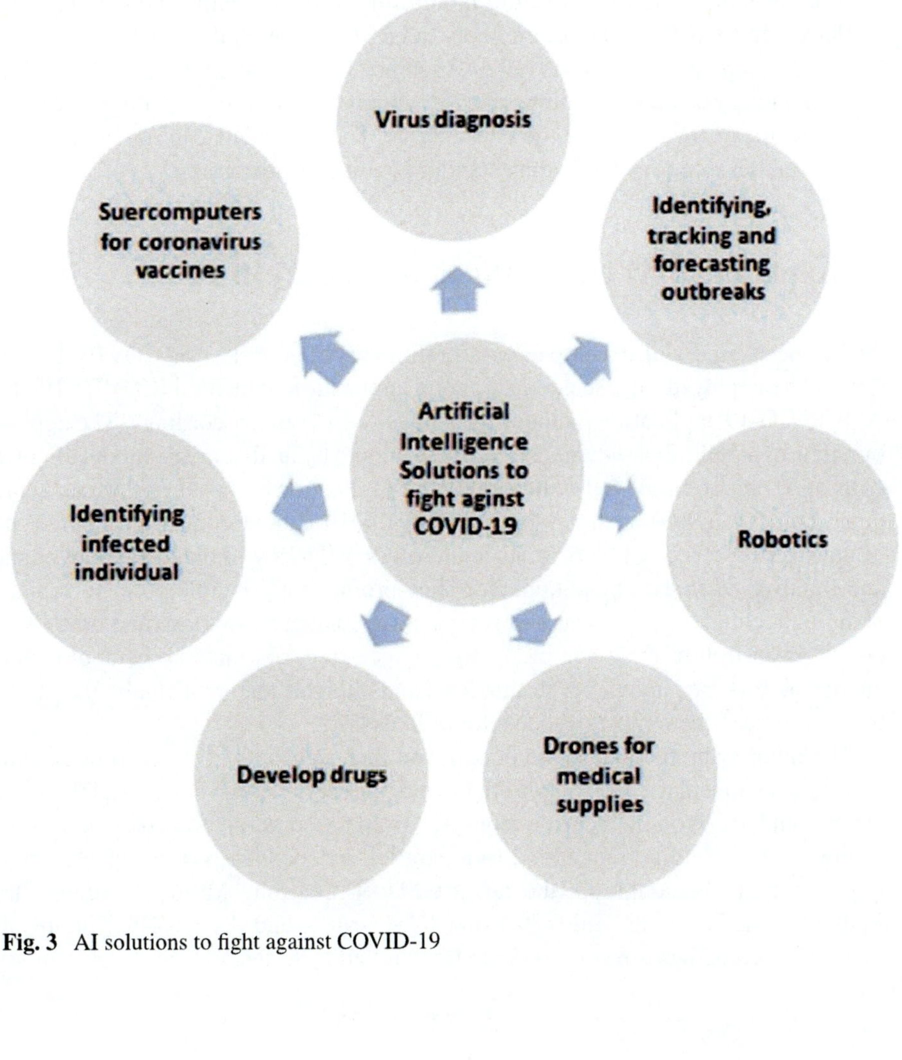

Fig. 3 AI solutions to fight against COVID-19

3. Drones for medical supplies. Drone delivery is one of the easiest and best delivery methods for emergency supplies where they need to be delivered during an outbreak of disease. Drones are also used to monitor public areas, document non-compliance with quarantine mandates and thermal imaging.
4. Robotics. Robots can be deployed with coronavirus to complete more tasks like cleaning, sterilizing, providing food and medicine to decrease the amount of contact among humans.
5. Develop drugs. Machine learning techniques can be used to structure drugs that can fight the most complicated diseases in this world, and now help support efforts to pleasure, coronaviruses, at the first time the company has to concentrate its production on infectious diseases. With the AI techniques and in a few months, his predictive abilities can be used to suggest existing drugs that may be useful.
6. Identifying infected individuals. While the use the technology and AI is definitely problematic, China's advanced monitoring program using facial recognition technology and temperature monitoring tools to identify people who may have a fever and who are more likely to be infected with the virus.
7. Supercomputers for coronavirus vaccine. The supercomputers and cloud computing resources are being used by researchers to rapid tracking to develop a vaccine or treatment for the coronavirus. These systems can run computation and model solutions faster than standard computer processing.

3 Application: AI for the Diagnosis of COVID 19

AI has been used and developed as a preliminary diagnostic for COVID-19. The test can be deployed on a scale via a phone application named AI4COVID-19. The AI4COVID-19 application includes the subject's 2 s cough recordings. The application returns a preliminary diagnosis within a minute by analysing the model of cough with an AI engine employed at the cloud. The application of the AI engine can distinguish more than 90% accuracy between COVID19 patient cough and many types of not COVID-19 cough. Performance of AI4COVID-19 will maybe get better a lot and improve when data be obtainable. They provided a principle proof to promote planning a clinical experiment and serves as an advocacy for classified data on the cough. Instead, it offers a complementary telecasting tool that can be deployed by anyone at anytime, anywhere, so that tests and clinical treatment can be directed to those who need them most, rescue a lot of lives [18].

The mobile app records sound/cough, and then transmits captured sounds to the server. In the first place, the sounds are fed into the server cough detector. If the sound is not found to be cough, it can be requested from the server by the user. As we show in Fig. 4, the sound is transmitted to a parallel, independent system of classifiers, if perceived to be coughing: the DL-MC, Deep Learning Multi-Classifier, Classical Machine Learning Multi-Classifier (CML-MC), and the DL-BC. The results of all three classifiers are subsequently transmitted to a mediator. Such three phases

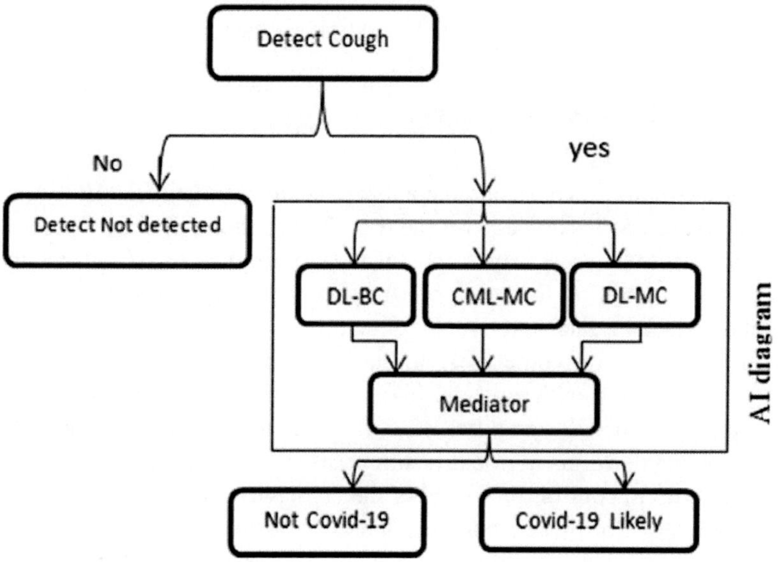

Fig. 4 AI application for AI4COVID-19

mediator-cantered novel architecture (DL-MC, DL-BC, and CML-MC) are designed to efficiently reduce the risk of misdiagnosis.

They used CNN structure like the insert Mel spectrogram images are of a lot of dimensions, to minimize the total complexity of the process before continuing it is first passed through a 2 * 2 max-pooling sheet. This comparatively smaller picture are then transferred to two co-evolutionary layers composed of 32 filters and 5 × 5 kernel sizes With the extreme grouping layer between them to learn about the complex features that are then transferred to the maximum 2 × 2 pools to reflect the acquired features in fewer dimensions. The images it is then settled to one dimension moved to two completely linked layers of 128 neurons each. For each of these, 0.5 dropouts are used to prevent over-fitting. The final layer is the neuronal recognition layer consist of two nerves to distinguish between a cough and not cough to data. The amount of co-evolutionary and completely entangled layers is held small to reduce possible over-fitting issues. Because ReLU is the existing norm for CNN networks, it is applied for the activations functions, while Adam's used such an optimizer because it's relatively improved performance and versatility. The detection model is completed by a binary cross-entropy loss function. Finally, they used the potential of AI to can fight COVID-19. So that can we have taken it as a case study to prove how researchers using AI against COVID-19.

4 Problems, Challenges and Future Perspectives

4.1 Problems and Challenges

AI is a powerful tool to possibly to combat pandemics of COVID-19. Even after the outbreak, there was a stampede for the use of AI. This chapter provides a review of early and selective necessarily, discusses the contribution of AI in the fight against COVID-19, in addition to the current restrictions on their contributions. Used to control by a lack of data, and by a lot of data.

Will require to overcome these limitations a delicate equilibrium among data privacy and public health, and the interaction strictly between human and AI. It is not likely that it will be addressed at the time to be of a lot of help during the current epidemic. So, as a global health crisis, looking to the medical industry to new techniques to control the spread of the epidemic COVID19 and controlled that disease. AI is one of these techniques that can easily path route the spread of that coronavirus, and identify patients at a large risk, and is useful at controlling that infection in a real-time. It can also predict the risk of death through an adequate analysis of the data of previous patients. AI can help the world in the fight against this virus through the population screening and medical assistance, notification, and proposals for infection control [16, 17]. That technology has the ability for improving designing, treatment, and reported the result to a COVID-19 patient, existence an evidence-based medical model.

We have specific several important applications of the AI of the COVID-19 epidemic. That technique plays a significant role in detecting the set of cases and predicting where will coronavirus be an effect in the future through the collection and analysis of all processing data.

4.2 Future Perspectives

Although the efforts from the whole world to understand COVID-19, there are many issues remain unsolved and cause problems and challenges. In the next, we explore such problems and challenges and present to the researchers and academics some future research points from the AI point of view.

1. The related and previous studies showed that coronavirus could survive on inanimate objects and surfaces however; the presence of coronavirus in the environment has not been discussed. Therefore, in the future, some emerging technologies such as drones, sensor networks and the Internet of Things can be deployed to deal with this problem.
2. The specific nature of the novel Coronavirus is still unknown, and new drugs for this disease have not been developed. In this regard, machine learning, data mining, deep learning approaches can be used to build classification, clustering

and predicting models of this virus to develop drugs and promote the research and development of vaccines.

3. There is a lack of using a simulation approach to study the behaviour of this virus. In the future, researchers should demonstrate how simulation-based methodology can be used to examine and predict the impacts of coronavirus COVID-19 and optimization software to uncover critical parameters and scenarios of positive and negative of the virus.
4. There is still a lack of accurate data sets and mathematical models to be used in model prediction implementation, except verified case data that can be used for model estimation model validation methods, including maximum likelihood methods and Bayesian inference methods. For modellers, the shortage of datasets is a more severe concern.
5. The option of a model that could represent the epidemic under study is a significant factor that can affect the model predictions. A complex model contains more biological and epidemiological knowledge about the epidemic; however, a drawback of a complex model is that it needs an estimate of more model parameters compared to a simpler one.
6. Medical imaging plays a significant role in the diagnosis and treatment of the disease. Imaging of the chest and lung can be used to diagnose and record the severity of the lesions and to allow objective evaluation of changes. The image processing techniques, for example, can detect lesions of potential coronavirus pneumonia, measure their length, shape, and intensity, and comparative changes in multiple lung lesions of the image, all of which provide a comprehensive determination to help doctors make fast decisions. Although a manual process for reading a CT scan can take a few minutes, the image can be read in a few seconds by AI techniques. There has been a shortage of image processing techniques to examine the results of chest imaging in patients with coronavirus disease on thin-section computed tomography (CT) scanning so far.
7. The PCR-based diagnostic method can take a long time (some days), whereas AI-based CT imaging may help medical staff more efficiently when rapid judgment is required.
8. AI techniques require high-quality input data for virus analysis. By gathering data relating to disease in sub-Saharan Africa, there aren't data this is not because of the absence of disease but because there isn't monitoring. This could reflect the type of harm that the virus can inflict if it spread to a country with a poorer health system. In this coronavirus scenario, AI can be used in societies for detection more efficient way of managing the effects of the new coronavirus, rather than attempt to close the borders. Governments are not yet entirely dependent on AI in this regard within the nations.

5 Concluding Remarks

The outbreak of the 2019 COVID-19 has caused around 1 million confirmed cases and thousands of deaths according to WHO statistics. At the moment of writing this chapter, the number is still increasing since COVID-19 is extremely contagious and easily spreadable between people via respiratory droplets. In this context, many efforts had been made to contain this disease by the early detection of COVID 19 to contribute to its containment before spreading.

Health care organizations are at the imperative need to decision-making techniques for the deal with this virus and help them to get the appropriate proposals at a real-time to avoid its dispersal. AI works at an effective method to simulate human intelligence. It might also plays a necessary role to understanding and proposing the developing of a vaccine to COVID-19. That results-based technology are applied to the examination, analysis, tracking and prediction of present and possibility patients in future. Large applications are implemented for track data from restored cases and deaths. Thus, many researchers have been developing specialized tools using Artificial Intelligent techniques and big data analytic tools for early prediction and diagnosis of COVID 19 cases to contribute to its containment before spreading. These models had been a great help to clinicians and decision-makers for making the appropriate recommendations for treatments. Finally, the value of integrating AI and big data tools strongly comes into play by providing clinicians and decision-makers with helpful information on COVID-19 disease, which will reduce their efforts.

References

1. McCall, Becky: COVID-19 and artificial intelligence: protecting health-care workers and curbing the spread. Lancet Digit. Health **2**(4), e166–e167 (2020)
2. Haleem, A., Javaid, M., Vaishya, R.: Effects of COVID 19 pandemic in daily life. Curr. Med. Res. Pract. (2020). https://doi.org/10.1016/j.cmrp.2020.03.011
3. Vaishya, R., Javaid, M., Khan, I.H., Haleem, A.: Artificial Intelligence (AI) applications for COVID-19 pandemic. Diab. Metab. Syndr. Clin. Res. Rev. **14**, 337–339 (2020)
4. Zhang, K., Liu, X., Shen, J., Li, Z., Sang, Y., Wu, X., Cha, Y., et al.: Clinically applicable AI system for accurate diagnosis, quantitative measurements and prognosis of COVID-19 pneumonia using computed tomography. Cell
5. Apostolopoulos, I.D., Mpesiana, T.A.: Covid-19: automatic detection from x-ray images utilizing transfer learning with convolutional neural networks Phys. Eng. Sci. Med. **43**, 635–640 (2020)
6. Cobb, J.S., Seale, M.A.: Examining the effect of social distancing on the compound growth rate of SARS-CoV-2 at the county level (United States) using statistical analyses and a random forest machine learning model. Public Health **185**, 27–29 (2020)
7. Sheikh, J.A., Singh, J., Singh, H., Jamal, S., Khubaib, M., Kohli, S., Dobrindt, U., Rahman, S.A., Ehtesham, N.Z., Hasnain, S.E.: Emerging genetic diversity among clinical isolates of SARS-CoV-2: lessons for today. Infect. Genet. Evol. **84**, 104330 (2020)
8. WHO: Coronavirus disease 2019 (COVID-19) Situation Report, 96 (2020). https://www.who.int/docs/default-source/coronaviruse/situation-reports/20200425-sitrep-96-covid-19.pdf?sfvrsn=a33836bb_2. Accessed 25 Apr 2020

9. LeCun, Y., Bengio, Y., Hinton, G.: Deep learning. Nature **521**, 436–444 (2015). https://doi.org/10.1038/nature14539
10. Ting, D.S.W., Liu, Y., Burlina, P., et al.: AI for medical imaging goes deep. Nat. Med. **24**, 539–540 (2018). https://doi.org/10.1038/s41591-018-0029-3
11. Narin, A., Kaya, C., Pamuk, Z.: Automatic Detection of Coronavirus Disease (Covid-19) Using X-Ray Images and Deep Convolutional Neural Networks. arXiv preprint arXiv:2003.10849 (2020)
12. Shi, F., Wang, J., Shi, J., Wu, Z., Wang, Q., Tang, Z., He, K., Shi, Y., Shen, D.: Review of artificial intelligence techniques in imaging data acquisition, segmentation and diagnosis for covid-19. IEEE Rev. Biomed. Eng. (2020)
13. Xu, X., Jiang, X., Ma, C., Du, P., Li, X., Lv, S., Yu, L., et al.: Deep Learning System to Screen Coronavirus Disease 2019 Pneumonia. arXiv preprint arXiv:2002.09334 (2020)
14. Shan, F., Gao, Y., Wang, J., Shi, W., Shi, N., Han, M., Xue, Z., Shi, Y.: Lung Infection Quantification of Covid-19 in CT Images with Deep Learning. arXiv preprint arXiv:2003.04655 (2020)
15. Shi, F., Xia, L., Shan, F., Wu, D., Wei, Y., Yuan, H., Jiang, H., Gao, Y., Sui, H., Shen, D.: Large-Scale Screening of Covid-19 from Community Acquired Pneumonia Using Infection Size-Aware Classification." arXiv preprint arXiv:2003.09860 (2020)
16. Bai, H.X., Hsieh, B., Xiong, Z., Halsey, K., Choi, J.W., Tran, T.M., Pan, I., Shi, L.B., Wang, D.C., Mei, J., Jiang, X.L.: Performance of radiologists in differentiating COVID-19 from viral pneumonia on chest CT. Radiology (2020). https://doi.org/10.1148/radiol.2020200823
17. Hu, Z., Ge, Q., Jin, L., Xiong, M.: Artificial Intelligence Forecasting of COVID-19 in China. arXiv preprint arXiv:2002.07112 (2020)
18. Imran, A., Posokhova, I., Qureshi, H.N., Masood, U., Riaz, S., Ali, K., John, C.N., Nabeel, M.: AI4COVID-19: AI Enabled Preliminary Diagnosis for COVID-19 from Cough Samples Via an App. arXiv preprint arXiv:2004.01275 (2020)

SAKHA: An Artificial Intelligence Enabled VisualBOT for Health and Mental Wellbeing During COVID'19 Pandemic

Jaideep Singh Sachdev, Arti Kamath, Nitu Bhatnagar, Roheet Bhatnagar, Arpana Rawal, Ashraf Darwish, and Aboul Ella Hassenian

Abstract COVID19 pandemic is playing havoc all around the world. Though the world is fighting this invisible enemy it has succumbed to the devastating potential of the Coronavirus. Largest of world economies and developed nations have been exposed and their health infrastructure has collapsed during this testing time. It is assessed and predicted that the novel coronavirus which is responsible for COVID19 pandemic, may turn into an endemic (just like HIV) and *will never go away*. It will become part and parcel of our life and humans have to learn to live with it even if the vaccine is developed. The government's world over is concerned with containment and eradication of this virus at the earliest and massive efforts are on at all front to contain it's spread. As of now (3rd week of May 2020), more than 4.4 million cases of the disease have been recorded worldwide and more than 300,000 have died. The world has also seen technological innovation during this time and mechanisms to tackle COVID19 patients. Innovations in carrying out quick testing using Rapid testing kits, Artificial Intelligence (AI) powered thermal scanning for temperature monitoring in the crowd, AI-enabled contact tracing, Mobile Apps, low-cost ventilators, and many other such similar solutions. All these pertain to

J. S. Sachdev · A. Kamath · N. Bhatnagar · R. Bhatnagar (✉)
Manipal University Jaipur, Jaipur, Rajasthan, India
e-mail: roheet.bhatnagar@jaipur.manipal.edu

J. S. Sachdev
URL: http://www.egyptscience.net

A. Rawal
Bhilai Institute of Technology, Durg, Chhattisgarh, India

A. Darwish
Faculty of Science, Helwan University, Helwan, Egypt

A. E. Hassenian
Faculty of Computer and Artificial Intelligence, Cairo University, Giza, Egypt

R. Bhatnagar · A. Darwish · A. E. Hassenian
Scientific Research Group in Egypt (SRGE), Cairo, Egypt

A. E. Hassanien and A. Darwish (eds.), *Digital Transformation and Emerging Technologies for Fighting COVID-19 Pandemic: Innovative Approaches*, Studies in Systems, Decision and Control 322, https://doi.org/10.1007/978-3-030-63307-3_6

checking for COVID19 symptoms and taking actions thereafter, but what about the stress, pain, and shock of a person who has been put under quarantine in a facility meant for the purpose or the person who is Corona positive? In this chapter, the authors have discussed briefly the pandemic and tried to provide a solution for the mental wellbeing of such people who are under quarantine and are isolated but heavily stressed or showing stress symptoms, by creating a VisualBOT which could understand the facial expression of the person and judge his mood, for providing suitable counseling and help.

Keywords Artificial intelligence · COVID19 · VisualBOT · Mobile app · Coronavirus

1 Introduction

COVID19 pandemic is the greatest global health crisis that humankind has ever faced. It has spread its roots to almost all the continents around the globe except for Antarctica. With this pandemic which is slated to become endemic, will cost more lives in the future and humans should learn to live with it in the post-Corona era—which is the new normal. Apart from the death toll that has resulted due to the pandemic, nearly 1/3rd of the world's population has been forced to live in lockdown or quarantine conditions. Man is a social being and loves to be always in touch with his near and dear ones, and even brief periods of isolation and loneliness can harm the mental state and physical state of humans [1]. The only solution to curtail the COVID-19 outbreak is lockdown and keeping persons with COVID symptoms in self or forced quarantine, which although is a temporary condition, can be very much demanding on the health of a person. If a person is tested COVID-19 positive, he needs to be put up under quarantine where his movement has to be restricted, practice social distancing and isolation as per prescribed guidelines, and set standards [2]. This is being practiced in all countries across the globe. In a way, it can arguably be considered as the largest psychological experiment being ever conducted on the human race. Social distancing, containment zones, lockdown, quarantine, and isolation are some of the measures being adopted by governments, but over some time these may start to have its side effects involving the psyche of people which may lead to psychological problems. The American Psychological Association reports that social isolation carries several health risks [3] which can lead to poor sleep, poor cardiovascular health, lower immunity, depressive symptoms, and many other similar psychological symptoms. There had been a lot of efforts which are underway throughout the world to contain the virus and the disease but the measures are not proving very much fruitful. The vaccine to such a condition like COVID'19 is the only solution but it will take some time. Such situations call for efforts from researchers and academicians to find solutions to cope up with the psychological condition of a person in quarantine and technology can play an important role. We have got advanced and trusted technologies that can be utilized for creating solutions to handle psychological stress-related problems among people. Computer Vision, AI, and NLP may be utilized to create a visual bot to effectively interact and analyze the state of mind of such persons.

Expressions are a very important part of a meaningful conversation because there are so many complex emotions, facial expressions, gestures, audio signals, and body postures that differ for different individuals which cannot be determined by spoken words. It is proven in the research of social psychology that facial expressions dominate the conversation than just simple words. Emotional Intelligence caters to all these requirements and is the latest platter provided in machine learning domains.

Automated identification of facial expressions by domain-specific assistant-bots has become a necessary and integral component for the subject of psychological and expression-of-thought phenomena with the increasing advancements in machine learning realm of human being-computing machine interaction (HCI) and also as a part of computer visual sensation research. As an example let us consider a situational scenario of online mode of delivering lectures through online platforms and taking online pen and paper examination, the use of bots seems very attractive as a solution. If the Conversational agents (Chatbots) are to become highly interactive in functionality, to mimic as strong alternatives to a vigilant teacher in classroom teaching environments or proctored examination environments, their design demands incorporating meta-cognitive skills along with intelligent communicative capabilities, provide instructions, motivation, mentoring and engagement among students, whether face-to-face or in remote execution modes. The only task that cannot be performed is the conduction of face-to-face end-course proctored examinations as can be seen in current times of COVID pandemic outbreak; so the design of emotion-sensing visual bots as substitutes of online human invigilators help us resolving the unprecedented crucial times.

In contrast to static images, the main Facial Expression Detection (FED) challenge in video images is to track facial features over time and spatial information separately, this is being resolved by using hybrid FED models, however, the results need further exploration and validation on a competitive note. Another major challenge that FED observed was from pose-variant faces and wild environments of both static image and video databases that caused the lowest performance of the design models and the prime reason to switch over to multimodal approaches as compared to unimodal approaches.

The concept of Visual bots is thus, deeply affined to the realm of Emotional AI. Emotional Intelligence is an interdisciplinary field of AI-enabled computer vision that uses neural network learning models to identify facial features of eyebrows, mouth, eyes, and to label these features with an emotion or feeling; this has been eventually enriched by including body language and voice modulation too. They use the basic underlying principle of emotion facial detection (also known as Facial Emotion Detection (FED) or Facial Coding principle), making use of components, namely, a camera or sensor to capture the facial expressions to detect a person's inner feelings and soulful expressions to be applied with effective computer vision algorithms; hence, forming an interdisciplinary realm with computer vision, machine learning (text mining, NLP, artificial intelligent techniques, human-computer interaction (HCI), virtual reality (VR) and augmented reality (AR).

2 Brief History of Chatbots and Emergence of Visual-Bots

An artificial intelligence (AI) program which could simulate interactive human conversation by using stored user phrases and audio signals is often called a Chatbot. These were developed for automating commonly used activities that need to be answered against user queries in a customer service and marketing industry. A chatbot is also known as an artificial conversational entity (ACE) as they do not have a physical presence and are created basically to automate the activities of repetitive nature e.g. user enquires about a process or a particular system, systems user manuals, etc.

The first bots were applied in the Finance industry for automatically doing online trading of equities based on market indicators. This was almost over a decade ago and now these have penetrated so deep in the business that the financial Robo-advice market is projected to grow to $7 trillion by 2025, according to CNBC [4].

Today, the application of Chatbots can be seen in almost all the domain including healthcare. They have become more intelligent and are capable of learning on their own with the help of Artificial Intelligence technology. These are called Conversational AI chatbots, but now the world is on the verge of another revolutionary step which combines the Computer Vision technology with the Artificial Intelligence Technology to build something which is known as *Visual Bots—the bots which have got Eyes and can see.* These visual bots are intelligent enough to understand the human behavior, do sentiment analysis of the person and judge the mood swings of person, thereby interacting with them in a more user-friendly way or more humane way. We, humans, make use of our eyes and brains to perceive objects. Our eyes focus on an object and send signals to our brain for its interpretation, and in this process, we learn, unlearn, and relearn many things. Humans have a great visual sense, are visually driven and this is the reason for having an increase in visual search engines, video tutorials, and visually based customer assistance. Visual engagement satisfies the customer rather than escalates dissatisfaction. Thus, Visual Bots are slowly going to bring a paradigm shift in bringing the much desired human touch to the customers, industries are well aware of these facts and we have technology that is making this a reality. But the fact remains, the chatbots are precursors to futuristic visual bots.

2.1 Evolution of Visual Bots

With the advancement in Computer Vision, AI, and NLP technology and the emergence of many other newer ones like Augmented Reality, the traditional conversational AI chatbots are changing into Visual bots. The business is likely to be benefitted by the use of visual bots that have evolved from the chatbots over time (Fig. 1).

Table 1, lists the journey of chatbots to the world of modern time visual bots. It shows the period of development of bot and its brief remark highlighting its purpose and features. The table lists the development from the 1950 era till 2018—the dawn of an era of visual bots. Almost all the top companies of the world are carrying extensive

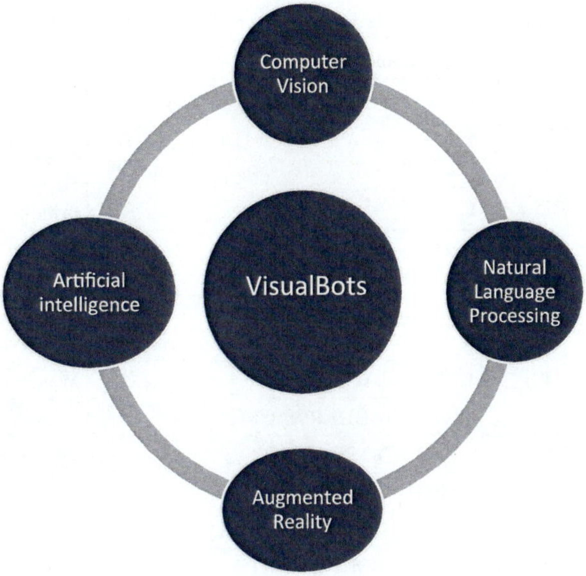

Fig. 1 Enabling technologies for visual bots

research into this realm of intelligent visual bots as they offer immense benefits and finds tremendous applications that we may not be able to even visualize now but are sure to unfold in times to come. This domain has attracted both academics as well as industry alike and we will surely see many interesting applications when it becomes more ubiquitous in our lives.

The evolution of chatbots can be seen as a result of the advancement and enrichment of technologies that keep on changing at a very fast pace. Figure 2, summarizes the different phases in the evolution of chatbots and the emergence of visual bots.

2.2 *Facial Emotion Recognition (FER) Classification*

In preliminary studies, it was found that the emotional intelligence community used to use hand-crafted facial features in conventional Facial Emotion Recognition (FER) approaches; however, these could be fetched from selected peak expression frames of image sequences, hence, could yield high performance only for image databases. On the other hand, (video-based) FER approaches began utilizing Spatio-temporal features to capture the expression dynamics in facial expression sequences, yielding a higher recognition rate than static FER approaches. In conventional approaches, pre-trained FE classifiers, namely, support vector machines, AdaBoost and Random Forests used relatively lower computing power and memory than deep learning-based approaches and earned the potentiality to be modeled as real-time embedded systems. Yet it does suffer from a few drawbacks. Deep-learning algorithms have provided a breakthrough in this research direction by enabling "end-to-end" learning

Table 1 Chatbots development timeline [5]

S. No.	Period and chatbots	Remarks
1	1950—TURING TEST	Developed by Alan Turing and proposes that a machine/computer is indistinguishable from a human during a text-only conversation. It proved to be the working principle for chatbot development
2	1966—ELIZA	Developed by Joseph Weizenbaum and could simulate human conversation
3	1972—PARRY	Its conversational strategy was better than ELIZA
4	1984—RACTER	Was better but not released publicly
5	1988—JABBERWACKY	Simulated natural human chat
6	1992—Dr. SBAITSO	Meant for MS-DOS based systems
7	1995—ALICE	a language processing robot
8	2001—SMARTER CHILD	A chatbot for an interactive chat
9	2006—IBM's WATSON	Was developed by IBM and capable of doing data analysis on a massive amount of data using machine learning and natural language processing capabilities
10	2010—SIRI	Offering from Apple
11	2012—GOOGLE NOW	Developed by Google for mobile apps
12	2015—ALEXA/CORTANA	Alexa is an Amazon chatbot while Cortana is a Microsoft bot
13	2016—MESSENGER BOTS	Introduced by FB
14	2018 onwards—the dawn of an era of Visual Bots	e.g. *Levi's AI-powered virtual stylist* and *Amazon Echo Look*

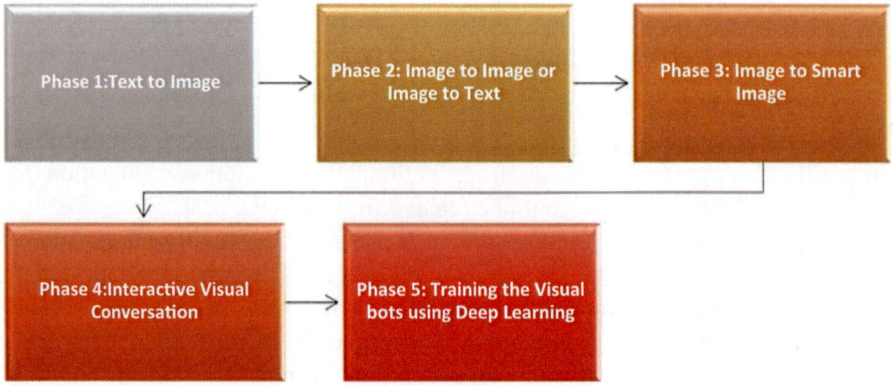

Fig. 2 Key phases in the development of a visual bot

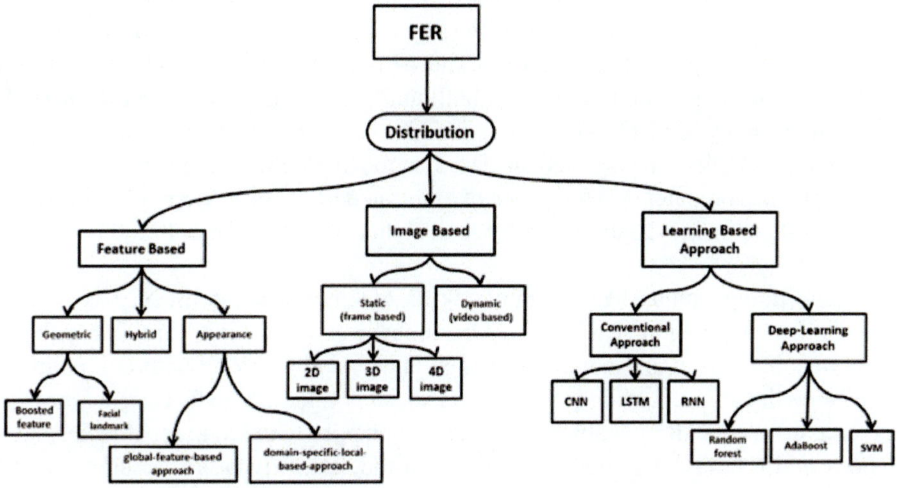

Fig. 3 Facial emotion recognition classification hierarchy

in feature extraction, classification, and recognition stages. With the implementation of deep learning models, the ML researchers have been able to manage completely removing or reducing to a great extent, the dependence on physics-based models and/or other pre-processing techniques. The deep learning models comprised a Convolutional Neural Network (CNN), recurrent neural network (RNN) models, and a combination of hybrid CNN-RNN or CNN-LSTM deep-learning models. Some other deep learning models used deep belief networks, deep auto-encoders, Generative adversarial Networks, Cascaded Networks, and Network ensembles [6–8]. The detailed classification hierarchy of facial emotion recognition/detection approaches implemented to date can be best illustrated in Fig. 3.

Much of the work has been carried out in FER domain and the common FED methods, emotion types, and databases since the past two decades from the available literature are as mentioned below:

a. The most used methods and algorithms for face detection are Viola-Jones face detector & Multi-task Cascade Convolution Neural Network (MTCNN) [9, 10]
b. The commonly used feature extraction approaches for FED are:

   facial landmarks and Local Binary Patterns (LBP);

c. The popular classifiers chosen for labeling emotions are

   LSTMs, SVMs, HMMs, KNNs, and Support Vector Regressions (SVR);

d. Most FER databases were experimented for being labeled with the six basic emotions: (anger, disgust, fear, happiness, sadness, and surprise), plus the neutral expression.

A thorough analysis was done on initially designed conventional FED models and was compared with some of the highly competent FED models. The attributes of comparison included the mandatory stages of emotion classification starting from feature vector representation to tools and APIs used and then to machine learning methods for classifying emotions in both uni-modal and multi-modal formats. The literature survey of FED models is divided into two sets; the first set of the survey is tabulated in Table 2 and is based on classification hierarchy, while second is tabulated owing to the competent performance of high-rated FED models in wild environments of both static image databases, video, and audio-video clip datasets and is presented in Table 3.

Embedding human-like cognitive skills of emotional intelligence is indeed a dynamic phenomenon. This requires a culmination of machine learning capabilities of computer vision and natural language understanding along with computing the gesture selection according to the rules of social psychology so that a set of messages conveyed in meaningful conversations can be mapped to the right set of facial expressions and (or) also by spoken words from the set of dynamic facial expressions changing in the set of video or audio-video images. Thus, a highly modular and scalable framework was felt to be designed to accommodate a vast set of multi-modal input attributes to decide the final gesture of emotion being expressed, both by facial parameters and word of mouth (natural language text). Hence, FED system development literature has to shift its focus from testing challenging in-the-wild environmental conditions as against static image frames.

The *EmotiW Challenge* stimulates the research done in emotion recognition in the wild which correlates with better results over the years. Face frontalization is used to overcome the pose-variant faces problem. The solution was appropriate as the *Acted Facial Expressions in The Wild* (AFEW) database is 2D. It is one of the most popular competitions in the domain.

In one of the latest developments, it was realized that considering audio features are far safer than text features that could be confusing if the interacting entities communicate in different context and language mediums. Features extracted in time-sequence have been proved to be useful in video-based applications, hence, the combination of acoustic features and facial features in both non-temporal and temporal mode is the latest breakthrough in the field of Facial Emotion detection that can build human-like mimicking capabilities in visual-bots.

3 Proposed Architecture and Framework for '*Smart* and *Adaptive Knowledge-driven Human Assistant*' (SAKHA)

The global COVID outbreak has compelled the human race to make a huge twist in their living-styles, thinking patterns, and behavioral patterns. Man has always tried to sustain his existence on this planet by evolving himself through the 'survival of fittest' principle. The proposed architecture of our visual bot (which has been named as, "*Sakha*": a Hindi word meaning Friend), is too proposed on design principle

Table 2 Survey of popular FED models according to FED classification hierarchy

Paper-ID: (Year)	Bot name/type	FED principle/feature space	Supporting architecture/tools/APIs	Learning mechanism	Data types	Emotions handled
(2018) [9]	Empathy-Bot	Conventional approach (FED algorithm)	Google cloud vision API, Raspberry Pi & android phone	A 3-layered convolutional neural network, Keras framework	Static image datasets	[Anger, fear, disgust, surprise, joy, and sadness]
(2016) [11]	ChatBot	Geometric-feature based (Haar features); Viola-Jones face detector	**Python**, OpenCV library	AdaBoost classifier	Static image datasets	Smiling gesture using mouth portion of images
(2017) [12]	HCI agent	Principal component analysis (PCA); Fisher linear discriminant analysis (FLD); Fast pixel based matching (FPBM)	**Python**, OpenCV library	These include gabor wavelets, linear discriminant analysis, and support vector machine and Adaboost algorithm	Static facial image (frames)	Human sadness, smile and human expressions
(2013) [13]	FORMAS: multimodal communication for participatory planning and decision analysis: tools and process models	Geometric features such as the shape of the facial components (e.g. eyes, mouth) and the locations of facial points (corners of the eyes, mouth)	Back-propagation of a feed-forward architecture	Feed-forward error back-propagation learning (ANN classifier)	Cohn-Kanade (CK) database	[Anger, disgust, surprise, happiness, sadness and fear]
(2014) [14]	*Geranium* (educative conversational agent)	Natural language generation and text-to-speech synthesizer; multimodal answer generation modules	GUI +ASA + NLU; Voice Extensible Mark-up Language (VoiceXML)	No information (black box)	Event-generated question-answers dialogues	(Neutral, happiness, sadness)

Table 3 Literature survey of competent FED models designed for real-time wild environments (courtesy *Emoti W challenge series*)

Bot name/type	FED principle/feature space	Supporting architecture/tools/APIs	Learning mechanism	Data types	Emotions handled
Emoti W-2018 (Winner) [15]	MTCNN feature extraction: 3D facial landmarks movements using Euclidean distances	Open Smile toolkit, 7-convolution layered deep neural architecture	SoundNet, Inception Net and DenseNet models	RAF [4] and FER2013 [20] datasets	[Angry, disgust, fear, happy, neutral, sad and surprise]
Emoti W-2017 (Winner) [16]	Supervised descent method (SDM): basic Local binary patterns (LBP) and mean LBP feature maps	Ad-a boost based multi-view face detector	Supervised scoring ensemble (SSE): ResNet, DenseNet and HoloNet models	Video clips (EmotiW 2017 dataset)	[Anger, disgust, fear, happy, sad and surprise]
Emoti W-2016 (Winner) [17]	CNN-features-based spatio-temporal RNN model	Open Smile toolkit, Hybrid CNN-RNN, and C3D Networks	HYBRID classifier on LSTMs and C3Ds	Video clips (AFEW 6.0 database); audio-video clips (APPEI database)	[Anger, disgust, fear, happiness, sad, surprise, and neutral]
Emoti W-2015 (Winner) [18]	AU-aware facial features, task-specific graph-structured facial feature relations	AU-aware facial features (open smile toolkit), 5 5-layer CNN models (Caffe toolkit)	Linear support vector machine (SVM) classifier, SVM with radius basis function (RBF) kernel	Annotated static images and short-lived video clips (EmotiW 2015 dataset)	[Angry, disgust, fear, happiness, sadness, surprise and neutral]
Emoti W-2014 (Winner) [19]	Image set models (i.e. linear subspace, covariance matrix, and Gaussian distribution)	Video-audio fusion scheme (OpenSMILE toolkit)	An optimal fusion of (kernel SVM, logistic regression, and partial least-squares) classifiers	Aligned face images (Emoti-W 2014 dataset)	[Angry, disgust, fear, happy, neutral, sad, and surprise]

(continued)

Table 3 (continued)

Bot name/type	FED principle/feature space	Supporting architecture/tools/APIs	Learning mechanism	Data types	Emotions handled
Emoti W-2013 (Winner) [20]	Convolutional neural network (visual data), deep belief net (audio data), K-Means based “bag-of-mouths” model (mouth features), relational autoencoder (Spatio-temporal visual features)	Our ConvNet uses the C ++ and CUDA implementation written by Alex Krizhevsky (Krizhevsky, 2012) interfaced in Python	SVM Classifier	Google dataset	[Angry, disgust, fear, happy, sad, surprise and neutral]

for sustained existence of humans, rescuing them out of mental stress, panic and crucial times of observing COVID prevention and treatment (quarantine and isolation) protocols and more. It is also shortened form of Artificial Intelligence enabled '***S****mart* and ***A****daptive* ***K****nowledge-driven* ***H****uman* ***A****ssistant*' At the conceptual level, *Sakha* is proposed to be designed as advice-seeker in two prominent situations, that can be faced by a person surrounded by the pandemic situation; one while fetching the news updates on pandemic effects and statistics that come from reliable sources like Government bodies and World Health Organization (WHO), and secondly, in building a constructive near-human dialogue during an interactive session, the generation of this very emotion also acts as one of the inputs for the generative functionality of bot (*Sakha*).

Sakha functions for his master (that human who is running an instance of '*Sakha*') by scanning for COVID patients in his proximity 24 $\times$ 7. It is designed to pass on the situation-updates, news-alerts, and statistics in the form of speech signals, also waiting for his master to respond in speech (natural language) formats. This visual bot would also gather his master's mental frame of mind, mood swings, speech, and facial image frames throughout the interactive session and temperaments at some external stimuli. The input to Sakha is mainly audio-video signals of his master through an appropriately designed user interface along with human gestures and emotions, traces of symptoms (if any), both of physical (COVID) and mental ailments, while the output from Sakha is intended to be conversational and counseling dialogue sessions with its master. The last component is proposed to be a recommendation framework, where prompt signals are sent back to the master, which is sure to elevate the man's positive psychology (thinking) and encouraging spirits.

User interface: The user interface is constructed with add-on features to a conventional human-computer interaction model. It provides multi-modal forms of input signals (both voice and image data streams). In turn, the high degree of interaction between the bot and the human component demands intelligently generated human-like dialogues where it is expected to generate new (NL) responses rather than just retrieve from predefined responses, echoing their raised IQ levels.

Automatic Speech Recognizer: This unit has an acoustic model that handles the speech portion of multi-modal input received from the user interface. In turn, it retrieves the free text form of the output in the form of words, forwarding them to the Natural language preprocessing unit; this unit is further transformed into Visual Bot (feature vector) input augmented by Sakha's feature space.

Sakha's Feature Space: One can simulate any of the popularly competent Facial Emotion Detection (FED) feature extraction models described in Table 3 say, for instance, a convolutional neural network for visual data streams, deep belief net for audio data streams, K-Means based "bag-of-mouths" model for extraction of mouth features, the relational autoencoder (Spatio-temporal visual features) for multi-modal feature space of the bot [20].

Multi-Modal Learning: The proposed machine learning model of Sakha rests upon the state-of-art deep learning (DL) models that have been implemented recently in Woe-BOT designed to feasibly engage their human counterparts showing depression symptoms with effective cognitive-behavioral therapies [21]. These DL models are destined to make a sense of raw data representations fed by Sakha's FED feature space through multi-layered algorithms on inter-connected neurons to infer responses to new data streams (facial gestures, spoken dialogues, and derived sentiments and opinions) of human counterparts [22, 23].

Response Selector: functions when the human ignites a thematic interactive dialogue with the bot (treated as test vectors), eventually computes and generates an optimal response (speech and/or text) by learning from each of the three feature categories trained through Multi-modal training step as shown in Fig. 4. In this way, three types of learning tasks can be accomplished, namely, close monitoring of COVID-infection symptoms, interactive dialogue generation, and recommending counseling dialogues. The first task may be a conglomerate of statistical correlations over recorded body temperatures, levels of congestion over screened hoarse-voice data streams sensed from the master, and also gestures classified over generated facial feature vectors lying in Sakha's feature space.

The proposed functionality of Sakha through its graphical user interface is illustrated in steps as shown in Fig. 5. It illustrates the expected sequel of conversations between the visual bot and the human when putting to testing in real-time domain environments.

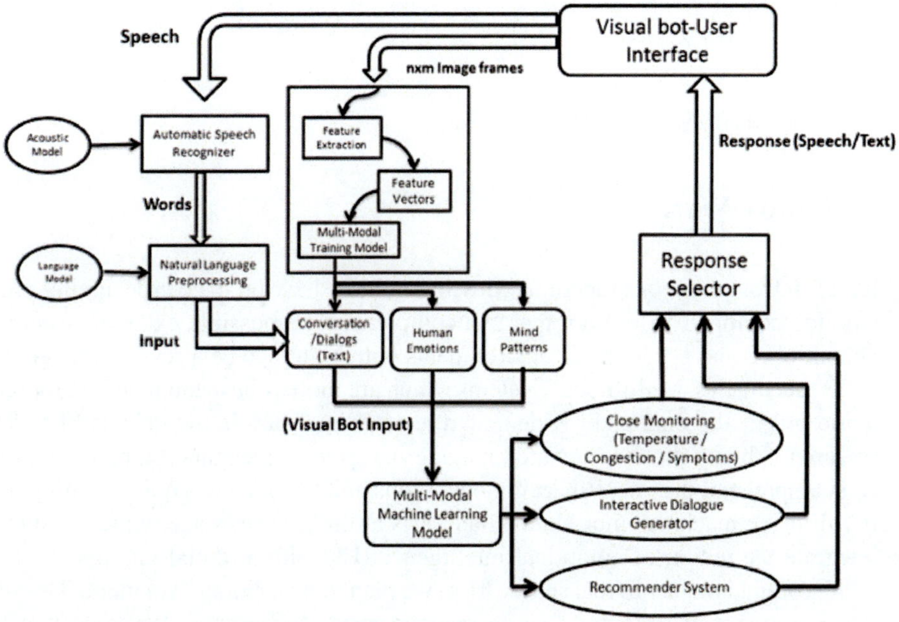

Fig. 4 Proposed architecture of "SAKHA" a Visual-bot

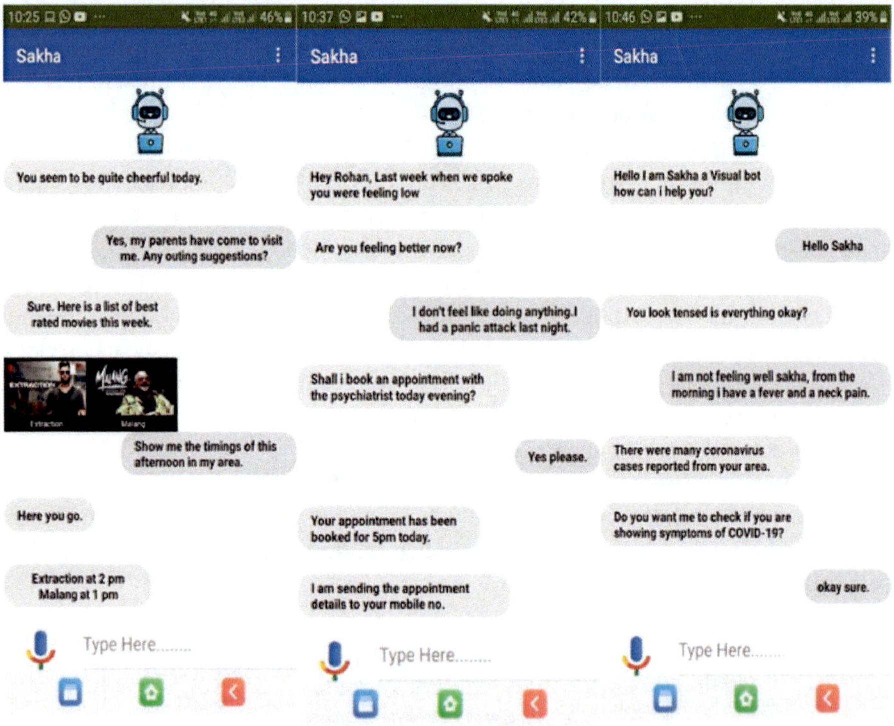

Fig. 5 Few snapshots of "SAKHA" visual bot

4 Future Work

It is clear from the above functionalities that, still, we have far to go in designing a full-emotion-equipped Visual-bot that can comprehend all possible external stimuli of uncontrolled and pose-variant environments with high facial expression recognition (FER) accuracies. Multimodal Systems which are more robust and practical for real-world practical scenarios provide new research dimension in the field of FER. The research is being directed towards making machines understand human emotions, can comprehend the vocal/speech modulations and this idea is a quite thrilling one. It will make machines more human and it is at the nascent stage, which is sure to integrate the notion of emotional intelligence (EI) with artificial intelligence (AI) with profound impact in real sense. Also, we plan to incorporate Augmented Reality into the proposed BOT for a more realistic visual experience. We hope that the proposed BOT will be able to uplift the mood of the people and would prove to be of great help in dealing with people under stress. It can easily be customized for other applications apart from healthcare.

References

1. Cherry, K.: How to Cope With Quarantine. Updated on 18 Mar 2020
2. Centers for Disease Control and Prevention: About Quarantine and Isolation. Updated 27 Jan 2020
3. Novotney, A.: Social isolation: it could kill you. Monit. Psychol. Am. Psychol. Assoc. **50**(5), 32 (2019)
4. https://www.cnbc.com/2018/04/06/how-to-make-sure-youre-investing-with-the-right-robo-advisor.html
5. https://www.valuecoders.com/blog/technology-and-apps/history-and-evolution-of-chatbots/
6. Ko, B.C.: A brief review of facial emotion recognition based on visual information. Sensors **18**(2), 401 (2018)
7. Li, S., Deng, W.: Deep facial expression recognition: a survey. IEEE Trans. Affect. Comput. (2020)
8. Minaee, S., Abdolrashidi, A.: Deep-Emotion: Facial Expression Recognition Using the Attentional Convolutional Network (2019). arXiv preprint arXiv:1902.01019
9. Sridhar, R., Wang, H., McAllister, P., Zheng, H.: E-Bot: a facial recognition-based human-robot emotion detection system. In: Proceedings of the 32nd International BCS Human-Computer Interaction Conference, vol. 32, pp. 1–5 (2018
10. Canedo, D., Neves, A.J.: Facial expression recognition using computer vision: a systematic review. Appl. Sci. **9**(21), 4678 (2019)
11. Gudipati, V.K., Barman, O.R., Gaffoor, M., Abuzneid, A.: Efficient facial expression recognition using AdaBoost and haar cascade classifiers. In: 2016 Annual Connecticut Conference on Industrial Electronics, Technology & Automation (CT-IETA), pp. 1–4. IEEE (2016)
12. Jayalekshmi, J., Mathew, T.: Facial expression recognition and emotion classification system for sentiment analysis. In: 2017 International Conference on Networks & Advances in Computational Technologies (NetACT), pp. 1–8. IEEE (2017)
13. Rázuri, J.G., Sundgren, D., Rahmani, R., Cardenas, A.M.: Automatic emotion recognition through facial expression analysis in merged images based on an artificial neural network. In: 2013 12th Mexican International Conference on Artificial Intelligence, pp. 85–96. IEEE (2013)
14. Griol, D., Molina, J.M., de Miguel, A.S.: Developing multimodal conversational agents for an enhanced e-learning experience. ADCAIJ Adv. Distrib. Comput. Artif. Intell. J. **3**(1), 13–26 (2014)
15. Liu, C., Tang, T., Lv, K., Wang, M.: Multi-feature based emotion recognition for video clips. In: Proceedings of the 20th ACM International Conference on Multimodal Interaction, pp. 630–634 (2018)
16. Hu, P., Cai, D., Wang, S., Yao, A., Chen, Y.: Learning supervised scoring ensemble for emotion recognition in the wild. In: Proceedings of the 19th ACM International Conference on Multimodal Interaction, pp. 553–560 (2017)
17. Fan, Y., Lu, X., Li, D., Liu, Y.: Video-based emotion recognition using CNN-RNN and C3D hybrid networks. In: Proceedings of the 18th ACM International Conference on Multimodal Interaction, pp. 445–450 (2017)
18. Yao, A., Shao, J., Ma, N., Chen, Y.: Capturing au-aware facial features and their latent relations for emotion recognition in the wild. In: Proceedings of the 2015 ACM on International Conference on Multimodal Interaction, pp. 451–458 (2015)
19. Liu, M., Wang, R., Li, S., Shan, S., Huang, Z., Chen, X.: Combining multiple kernel methods on Riemannian manifold for emotion recognition in the wild. In: Proceedings of the 16th International Conference on Multimodal Interaction, pp. 494–501 (2014)
20. Ebrahimi Kahou, S.: Emotion Recognition with Deep Neural Networks. Doctoral dissertation, École Polytechnique de Montréal (2016)
21. Fitzpatrick, K.K., Darcy, A., Vierhile, M.: Delivering cognitive behavior therapy to young adults with symptoms of depression and anxiety using a fully automated conversational agent (Woebot): a randomized controlled trial. JMIR Mental Health **4**(2), e19 (2017)

22. Lee, J.: How chatbots Use AI, Machine Learning, and NLP to Transform Marketing and Sales (2018) https://blog.growthbot.org/how-chatbots-use-ai-machine-learning-and-nlp-to-transform-marketing-and-sales
23. Noda, K., Arie, H., Suga, Y., Ogata, T.: Multimodal integration learning of robot behavior using deep neural networks. Robot. Auton. Syst. **62**(6), 721–736 (2014)

Classification of COVID19 X-ray Images Based on Transfer Learning InceptionV3 Deep Learning Model

Hassan Amin, Ashraf Darwish, and Aboul Ella Hassanien

Abstract The World Health Organization (WHO) has recently announced the novel coronavirus 2019 as a pandemic. Many preventative plans and non-pharmaceutical efforts have emerged and been in use to manage and control the spread of the disease which includes infection control, proper isolation of patients, and social distancing. The main test used to confirm a COVID-19 case is the RT-PCR test. However, this approach needs analysis time and specimen collection. Therefore, the importance of medical imaging is increased to screen COVID-19 cases. Hence radiology has a pivotal role in managing COVID-19 infection using CT scans and chest X-ray (CXR) throughout the screening, diagnosis, and prognostication processes of the disease. In this paper, a new model using the transfer learning method and InceptionV3 algorithm has been presented to classify the X-ray images into COVID-19, Normal, and Pneumonia classes. The experimental results show that the proposed model achieved 98% Accuracy on the test set for classifying the images from the 3 different classes.

Keywords COVID19 X-ray images · Transfer learning InceptionV3 algorithm

H. Amin (✉) · A. Darwish
Faculty of Science, Helwan University, Cairo, Egypt
e-mail: hassanmamin994@gmail.com
URL: http://www.egyptscience.net

A. Darwish
e-mail: ashraf.darwish.eg@ieee.org
URL: http://www.egyptscience.net

A. E. Hassanien
Faculty of Computers and Artificial Intelligence, Cairo University, Giza, Egypt
e-mail: aboitcairo@cu.edu.eg
URL: http://www.egyptscience.net

H. Amin · A. Darwish · A. E. Hassanien
Scientific Research Group in Egypt (SRGE), Giza, Egypt

A. E. Hassanien and A. Darwish (eds.), *Digital Transformation and Emerging Technologies for Fighting COVID-19 Pandemic: Innovative Approaches*, Studies in Systems, Decision and Control 322, https://doi.org/10.1007/978-3-030-63307-3_7

1 Introduction and Related Work

COVID-19 infection has been spreading quickly around the world. In the literature, numerous publications have been published to help radiologists identify COVID-19's effects on the lungs in the X-ray and CT scans [1–4]. With the anticipation of a worldwide increase in the number of cases of COVID-19, we will inevitably witness a rise in demand for chest imaging along with an increased risk of the radiology staff catching the disease due to exposure to infected patients. Severe symptoms of the disease show the same symptoms of pneumonia/lower respiratory tract infection (e.g., low levels of blood oxygen; difficulty in breathing and coughing episodes; CT, chest X-ray, or lung ultrasound changes consistent with consolidation) [5, 6]. One of the main challenges is COVID-19 being a new virus that poses a global threat, so it presents novel challenges in primary prevention which require an effective vaccine and secondary prevention which requires an effective therapeutic compound to treat infected patients. It can also introduce challenges to tertiary prevention that aims at remedying a stable rate of infection or even prevent its recurrence. Although COVID-19 is the most severe SARS outbreak, it is not the first that concerned humanity, and for decades, coronaviruses were one of the serious concerns in veterinary science. However, it has not, but recently, impacted humans rather seriously. Some of the techniques used by radiologists in their diagnosis are the use of X-ray images and CT scans. Which is being employed to effectively detect COVID-19 symptoms? However, they are exhaustive and time-consuming even for expert radiologists. Computer vision and deep learning algorithms such as convolutional neural networks (CNN) can effectively help radiologists in detecting COVID-19 cases [7]. Using chest X-ray images and CT scans, a radiologist can accurately detect if the patient is COVID-19 infected and identify his stage of deterioration or recovery. One of the major effects of COVID-19 on the patient is damaging their lungs, which can be visible in the chest X-ray and CT scans. AI models can accurately detect the lung damage signs caused by COVID-19 in its early stages. The inception of artificial neural networks (ANNs) was applied for binary classification of people infected with COVID-19 and healthy people using 1119 CT scan images [8]. The modified inception model based on transfer-learning concluded an accuracy of 89.5% and 0.87, 0.88, for sensitivity and specificity respectively on the used validation dataset. Adding to that, the model was able to produce an accuracy of 79.3% and 0.67, 0.83 for sensitivity and specificity respectively on an external validation dataset. Hence the findings show that deep learning techniques are reliable in finding COVID-19 radiological signs. A brief report on the epidemiology and imagery of coronavirus syndromes was provided in [9]. The authors described the key characteristics and varieties in chest radiography (Chest CT) for SARS-infected MERS coronavirus (which got reported for the first time in Saudi Arabia in 2012), as well as for COVID-19 (a recent study in The Lancet). Overall, the findings of the imaging are highly unspecific and may overlap with the symptoms of influenza H1N1, pneumonitis cytomegalovirus, or atypical pneumonitis. Moreover, the authors presented the robust, safety measures, and anticipations for the radiology medical staff and the employees to take care of the patients even he is known COVID-19 or just with known or just suspected COVID-19. In [10] Authors focus on the important role of chest imaging whether in diagnosing stage or in the following up the patients. Similar to other coronavirus Chest imaging for COVID-19 patient shows widespread asymmetric airspace opacities. For 40 of 41 COVID-19 patients, the first report described that on the initial chest CT there is bilateral lung involvement. The report describes also for a patient who in the ICU there is also a consolidative pattern. While for patients who were not in the ICU there is also a predominantly ground-glass pattern [11].

2 Basics and Background

2.1 COVID-19 Pandemic

In December 2019 in Wuhan, China, the first incident of the novel coronavirus was reported [12]. On January 30th, 2020 the World Health Organization (WHO) stated the COVID-19 outbreak as a public health emergency of global concern.

Over the ensuing months, reports have emerged of the worldwide spread of COVID-19 in each inhabited continent. As the data we have for COVID-19 continue to give us more insights about this disease, we still have limited information regarding its effects on pregnancy [13]. Management is mainly focused on providing supportive care, having oxygen therapy to be the biggest treatment intervention. Antivirals and Corticosteroids have been advised to be used in critical scenarios. However, so far no antiviral has been clinically proven to treat COVID-19 patients and no vaccine is yet produced to help with the disease.

Radiological findings can also help to diagnose pneumonia in virally infected patients. Bilateral and Multi lobe lung involvement was common in over 75 and 71% of adult patients. Although the virus can spread through the air, disinfecting the air is not proven to yield any benefits for stopping the spread of the virus. Human to human infection must be the main focus to limit the spread of the virus. Since personal protective equipment resources are currently in short supply, their use should be carefully considered. In particular, surgical masks are being widely used among communities although no clinical proof shows that they have a role in reducing or preventing the acquisition of COVID-19. Infected individuals that have been identified should rapidly be isolated and provided with optimized care. Patients suspected of infection should also be given a medial mask isolated in a room whenever possible. Employing properly ventilated single rooms while implementing aerosol-generating procedures (AGPs) is critical to limit the spread among suspected patients and the health care workers. Patients should get clear instructions to cover their mouth and nose during sneezing or coughing using tissues. Enforcing hand hygiene after contact with respiratory secretions must get employed, and whenever possible, having dedicated equipment or using disposable ones (e.g. thermometers and stethoscopes) for each suspected case.

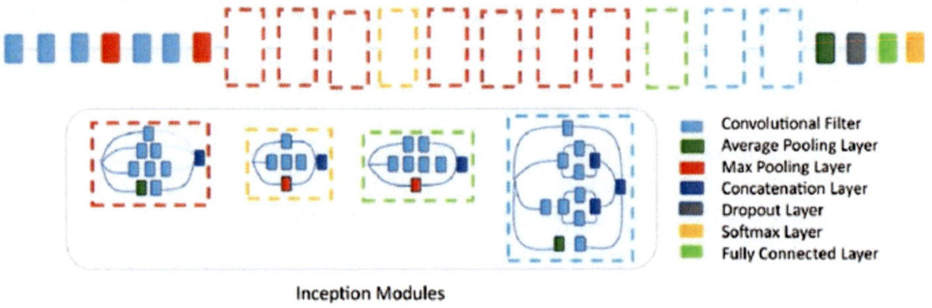

Fig. 1 Convolutional neural network architecture and Inception V3

2.2 InceptionV3 Algorithm

InceptionV3 [14] is a convolutional neural network that is based on the idea of an inception module as shown in Fig. 1. In a traditional convolutional neural network layer, we take the output of one layer and that would be the input to the next layer, and this pattern is followed until the prediction. While in an inception module it takes the previous layer input and passes it to four different operations in parallel and concatenates the outputs from all of these different layers to be able to learn better representations of the data. It was trained on more than 1000 classes and more than 1.4 million images from the ImageNet dataset [15] giving the state of the art performance. The Inception algorithm is used to extract features from images that eventually gets passed to a classifier that gets trained to learn mappings from the extracted features and the target classes of the images.

3 The Proposed Classification Model

In this section, the proposed classification model is discussed. The dataset description that has been used in this paper and the different phases of the proposed model are presented. Figure 2 describes the proposed model phases.

3.1 Dataset Description

In this chapter, a new dataset was generated based on two other datasets:

- Chest X-ray Images (Pneumonia) dataset available on Kaggle [16] which is a collection of 5863 X-ray images of 2 classes (Normal/Pneumonia).
- COVID-19 images dataset available on GitHub [17], the images were extracted from various publications and stored on GitHub as centralized storage for

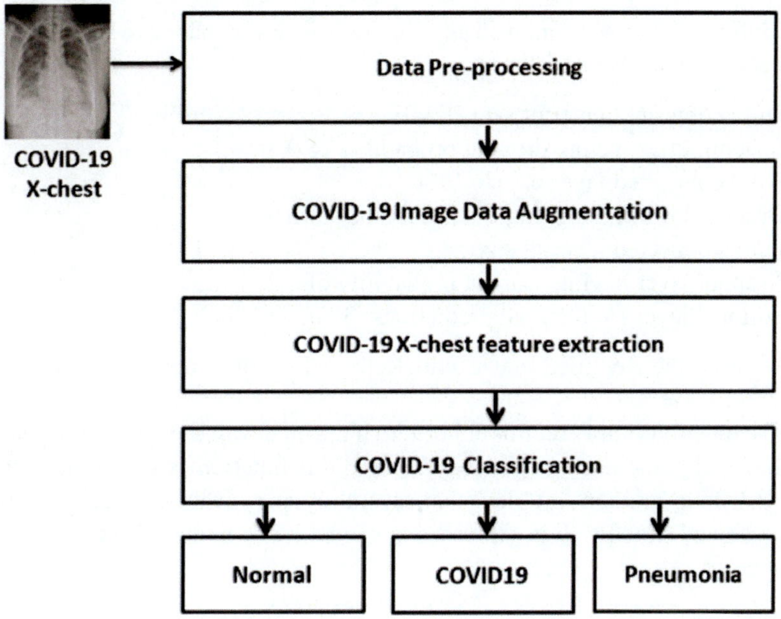

Fig. 2 Framework for COVID-19 classification

them. The dataset as of the current time has 162 X-ray images for COVID-19 confirmed cases.

3.2 *Experimental Implementation*

Image classification tasks have two main steps; feature extraction and classification. The diagram representing our framework is shown in Fig. 2. We first collected the images from the 2 datasets [16, 17] to generate the used dataset for training and testing. 194 images from the original Pneumonia dataset has been randomly selected for both the Normal and Pneumonia categories, and the full 163 images from the COVID-19 dataset has been selected. The data was split into 80:20 train: test splits, totaling 130, 156, 156 images for COVID19, Pneumonia, and Normal respectively in the training set. While in the testing set the images counts were 32, 38, 38 for COVID19, Pneumonia, and Normal respectively.

Data pre-processing is performed by resizing the images to 299 $\times$ 299 pixels and Data augmentation is performed on the training set in the form of image shear with 20% and 10-degree rotation. This data is fed into a feature extractor algorithm to extract the feature descriptor values. The Inception V3 model is used for the feature extraction stage, it is a model that was trained on more than 1000 classes and more than 1.4 million images from the ImageNet dataset, it is an image recognition model that extract features from images using Convolutional Neural Networks. Its last Softmax layer was removed and 7 more layers were added to better fine-tune the model on the dataset as follows:

- Fully connected layer of size 1024
- Dropout layer having dropout probability of 0.5
- Fully connected layer of size 1024
- Dropout layer having dropout probability of 0.5
- Fully connected layer of size 512
- Dropout layer having dropout probability of 0.2
- Softmax layer for the final predictions (3 output classes).

We used the Rectified linear unit (Relu) activation function for the added fully connected layers.

The function simply returns either zero if the input value less than or equals to zero, or the input value itself otherwise. We used this function as it does not suffer from the vanishing gradient problem that several other activation functions suffer from like tanh and sigmoid. It is expressed mathematically in the following equation:

$$G(z) = \max(0, z) \tag{1}$$

Here, G() is the Relu function and z is the input to the function, the function returns zero if the input value is negative otherwise it returns the input value.

The dataset is divided with ratios of 80 and 20, 80% is used in the training process and 20% to test the results. From the 80% of the training data, a split of 10% was used to validate the results of the model during the training process. Keras python framework was used for constructing the model, Epochs are set to 150 and we used ReduceLROnPlateau callback function to reduce the value of the learning rate when the model stops improving based on monitoring the validation loss value.

3.3 *Results, Discussion, and Analysis*

The proposed model is trained on a local machine with configurations as Ryzen 5 3600 CPU @ 3.6 × 6 GHz, 32 GB of DDR4 RAM and an NVIDIA RTX 2080 SUPER Graphics card @ 1815 MHz Boost clock and 8 GB of VRAM, images used for the analysis were resized to 299 × 299 pixel for all the 3 categories. Data augmentation is performed in the form of image shear with 20% and 10-degree rotation. Samples for the three classes of images are shown in Fig. 3.

The following Fig. 4 shows the accuracy, loss, validation accuracy, and validation loss during the training process.

After training, the proposed model was able to achieve 98% accuracy on the test set for classifying the images from the 3 different classes (COVID19, Normal, and Pneumonia).

The generated confusion matrix is in the following Table 1.

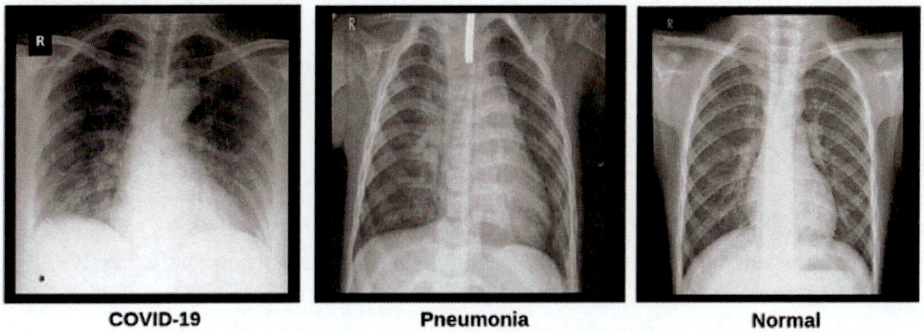

Fig. 3 Example of X-ray images for COVID19, Pneumonia, and normal classes

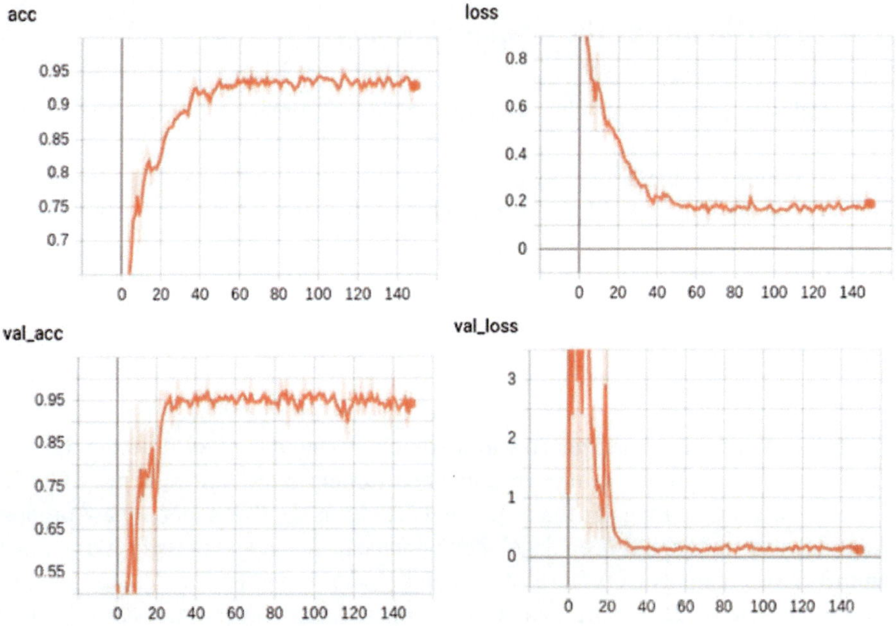

Fig. 4 Accuracy, loss, validation accuracy, and validation loss

Table 1 The confusion matrix

Item	COVID19	NORMAL	Pneumonia
COVID19	32	0	0
NORMAL	0	37	1
Pneumonia	0	1	37

Table 2 Precision, recall and F1 score for the proposed model

Item	Precision	Recall	F1-score
COVID-19	1.00	1.00	1.00
Pneumonia	0.97	0.97	0.97
Normal	0.97	0.97	0.97

As shown by the confusion matrix, the model was able to successfully classify all 32 images of COVID19 in the test set and misclassified only 1 image of the NORMAL class as Pneumonia and 1 Pneumonia class as NORMAL.

3.4 The Validity of the Proposed Model

The following table shows the experimental results, precision, recall, and F1 score on the test set in Table 2.

As shown in Table 2, the model has 100% for the COVID-19 class for precision, recall, and F1-score, and having 97% across all metrics for both Pneumonia and Normal classes.

4 Conclusion

COVID-19 infection has been spreading quickly around the world. In the literature, numerous publications have been published to help radiologists identify COVID-19's effects on the lungs in the X-ray and CT scans. In this paper, a new proposed classification model has been developed which can detect and classify X-ray images into 3 classes, namely COVID19, Pneumonia, and Normal. The model was developed based on InceptionV3 using transfer learning and extra layers were added to better fine-tune the model on the dataset. Our experimental results show that the proposed model was able to accurately classify the three classes with 98% accuracy on the test set used.

References

1. Zu, Z.Y., Jiang, M.D., Xu, P.P., et al.: Coronavirus disease 2019 (COVID19): a perspective from China. Radiology (2020). https://doi.org/10.1148/radiol.2020200490
2. Kanne, J.P., Little, B.P., Chung, J.H., et al.: Essentials for radiologists on COVID-19: an update—radiology scientific expert panel. Radiology (2020). https://doi.org/10.1148/radiol.2020200527
3. Kim, H.: Outbreak of novel coronavirus (COVID-19): What is the role of radiologists? Eur. Radiol. **30**, 3266–3267 (2020). https://doi.org/10.1007/s00330-020-06748-2

4. Hosseiny, M., Kooraki, S., Gholamrezanezhad, A., et al.: Radiology perspective of coronavirus disease 2019 (COVID-19): lessons from severe acute respiratory syndrome and the Middle East respiratory syndrome. Ame. J. Roentgenol. **214**(5), 1078–1082
5. Pan, F., Ye, T., Sun, P., et al.: Time course of lung changes on chest CT during recovery from 2019 novel coronavirus (COVID-19) pneumonia. Radiology **295**(3), (2020) https://doi.org/10.1148/radiol.2020200370
6. Shi, H., Han, X., Jiang, N., et al.: Radiological findings from 81 patients with COVID-19 pneumonia in Wuhan, China: a descriptive study. Lancet. Infect. Dis. **20**(4), 425–434 (2020)
7. Brunese, L., Mercaldo, F., Reginelli, A., Santone, A.: Explainable deep learning for pulmonary disease and coronavirus COVID-19 detection from X-rays. Comput. Methods Programs Biomed. **196** (2020). Article#105608
8. Wang, S., Kang, B., Ma, L., Zeng, X., Xiao, X., Guo, J., Cai, M., Yang, J., Li, Y., Meng, X., et al.: A deep learning algorithm using CT images to screen for coronavirus disease (COVID-19). (2020) medRxiv preprint medRxiv:2020.02.14.20023028
9. Tan, L., et al.: Lymphopenia predicts disease severity of COVID-19: a descriptive and predictive study. Signal Transduct. Targeted Ther. **5**(1), 1–3 (2020)
10. Wang, D., Hu, B., Hu, C., et al.: Clinical characteristics of 138 hospitalized patients with 2019 novel coronavirus-infected pneumonia in Wuhan. China. Jama **323**(11), 1061–1069 (2020)
11. Bao, C., Liu, X., Zhang, H., Li, Y., Liu, J.: COVID-19 computed tomography findings: a systematic review and meta-analysis. J. Am. Coll. Radiol. (2020). https://doi.org/10.1016/j.jacr.2020.03.006
12. Berghella, V.: Now! protection for obstetrical providers and patients. AJOG MFM 2020, in press (2020). Available at: https://www.sciencedirect.com/science/article/pii/S2589933320300392
13. Li, N., Han, L., Peng, M., Lv, Y., Ouyangg, Y., Liu, K., et al.: Maternal and neonatal outcomes of pregnant women with COVID-19 pneumonia: a case-control study. Clin. Infect. Dis. (2020). https://www.medrxiv.org/content/10.1101/2020.03.10.20033605v1. Last accessed: 2 Apr 2020
14. Dong, N., Zhao, L., Wu, C.H., Chang, J.F.: Inception v3 based cervical cell classification combined with artificially extracted features. Appl. Soft Comput. **93** (2020). Article#106311
15. Russakovsky, O., Deng, J., Su, H., Krause, J., Satheesh, S., Ma, S., Berg, A.C.: Imagenet large scale visual recognition challenge. Int. J. Comput. Vision **115**(3), 211–252 (2015)
16. Chest x-ray Pneumonia dataset. https://www.kaggle.com/paultimothymooney/chest-xray-pneumonia
17. COVID-19 chest x-ray dataset. https://github.com/ieee8023/covid-chestxray-dataset

Artificial Intelligence Approach to Predict the COVID-19 Patient's Recovery

Aya Salama, Ashraf Darwsih, and Aboul Ella Hassanien

Abstract Coronavirus is the new pandemic hitting all over the world. Patients all over the world are facing different symptoms. Most of the patients with severe symptoms die especially the elderly. In this chapter, three machine learning techniques have been chosen and tested to predict the patient's recovery of Coronavirus disease. The support vector machine has been tested on the given data with a mean absolute error of 0.2155. The Epidemiological data set is prepared by researchers from many health reports of real-time cases to represent the different attributes that contribute as the main factors for recovery prediction. Deep analysis with other machine learning algorithms including artificial neural networks and regression models has been tested and compared with the SVM results. The experimental results show that most of the patients who could not recover had a fever, cough, general fatigue, and most probably malaise.

Keywords Coronavirus · COVID-19 · Multi-layer perception · Support vector machine · Regression · Artificial intelligence

A. Salama · A. E. Hassanien (✉)
Faculty of Computers and Artificial Intelligence, Cairo University, Cairo, Egypt
e-mail: aboitcairo@gmail.com
URL: http://www.egyptscience.net/

A. Salama
e-mail: aya.salama@aucegypt.edu
URL: http://www.egyptscience.net/

A. Salama · A. Darwsih · A. E. Hassanien
Scientific Research Group in Egypt (SRGE), Cairo, Egypt
e-mail: Ashraf.darwish.eg@ieee.org
URL: http://www.egyptscience.net/

A. Darwsih
Faculty of Science, Helwan University, Cairo, Egypt

A. E. Hassanien and A. Darwish (eds.), *Digital Transformation and Emerging Technologies for Fighting COVID-19 Pandemic: Innovative Approaches*, Studies in Systems, Decision and Control 322, https://doi.org/10.1007/978-3-030-63307-3_8

1 Introduction

Coronavirus or what the so-called COVID-19 is the most booming topic now due to the enormous spread in a very short duration, causing a significant number of deaths [1, 2]. Corona is an infectious decease created by severe acute respiratory syndrome coronavirus 2. This virus first appeared in China and then it spread to the whole world killing people everywhere. Most of the patients cannot overcome this decease. Hospitals do not accept all cases due to the limited number of beds especially in developing countries like Egypt. However, hospitals admit cases based on their severity. Corona is now the common enemy to the whole world [3, 4]. People in the front line need all the possible ways of technology that could help. Artificial intelligence and machine learning already proved their significance much similar decease. Even now with corona, machine learning, and technology, in general, is one of the main weapons enabling us to face that pandemic virus [5, 6]. Artificial Intelligence has been used in [7] to detect the virus where the possible next outbreak is. However, in this work, we are trying to conduct a more helpful tool for the white army in frontlines to be able to predict the severity of the cases by estimating the recovery possibility of patients. We trained our models on the data set given by online [8].

The data set given by BoXu et al. showed that in some cases, patients with no symptoms at all died by the Coronavirus. For this reason, in this chapter, we provide preliminary experiments to predict the severity of the case as it may be manipulating to just depend on the obvious symptoms. The number of patients increases exponentially daily. However, the symptoms of most of the patients were not recorded in the Data set. Moreover, some records had symptoms but the final state of the patients of weather they recovered or died was not submitted yet. For this reason, we worked only on 108 records with symptoms. In addition to symptoms, age was also given. Moreover, we found that the factor of traveling to one of the cities where the virus is spread is also an important factor. Therefore, we added this factor to the inputs to the implemented classifier. These cities are China, France, and Italy.

The remainder of this chapter is organized as follows. Section 2 presents the COVID-19 overview. Section 3 presents some machine learning classifiers with the implementation and experimental results. Section 4 concludes this chapter.

2 COVID-19: Overview

At the beginning of April 2020, more than 1,000,000 confirmed cases of COVID 19 and more than 50,000 deaths have existed globally where humanity in our plant currently lives in COVID 19 pandemic. Infected cases data of COVID-19 can be detected using several medical means. These means of COVID-19 medical diagnosis include clinical characteristics and radiologist's diagnosis. Clinical characteristics involve human temperature monitoring and reverse transcription-polymerase chain reaction (RT-PCR). With the daily rapid growth in the number of newly confirmed and suspected COVID-19 cases, the rate of information evolution has become extremely fast and it became a critical issue to handle this huge amount of data every day. To the date (Mar 30th, 2020), there have been 769,826 confirmed cases and 36,934 deaths all around the world [9] and the rate is rapidly increasing. Big data analytics tools must handle and analyze the massive amount of big medical data quickly since time is a very significant issue in healthcare applications that affect the containment of the disease.

Advanced data analytics tools are needed to extract knowledge from patients' data to develop original methods to early diagnose cases with COVID-19. Big data analytics is the process of analyzing COVID-19 huge data collected from different sources (Chest CT scans and X-Ray images, RT-PCR data, CBC, sensors, social networks, and medical reports) [10]. COVID-19 big data Analytics is used to extract hidden patterns, valued perceptions relationships from the confirmed and suspected cases. Healthcare workers then take these results and conclude some findings that help in CONID-19 pandemic containment.

Based on the web of science database as in Table 1, the record count in each column is the total number of articles published in the selected year. The count includes Early Access articles that are fully peer-reviewed, citable, and published but have not been assigned a volume/issue/page number.

The main test to detect the COVID-19 confirmed cases is RT-PCR. However, this approach needs analysis time and specimen collection. Therefore, medical imaging is important for screening COVID-19 cases [11–13]. Two days after the onset of flu-like symptoms from COVID-19, about 50% of COVID-19 confirmed cases have normal CT. Based on chest CT scans, pneumonia patients due to COVID-19 have lung abnormalities [14]. However, COVID-19 RT-PCR may be initially negative due to the low sensitivity of COVID-19 RT-PCR. About 50–75% of COVID-19 patients have lung abnormalities such as multifocal ground-glass opacities or peripheral focal based on the early COVID-19 infection. During the COVID-19 progresses, significant consolidation and paving dominate the finding of the CT scans and reach its peak within 9–13 days.

Due to possible epidemics and potential global risks, the outbreak of the latest coronavirus (COVID-19), detected in late 2019, needs particular attention. In addition to clinical practices and therapies where machine learning is a new paradigm for health care, several different Machine Learning Algorithms (ML)-based machine learning methods are used for data analysis and decision-making processes. That means machine learning-based systems help detect COVID-19 outbreaks and predict the existence of their worldwide spread as well. Using machine learning software, new coronavirus outbreaks can be detected and the extent of their spread worldwide awaited. The basic theory behind machine learning-based devices, however, is that they need sufficient training data (for all possible cases. Often, traditional machine learning requires a clean set of annotated data so that classifiers can be well trained which falls within the scope of supervised learning.

Table 1 COVID related sciences from the web of science

No.	Field: web of science categories	Record count	% of 107
1	Medicine general internal	42	39.252
2	Infectious diseases	9	8.411
3	Public environmental occupational health	6	5.607
4	Critical care medicine	5	4.673
5	Medicine research experimental	5	4.673
6	Virology	5	4.673
7	Mathematical computational biology	4	3.738
8	Radiology nuclear medicine medical imaging	4	3.738
9	Biology	3	2.804
10	Multidisciplinary sciences	3	2.804
11	Pediatrics	3	2.804
12	Pharmacology pharmacy	3	2.804
13	Cell biology	2	1.869
14	Immunology	2	1.869
15	Oncology	2	1.869
16	Tropical medicine	2	1.869
17	Anesthesiology	2	0.935
18	Biochemical research methods	1	0.935
19	Biochemistry molecular biology	1	0.935
20	Biotechnology applied microbiology	1	0.935
21	Business finance	1	0.935
22	Cardiac cardiovascular systems	1	0.935
23	Chemistry multidisciplinary	1	0.935
24	Dentistry oral surgery medicine	1	0.935
25	Dermatology	1	0.935

3 Some Machine Learning Classifiers for COVID-19

Machine learning is the main branch of AI [15–17]. Artificial Neural Network (ANN) and Support Vector Machine (SVM) is the most known machine learning approaches used in multiple domains. ANN simulates the neurons in human brains, where nodes in the neural network represent the neurons. For non-linearly separable problems, more than one hidden layer is used and each layer consists of one or more nodes [18]. ASVM is very similar to ANN. However, ASVM depends on the kernel trick in which the features are converted into higher dimensional space to be linearly separable. Then, optimization techniques are used to maximize the hyperplane between classes to obtain the best classification results [19–21].

3.1 Artificial Neural Networks (ANNs)

ANNs are inspired by the biological behavior of brain networks. It has obtained an interest and used in various applications such as pattern and speech recognition and disease diagnosis. The backbone of the ANNs development is the neural network model. Recently, ANNs have become well known and helpful models in some approaches such as classification, clustering, prediction, and in many disciplines.

ANN works are based on layers of hidden nodes as mentioned before. ANN learns by training on labeled data sets. This way of learning is called supervised learning. Training usually takes several iterations. In each iteration, classification error is computed. This error is used to update weights on outputs nodes. Then these errors are backpropagated to update the weights on hidden nodes using the chain of first derivatives. This process takes place until the minimum classification error is reached or the maximum number of iterations [22, 23].

The data analysis factors explain the importance of ANNs which is efficient and successful in providing a high level of capability to be used with complex and non-complex problems. ANNs have an advantage is that it can make models more accurate and easy to use from complex problems. The ANN is a good model that can be used with medical applications.

3.2 Support Vector Machine (SVM)

SVM is a computational power kernel-based tool for data regression and classification. Compared with other machine learning techniques, SVM has better generalization performance. Therefore, SVM has achieved a high level of performance in many real-world applications such as image processing of medical applications.

SVM can be considered as ANN if sigmoid function as activation functions to update weights. However, instead of using multiple layers to solve the nonlinear problem, SVM tends to move the features to higher dimensions that it can be separable by a hyperplane. Then the goal is to maximize this hyperplane because the more features are separated, the more accurate classification becomes [24, 25]. Equations from 1 express the SVM equations.

Consider that there are a series of data points $D = [x_i, d_i]_i^n$ where n is the size of data and d_i represents the target value and x_i represents the input space vector of the sample. The SVM estimates the function as given in the following two equations:

$$f(x) = \omega\varphi(x) + b \tag{1}$$

$$R_{SVMs}(C) = \frac{1}{2}\omega^2 + C\left(\frac{1}{n}\right)\sum_{i=1}^{n} L(x_i, d_i) \tag{2}$$

where $\varphi(x)$ is the high-dimensional space feature, b is a scalar, ω is a normal vector and $C(\frac{1}{n})\sum_{i=1}^{n} L(x_i, d_i)$ signifies the empirical error. *b* and *w* can be assessed by Eq. (2).

3.3 Data Sets Characteristics and Analysis

The data set from the health report all over the world gave the following information: ID, age, sex, city, province, country, Wuhan-0, not-Wuhan-1, latitude, longitude, geo-resolution, date-onset-symptoms, date-admission-hospital, date-confirmation, symptoms, lives-in-Wuhan, travel-history-dates, travel-history-location, reported-market-exposure, additional-information, chronic-disease-binary, chronic-disease, source sequence-available, outcome, date-death-or-discharge, notes-for-discussion location, travel-history-binary. The sample of the available data online is shown in Table 2.

After processing the data, we found that most of the patients who couldn't recover had a fever, cough, general fatigue, and most probably malaise. Besides, most of the patients who died live in Wuhan in china or visited Wuhan, France, Italy, or Iran. For this reason, we used this data to classify the patients. Also, patients who couldn't recover from the Coronavirus are above 50. Therefore after processing the data, the attributes that contributed in classification are the following: age, lives in China (Boolean), visited recently Italy (Boolean), china, France or Itan (Boolean), has fever (Boolean) has cough (Boolean), has a sore throat (Boolean), has diarrhea (Boolean), has a general weakness (Boolean), has nosal problem (Boolean), has a headache (Boolean), has malaise (Boolean), has penomedia (Boolean). The sample of the processed data is shown in Table 3.

The visualization of data is also given in Fig. 1. This figure represents a histogram for each attribute. As obvious in this chart that patients with age above 50 are more vulnerable to get infected. It also shows that most of the studied cases didn't suffer from sneezing o diarrhea or nasal problems. However, the more common symptoms are fever and cough.

All the data set given by the WHO organization or any other responsible committee concerned about patients diagnosed positively with Corona only. For this reason, we could not use this data for deciding if patients have Corona or not.

3.4 Methodology and Results

In this work, we tested the artificial neural network, support vector machine and linear regression model in recovery estimation of Corona patients as illustrated in Fig. 2.

Table 2 The collected data from hospital reports

ID	Age	Sex	City	Province	Country	Wuhan 0	Latitude	Longitude	Geo_resol	Date_onset	Date_adm
				Tucuman	Argentina		−26.94	−65.34	Admin1		
				Buenos Ai	Argentina		−34.6033	−58.3817	Admin1		
				Buenos Ai	Argentina		−34.6033	−58.3817	Admin1		
				Buenos Ai	Argentina		−34.6033	−58.3817	Admin1		
				Cordoba	Argentina		−31.4167	−64.1833	Admin1		
				Cordoba	Argentina		−31.4167	−64.1833	Admin1		
				Cordoba	Argentina		−31.4167	−64.1833	Admin1		
				Cordoba	Argentina		−31.4167	−64.1833	Admin1		
				Cordoba	Argentina		−31.4167	−64.1833	Admin1		
				Cordoba	Argentina		−31.4167	−64.1833	Admin1		

Table 3 The processed data set

Throat	Pneumon	Weakness	Sneezing	Nosal	Diarrhea	Breath	Head	Traveled	Recovered
0	0	0	0	0	0	0	1	1	1
1	0	0	0	0	0	0	0	1	1
0	0	1	0	0	0	0	0	1	1
0	0	0	1	0	0	0	0	1	1
0	1	0	0	0	0	0	0	1	1
0	1	0	0	0	0	0	0	0	1

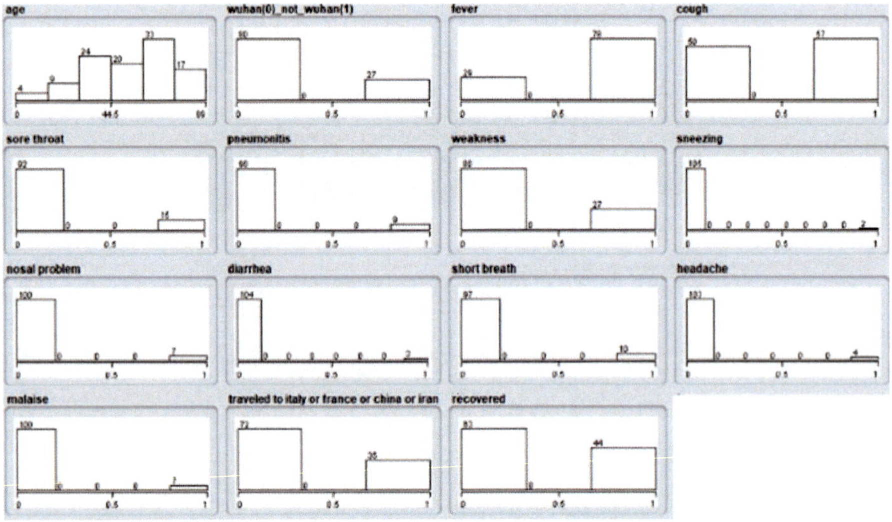

Fig. 1 Data visualization

For the multi-layer perception, the learning rate was 0.3, momentum was 0.2, and the sigmoid function was used with 10-fold cross-validation. All these experiments were carried out using the Weka classifier.

The mean square error and the absolute error of the different conducted experiments are elaborated in Table 4.

As obvious, SVM resulted in the minimum absolute error. However, Linear regression resulted in the minimum root mean squared error which is slightly less than resulted from the SVM. For this reason, SVM can be considered as the best model that can be used for recovery prediction. On the other side, Multi-layer perception was dramatically below the other methodologies tested for prediction.

Since most of the patients now are not from China. These experiments were also carried out by removing the Boolean attribute of living in China. As elaborated in Table 5, SVM still results in the minimum classification error among all the other classifiers. However, ANN performed better after removing the Boolean attribute of living in China.

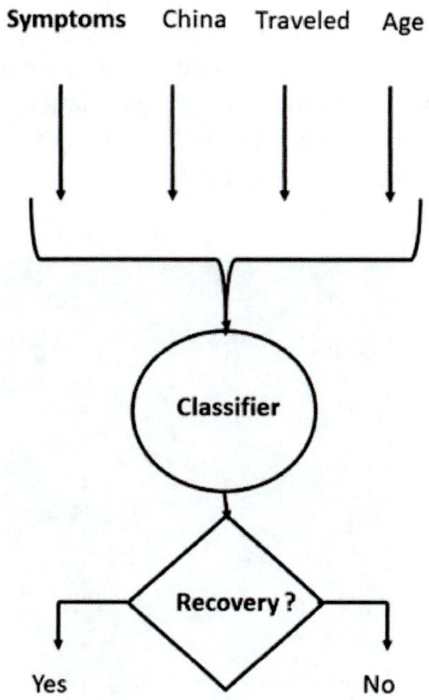

Fig. 2 Summarized chart of the proposed approach

Table 4 Experimental results

The used technique	Mean absolute error	Root mean squared error
SVM	0.21	0.46
ANN	0.53	0.702
Regression	0.3001	0.3894

Table 5 Results after removing China attribute

	Mean absolute error	Root mean squared error
SVM	0.21	0.46
ANN	0.46	0.61
Regression	0.34	0.41

The experiments were also conducted on the symptoms only to cover records without traveling history. However, as illustrated in Table 5, there was a significant increase in the classification error as shown in Table 6. Therefore, symptoms alone are not enough to decide the severity of Corona cases.

Table 6 Results after working on symptoms only

	Mean absolute error	Root mean squared error
SVM	0.38	0.48
ANN	0.48	0.61
Regression	0.36	0.43

Graphical representations of the impact of using different combinations of attributes on classification error are given in Figs. 3 and 4.

As obvious in the correlation coefficient matrix given in Fig. 5. Data are weakly correlated. So we cannot say that if the patient has a fever, then he should have a cough for example.

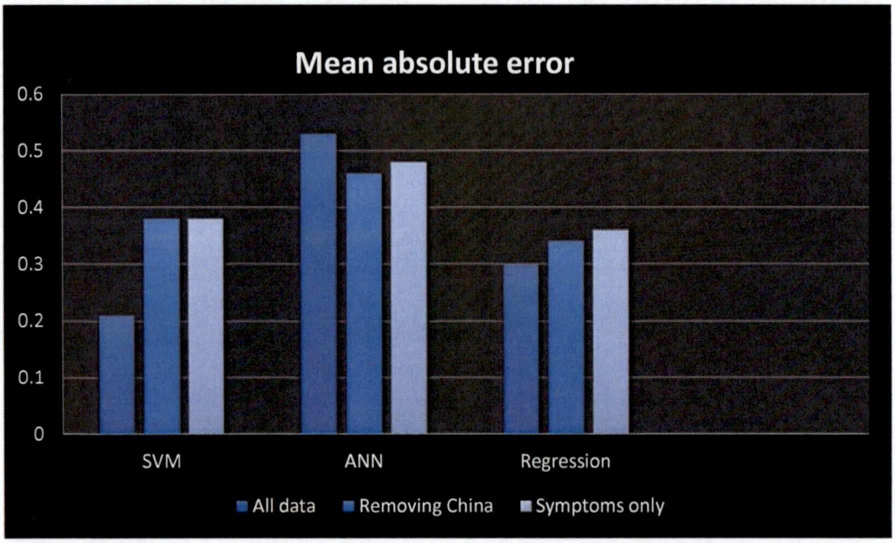

Fig. 3 Graphical representation of the impact of different attributes set on mean absolute error

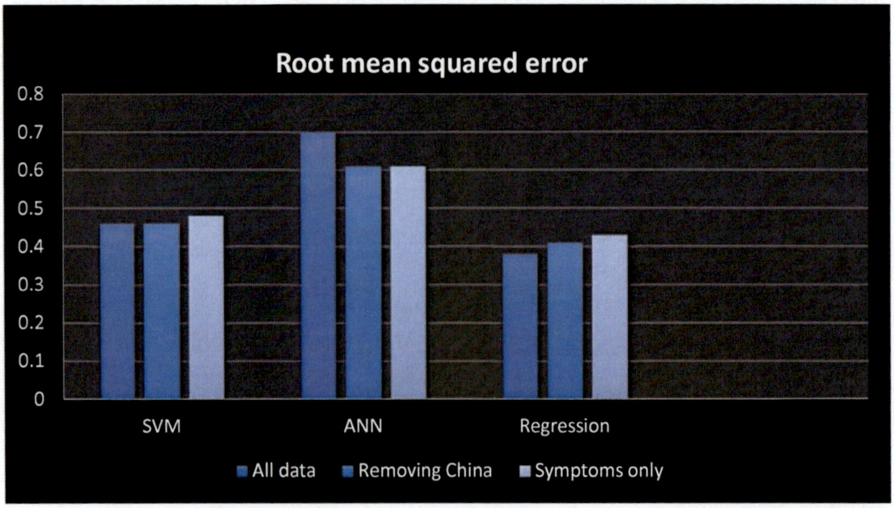

Fig. 4 The impact of different attributes set on the root mean squared error

	age	China	fever	cough	throat	pneumon	weak	sneez	nosal	diarrhea	short brea	heada	malais	travelec
age	1													
China	-0.2	1												
fever	-0.1	-0.08	1											
cough	0.01	-0.23	0.27	1										
throat	-0.2	-0.17	-0.06	0.05	1									
pneun	0.15	-0.02	-0.42	-0.26	0.072	1								
weekr	0.11	0.059	-0.13	-0.06	-0.11	-0.09853	1							
sneez	-0.1	-0.08	0.08	0.13	0.143	-0.04182	0.08	1						
nosal	-0.1	-0.07	0.08	0.02	0.22	-0.08018	0.02	0.24	1					
diarrh	-0.1	0.032	-0.02	-0.07	-0.07	-0.05147	-0.1	-0.02	-0.04	1				
breath	0.17	0.035	-0.09	-0.09	-0.04	-0.0973	0.04	-0.04	0.04	-0.0545	1			
heada	-0.1	-0.11	0.12	0.09	0.204	-0.05972	-0	-0.03	0.15	-0.0335	-0.06327	1		
malaia	-0.1	-0.07	0.16	0.17	0.111	-0.08018	-0.2	-0.04	-0.07	-0.0449	-0.08495	0.15	1	
travel	-0.4	0.191	0.11	0.01	0.235	0.147567	-0.1	0.05	0.14	-0.1184	-0.08699	0.07	-0.02	1
recove	-0.5	0.214	0.04	-0.09	0.21	0.157327	-0.1	0.02	0.09	-0.0269	-0.07257	-0.06	0.009	0.5509

Fig. 5 Correlation coefficient matrixes

4 Conclusion and Future Work

In this chapter, different classification models were tested to make the best use of the clinical data provided online to be able to predict the severity of the corona cases. SVM on 15 attributes of symptoms and other relevant information of patients achieved minimum classification error of 0.21 which proves the feasibility of the proposed approach. We also proved that symptoms alone cannot help in deciding the severity of the cases. For future work, if data sets can be gathered by researchers or WHO organization or based on personal efforts to include symptoms and other information of suspects of Corona to be able to diagnose that new coronavirus. Moreover, data is changing and is added every minute. As a result, more records can be usable by our model.

References

1. Bogoch, I.I., et al.: Pneumonia of unknown etiology in Wuhan, China: potential for international spread via commercial air travel. J. Travel Med. (2020). https://doi.org/10.1093/jtm/taaa008

2. Elmousalami, H.H., Hassanien, A.E.: Day level forecasting for coronavirus disease (COVID-19) spread: analysis, modeling, and recommendations. arXiv preprint arXiv:2003.07778 (2020)
3. Xu, B., et al.: Epidemiological data from the COVID-19 outbreak, real-time case information. Sci. Data Nature **7** (2020)
4. Novel, C.P.E.R.E.: The epidemiological characteristics of an outbreak of 2019 novel coronavirus diseases (COVID-19) in China. Zhonghua liu xing bing xue za zhi = Zhonghua liuxingbingxue zazhi **41**(2), 145 (2020)
5. Dong, E., Du, H., Gardner, L.: An interactive web-based dashboard to track COVID-19 in real time. Lancet Infect. Dis. **3099**, 19–20 (2020)
6. Hutter, F., Kotthoff, L., Vanschoren, J.: Automated machine learning: Springer (2019)
7. Brownstein, J.S., Freifeld, C., Madof, L.C.: Digital disease detection—harnessing the web for public health surveillance. N. Engl. J. Med. **360**, 2153–2157 (2009)
8. Xu, B., Gutierrez, B., Mekaru, S., Sewalk, K., Goodwin, L., Loskill, A., Cohn, E.L., Hswen, Y., Hill, S.C., Cob, M.M., Zarebsk, A.E., Li, S., Wu, C.-H., Hulland, E., Julia D., O'Brien, K., Scarpino, S, Brownstein, J.S., Pybus, O.G., Pigott, D.M., Kraemer, M.U.G.: Epidemiological data from the COVID-19 outbreak, real-time case information. Available at: https://www.nature.com/articles/s41597-020-0448-0.pdf
9. WHO.: Coronavirus disease 2019 (COVID-19) Situation Report—66. 2020. https://www.who.int/docs/default-source/coronaviruse/situation-reports/20200330-sitrep-70-covid-19.pdf?sfvrsn=7e0fe3f8_2. Accessed 31 Mar 2020
10. Pan, F., et al.: Time course of lung changes on chest CT during recovery from 2019 novel coronavirus (COVID-19) pneumonia. Radiology 200370 (2020)
11. Kanne, J.P., Little, B.P., Chung, J.H., Elicker, B.M., Ketai, L.H.: Essentials for radiologists on COVID-19: an updateradiology scientific expert panel. Radiology 200527 (2020)
12. Fang, Y., et al.: Sensitivity of chest CT for COVID-19: comparison to RT-PCR. Radiology 200432 (2020)
13. Lan, L., et al.: Positive RT-PCR test results in patients recovered from COVID-19. Jama 4 (2020)
14. Shan, F., Gao, Y., Wang, J., Shi, W., Shi, N., Han, M., Xue, Z., Shi, Y.: Lung infection quantification of COVID-19 in CT images with deep learning. arXiv preprint arXiv:2003.04655 (2020)
15. Gozes, O., Frid-Adar, M., Greenspan, H., Browning, P.D., Zhang, H., Ji, W., Bernheim, A., Siegel, E.: Rapid AI development cycle for the coronavirus (COVID-19) pandemic: initial results for automated detection and patient monitoring using deep learning CT image analysis. arXiv preprint arXiv:2003.05037 (2020)
16. Long, J.B., Ehrenfeld, J.M.: The role of augmented intelligence (ai) in detecting and preventing the spread of novel coronavirus (2020)
17. Marcello Lenca, M., Vayena, E.: On the responsible use of digital data to tackle the COVID-19 pandemic. Nat. Med. (2020). https://doi.org/10.1038/s41591-020-0832-5
18. Kosko, B.: Neural Networks and Fuzzy Systems. Prentice Hall, Englewood Cliffs, NJ (1992)
19. Bhavsar, Hetal, Ganatra, Amit: Variations of support vector machine classification technique: a survey. Int. J. Adv. Comput. Res. **2**(6), 230–236 (2012)
20. Xu, X., Jiang, X, Ma, C., Du, P., Li, X., Lv, S., Yu, L., Chen, Y., Su, J., Lang, G., et al.: Deep learning system to screen coronavirus disease 2019pneumonia. arXiv preprint arXiv:2002.09334 (2020)
21. Song, Y., Zheng, S., Li, L., Zhang, X, Zhang, X., Huang, Z., Chen, J., Zhao, H., Jie, Y., Wang, R., et al.: Deep learning enables accurate diagnosis of novel coronavirus (COVID-19) with CT images. medRxiv preprint. medRxiv:2020.02.23.20026930 (2020)
22. Hopfield, J.J.: Neural networks and physical systems with emergent collective computational properties. Proc. Nat. Acad. Sci. USA **79**, 2554–2588 (1982)
23. Wang, S., Kang, B., Ma, J., Zeng, X., Xiao, M., Guo, J., Cai, M., Yang, J., Li, Y., Meng, X., et al.: A deep learning algorithm using CT images to screen for corona virus disease (COVID-19). medRxiv preprint. medRxiv:2020.02.14.20023028 (2020a)

24. Dawkins, P.: Paul's online math notes. Accessed 1 Septe 1 2017. http://tutorial.math.lamar.edu/Classes/CalcII/EqnsOfPlanes.aspx
25. Hassanien, A.E., Mahdy, L.N., Ezzat, K.A., Elmousalami, H.H., Ella, H.A.: (2020) Automatic X-ray COVID-19 lung image classification system based on multi-level thresholding and support vector machine. https://doi.org/10.1101/2020.03.30.20047787

Deep Learning Technology for Tackling COVID-19 Pandemic

Mona Soliman, Asahraf Darwish, and Aboul Ella Hassanien

Abstract Although the COVID-19 pandemic continues to expand, researchers around the world are working to understand, diminish, and curtail its spread. The primary fields of research include investigating transmission of COVID-19, promoting its identification, designing potential vaccines and therapies, and recognizing the pandemic's socio-economic impacts. Deep Learning (DL), which uses either deep learning architectures or hierarchical approaches to learning, is developed a machine learning class since 2006. The exponential growth and availability of data and groundbreaking developments in hardware technology have led to the rise of new distributed and learning studies. Throughout this chapter, we discuss how deep learning can contribute to these goals by stepping up ongoing research activities, improving the efficiency and speed of existing methods, and proposing original lines of research.

Keywords Deep learning · Supervised learning · Unsupervised learning · Health care · COVID-19

1 Introduction

The novel corona virus disease pneumonia (COVID-19) occurred in Wuhan city, China, at the end 2019. A novel corona virus has been reported as the causative agent on 7th January 2020. The virus was originally called 2019-nCoV and emerged from the same family as the agents that triggered previous extreme respiratory outbreaks: SARS in Asia in 2002 and MERS in the Middle East in 2012 [1]. On 24 January 2020, Huang et al. summarized the clinical characteristics of 41 COVID-19 patients, suggesting that the severe onset symptoms vary from fever and trouble breathing to

M. Soliman (✉) · A. E. Hassanien
Faculty of Computers and Artificial Intelligence, Cairo University, Giza, Egypt
e-mail: mona.solyman@fci-cu.edu.eg
URL: http://www.egyptscience.net

M. Soliman · A. Darwish · A. E. Hassanien
Scientific Research Group in Egypt (SRGE), Giza, Egypt

A. E. Hassanien and A. Darwish (eds.), *Digital Transformation and Emerging Technologies for Fighting COVID-19 Pandemic: Innovative Approaches*, Studies in Systems, Decision and Control 322, https://doi.org/10.1007/978-3-030-63307-3_9

acute respiratory arrest and death [2]. All these 41 patients had pneumonia and their chest CT examination showed abnormalities. Acute respiratory distress syndrome, acute heart failure and secondary infections were among the complications. Since then the virus has spread exponentially in both rich and poor countries, leading to a Covid-19 global pandemic. We have no accepted medicines to alter the course of the disease, nor a vaccine to monitor its spread until this work is published [4]. It is too early to foresee any plausible scenario but it will have a clear global impact [3].

Deep Learning (DL), which uses either deep learning architectures or hierarchical approaches to learning, is a largely developed ML class since 2006. Learning is a method of estimating model parameters, such that a specific task can be performed by the qualified model (algorithm). The exponential growth and availability of data and groundbreaking developments in hardware technology have led to the rise of new distributed and deep learning studies. DL technology is based on the artificial neural network system (ANNs). These ANNs are constantly taking learning algorithms, and the efficiency of training processes can be improved by continuously increasing the amount of data. The quality depends on the huge volumes of data [5]. The training process is called deep because the number of levels of the neural network increases with time. The working of the deep learning process is purely dependent on two phases which are called the training phase and inferring phase [6].

DL utilizes graph technologies with transformations among neurons to develop many-layered learning models. Many of the latest deep learning techniques have been presented and demonstrated promising results across different kinds of applications such as Natural Language Processing (NLP), visual data processing, speech and audio processing, and accuracy rate on the image detection/classification competition task [7]. Deep learning technologies, such as convolutional neural networks (CNNs) with the strong ability of nonlinear modeling, have extensive applications in health care systems as well.

While the pandemic COVID-19 continues to grow, researchers around the world are working to understand, reduce, and curtail its spread. Primary areas of study include studying COVID-19 transmission, encouraging its detection, developing future vaccines and therapies, and understanding the socio-economic impacts of the pandemic. In this chapter, we explore how deep learning can contribute to these goals by stepping up ongoing research activities, improving the efficiency and pace of existing methods, and proposing original research lines. The aim of this review is not to evaluate the impact of the techniques mentioned, nor to suggest their use, but rather to show the reader the scope of current applications and to provide an initial picture and road map of how deep learning could support the global response to the COVID-19 pandemic.

The rest of this chapter organized as follows. The next section is about exploring some basic knowledge about COVID-19. Section 3 provides the basic concept of deep learning and its main terminologies. Existing researches with future directions of deep learning for COVID-19 is presented in Sect. 4. Section 5 concludes the main concepts and issues introduced in this chapter.

2 SARS-CoV-2(COVID-19: Strategies and Evidence

Corona viruses (CoV) were first detected in 1960s. Because of the specific appearance of crown, the virus was called Corona as sugary proteins that surround the particle. The CoV holds the longest genome of any RNA related virus [8]. CoV is commonly observed in animals, and some of the viruses can be transmitted to humans. Corona viruses are RNA viruses that are divided into four genera: alpha, beta, gamma, and delta [8]. Alpha corona viruses and beta corona viruses are known to infect humans. SARS-CoV-2 is related to bat corona viruses which enters human cells through the angiotensin converting enzyme 2 (ACE2) receptor [10]. Bats are a normal CoV host but not the only animal with the ability capability of transmitting the virus to humans [8]. COVID-19, the disease caused by the SARS-CoV-2 virus, has been declared a pandemic by the World Health Organization at march 12, 2020. No currently approved antiviral drug or vaccine for use against COVID-19 is available worldwide. Nevertheless, in clinical trials, few broad-spectrum antiviral drugs were assessed against COVID-19 which resulted in some clinical recovery cases [4].

2.1 *Transmission*

SARS-CoV-2 is transmitted by respiratory droplets from person to person, which are normally distributed when an infected person coughs or sneezes [1]. Since droplets typically fall within a few meters, if people stay at least 2 meters apart, the probability of transmission is common. Transmission is not thought to occur naturally by inhaling aerosols (virions suspended in air), although there are fears that the virus might be aerosolized during other behaviors (e.g. singing) or procedures (e.g. intubation or the use of nebulizers) [9]; And that it can last longer than 3 hours in aerosols. SARS-CoV-2 RNA has been found in blood and stool although there has been no evidence of fecal-oral spread. SARS-CoV-2 can survive for days on cardboard, plastic, and stainless steel. As a result, inanimate surface contamination may play a role in the propagation [8].

COVID-19's clinical scope ranges from asymptomatic or paucisymptomatic types to clinical conditions described by extreme respiratory failure requiring mechanical ventilation and treatment in an intensive care unit (ICU) to sepsis, septic shock, and multiple organ dysfunction syndromes(MODS) [3]. It has also been identified asymptomatic infections but their frequency is unclear.

2.2 *Diagnosis*

Pneumonia tends to be the most common extreme form of infection, usually characterized by fever, cough, dyspnea, and bilateral chest imaging infiltrations. Various reports propose that patients may be infectious 1 to 3 days before the onset of symptoms [3]. There are no clear clinical features that can yet allow a valid distinction between COVID-19 and other respiratory viral infections [9].

The risk factors for Covid-19 symptoms include the elderly (e.g., $\leq$65 years), cardiovascular disease, chronic lung disease, hypertension, diabetes, and obesity [8]. It is uncertain if such other factors (kidney disease, immunosuppression, cancer, and infection with unregulated human immunodeficiency virus [HIV]) impose an increased risk of infection. Laboratory reports can include lymphopenia and high levels of d-dimer, lactate dehydrogenase, C-reactive protein, and ferritin in hospitalized patients [9]. Covid-19 diagnosis is typically based on SARS-CoV-2 detection by polymerase-chain-reaction (PCR) assay. Soon after the onset of symptoms, the sensitivity of nasopharyngeal swabs testing with PCR seems to be significant, but with unknown frequency, false negatives may occur [3]. If a person is suspected of having Covid-19 but has negative testing of a nasopharyngeal swab, repeat testing is advisable, particularly if that person lives in an area with active society transmission.

2.3 *Evaluation and Management*

Primary prevention is the best method against COVID-19. It is best practice to avoid infected geographical areas and positive identified individuals for the virus. Like in any pandemic, could not be able to fully escape exposure [8]. In order to achieve this goal, the main steps are the following: (1) to use face masks; (2) to cover coughs and sneezes with tissues; (3) to wash hands frequently with soap or disinfection with hand sanitiser containing at least 60% alcohol; (4) to avoid contact with infected people; (5) to maintain an acceptable distance from people; and (6) Stop touching eyes, nose and mouth with unwashed hands [3]. The extent of the disease is driven by the diagnosis and treatment of Covid-19. Recent results from China show that 81% of people with Covid-19 had mild to moderate disease (including people without pneumonia and people with mild pneumonia), 14% had serious disease, and 5% had critical disease [9].

No specific antiviral treatment for COVID-19 is recommended, and there is currently no vaccine available. The diagnosis is symptomatic, and the main diagnosis is oxygen therapy [4]. Depending on the seriousness of the condition and local epidemiology, various approaches can be employed. Home treatment is ideal for asymptomatic or paucisintomatic patients. For around 14 days they require a regular assessment of body temperature, blood pressure, oxygen saturation and respiratory symptoms. Control of these patients should concentrate on preventing transmission to others and monitoring with prompt hospitalization for clinical status if appropriate [3]. When in the same room (or vehicle) as other people and when presenting to health care facilities, they will wear a face mask. It's also essential to disinfect frequently touched surfaces.

The laboratory tests will include a full blood count and a detailed metabolic panel in hospitalized patients. A baseline electrocardiogram should be obtained in most cases, and particularly if a drug that affects the corrected QT (QTc) interval is considered [9]. Chest X-ray is usually the primary type of imaging. Many centers use ultrasonography for the heart, too. The American College of Radiology advises against the use of computed tomography (CT) to diagnose Covid-19 as a screening or initial imaging test, recommending that it should be used "sparingly" and only in hospitalized patients where specific indications exist [3].

The COVID-19 outbreak is demonstrating to be an unprecedented disaster, especially in the most suffering countries, including China, Italy, Iran and the United States in all aspects, particularly health, social and economic. It is too early to foresee any plausible scenario but it will have a clear global impact [3].

3 Deep Learning: Basic Terminologies

Deep-learning approaches are multi-level representation-learning methods, generated by combining simple but non-linear modules which transform each representation to a higher, slightly more abstract level at a single level (starting with the raw input). Highly complex functions can be learned with adequate composition of these transformations. Deep learning allows for the feeding of raw data (pixels in the case of image data) into the learning algorithm without first choosing features or determining a feature vector. Deep learning algorithms can learn the proper set of features, and they do so much better than using hand-coding to get these features back. Instead of building a set of rules and methodologies to obtain features from raw data, deep learning includes learning these features automatically during the training process [37]. In the following subsections we will explore in more details the basics of deep learning.

3.1 Deep Learning History

The move from perceptron to basic neural networks is merely to bring together the perceptron. Through putting the perceptron side by side, we obtain a single one-layer neural network and by piling one-layer neural network on the other, we obtain a multi-layer neural network, also referred to as multi-layer perceptron (MLP) [37]. A remarkable property of neural networks, commonly known as the universal approximation property, essentially describes that an MLP can represent any function in three different ways: First, Boolean approximation of one hidden layer with MLP can represent exactly any boolean function. Second, Continuous Approximation where MLP of one hidden layer can guess with arbitrary accuracy any bounded continuous function. And finally, arbitrary Approximation where MLP of two hidden layers can approximate any function with arbitrary precision.

Neural network history can be traced back to 1943, when Walter Pitts and Warren McCulloch developed a computer model that was based on human brain neural networks [11]. To simulate the thinking process, they used a combination of algorithms and mathematics which they called threshold logic.

In 1960, Henry J. Kelley gets credit for creating the fundamentals of a continuous back propagation model. In 1962, Stuart Dreyfus built up a simpler version, based solely on the chain rule. Efficient backpropagation of errors (BP) in arbitrary, discrete, probably sparsely connected, apparently NN-like networks was first described in a 1970 master's thesis by Linnainmaa [13], without reference to NNs. Ivakhnenko [14] equipped an 8-layer neural network in 1971, using the Group Data Handling Method (GMDH) algorithm. The first NN-specific application of efficient BP was described in 80s [15]. NN can be trained effectively later and have been able to deliver decent results in visual object recognition tasks A number of years later related work was published.

Dataset were limited at the time and machines were not powerful enough, so the performance was either close or worse than the performance of modern computer vision algorithms. For years, the question of how to retain this expressive capacity of the network while raising the number of computing units was raised. Intuitively, Bengio and Delalleau [16] indicated that it is nature to follow deeper networks, because: biological neural system is a deep architecture, and humans prefer to replicate concepts as the composition of concepts at lower levels at one level of abstraction.

In 1992 the Cresceptron used Max-Pooling (MP) layers [17]. Here an activation array of two dimensional layers or units is divided into smaller rectangular arrays. In a down sampling layer, each is replaced by activating its maximally active unit. The benefits of this procedure include reducing parameters, increasing computational performance and controlling overfitting [18]. A later, more complicated version of the Cresceptron [19] also included "blurring" layers to increase object position tolerance.

Since 2009, several official international pattern recognition competitions have won supervised deep NNs, producing the first super-human visual pattern recognition results in specific domains.

When CNNs or convnets are merged with MP, they are MPCNNs similar to Cresceptron or HMAX with alternating convolutionary and max-pooling layers. However, unlike Cresceptron and HMAX, BP [20] trains MPCNNs. For many new, competitive, feed forward, visual Deep Learners, BP-trained MPCNNs have become central to this. A GPU-MPCNN ensemble accomplished best outcomes in 2012 on the ImageNet classification benchmark which is prominent in the computer vision society. In 2012, too, a broad NN (109 free parameters) has been trained on unlabeled data in unsupervised mode, then introduced to ImageNet. So, Deep Learners had obtained amazing results in image recognition and classification by 2011/2012 [39]. In 2012 DL [23] also won the first pure image segmentation contest, again via a GPU-MPCNN ensemble [22].

In 2014, algorithmic architectures of generative adversarial networks (GANs) was proposed using two neural networks, pitting one against the other (thus the "adversarial") to produce new, virtual data instances that can be passed on to real data. They are commonly used in generating image, generating video and generating voice [21]. In 2017 Hinton and his team implemented a dynamic routing mechanism for capsule networks. In 2017 Hinton and his team implemented a dynamic routing mechanism for capsule networks [35].

The creation of CUDA for Nvidia GPUs which allowed over 1000 GFLOPS per second and the publication of the ImageNet dataset, with 1.2 million images classified in 1000 categories, were important facts for the popularization of CNNs with several layers [24]. In the new millennium, deep NNs have finally attracted wide-spread interest, primarily by outperforming alternative machine learning methods such as kernel machines in numerous important applications. The next figure shows a brief progress of deep learning over time period.

These deep models show great performance not only in Computer Vision tasks but also in other tasks such as speech recognition, signal processing, natural language processing, Recommendation systems, and different health care applications like medical image analysis and Drug discovery and toxicology [38] (Fig. 1).

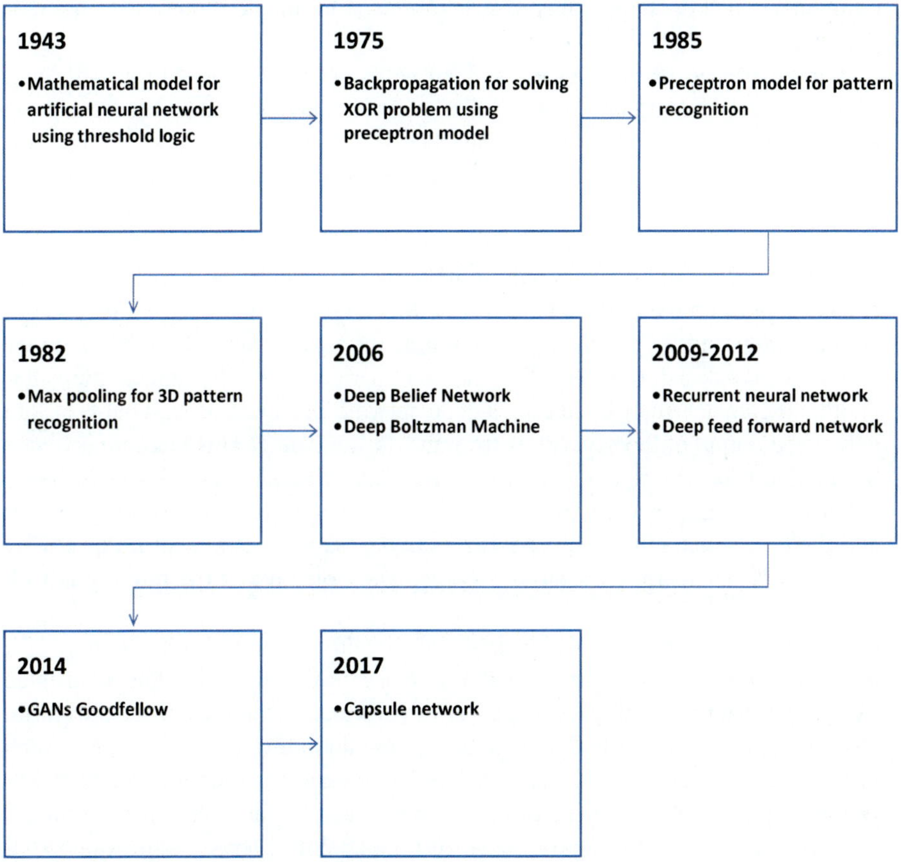

Fig. 1 Deep learning history time-line

3.2 Deep Learning: How It Works

Deep learning methods are a study of artificial neural networks and related machine learning algorithms that usually include learning representations at various levels of hierarchy such that complex concepts can be built from simpler ones [6]. Such deep nets are part of the broader area of data learning machine learning representations. The most widely accepted forms of learning include supervised learning, unsupervised learning, semi-supervised learning and reinforced learning.

3.2.1 Supervised Learning

Deep learning methods reduce feature engineering for supervised learning tasks by transforming the data into compact intermediary representations identical to major components and deducing layered structures that eliminate representational redundancy [25].

For supervised learning process, deep networks assign input to goal through a sequence of layered transformations and learn these layered transformations through exposure to the examples of training. The transformations a layer applies to its input will be decided by the weights of the layer, which are essentially a bunch of numbers. In other words, the weights of the transformations carried out by a layer are parameterised. In this scenario, learning can be described as the process of identifying the weight values of all layers within the network in such a way that input samples can be mapped correctly to their corresponding targets [25]. A deep learning network contains thousands of parameters, and finding the correct values for these parameters is not an easy decision, especially when one parameter's value affects the value of another parameter. In order to train a deep network one needs to find out how far the calculated output of the network is from the desired value. This measure is obtained by using a loss function, also called as objective function. The objective function calculates the difference between the predicted output obtained from the network and the true target value for a specific example. This gives a measure of how well the network has learnt a specific example. The objective of the training is to find the values for the weights that minimize the chosen error function. The difference obtained is then used as a feedback signal to adjust the weights of the network, in a way that loss score for the current example is lowered. This adjustment is done by the optimizer-back-propagation algorithm, the central algorithm in deep learning. Back-propagation algorithm involves assigning random values to the weight vectors initially, so that the network just implements a series of random transformations. Initially, the output obtained from the network can be far from what it should be, and accordingly the loss score may be very high. With every example that is fed to the network, the weights are adjusted in such a direction that makes the loss score to decrease. This process is repeated a number of times, until the weight values that minimize the loss function are obtained [37].

3.2.2 Unsupervised Learning

Deep learning algorithms can be applied to tasks of unsupervised learning. This is an significant advantage, because unlabeled data is more plentiful than the data labelled. Current unsupervised methods of deep learning typically fall into four groups [26]:

Clustering analysis: Study of clustering is a common approach to unsupervised learning by computers. The objective remains the discovery on the entire training data of multiple class consistent clusters (or groups). This is a challenging task with both the data and cluster size being exponentially proportional to the complexity and the solution space. This is especially so for clustering data in complex structures and representations such as images and videos Recent studies have tried to optimize clustering analysis and representation learning to maximize their complementary benefits along with the growth of deep learning techniques [27–29].

Sample specificity learning: Learning sample specificity goes to the other extreme by considering each sample as an individual class. The main point is that supervised deep neural network learning automatically discloses the correlation of visual similarity among various classes from end-to-end optimisation [26]. However, such supervision does not specifically model the limits of class judgment as an study of the clustering. Consequently, it is likely to generate more fluid class structures and less discriminatory feature representations.

Self-supervised learning: Rather than predicting human-annotated labels, such methods predict computed pseudo-labels from the raw data themselves. These approaches are typically based on one single monitoring signal and require some engineering effort [30]. The principle of learning representations from Self-supervised learning takes advantage of internal data structures and formulates predictive tasks for training a model. Specifically, given the rest [31] the model must predict either an omitted aspect or part of an case.

Generative models: is a principled way of learning the true data distribution of the training set in an unsupervised manner. The most commonly used and efficient generative models include Restricted Boltzmann Machines and Generative Adversarial Networks [25].

3.2.3 Semi-supervised Learning

The use of supervised and unsupervised models of deep learning has evolved rapidly due to their success in learning complicated issues. High-performance computing tools, the availability of vast quantities of data (labeled and unlabeled) and state-of-the-art open-source libraries are increasingly making deep learning feasible for different applications. Semi-supervised learning algorithms aim to learn the relationships between the data by using huge amounts of unlabeled data, which can often be more readily accessible than labeled data [32]. Among current methods, the simplest semi-supervised learning algorithm is based on a self-training scheme in which the model is bootstrapped with additional labeled data obtained from its own highly confident predictions; Repeat this cycle until some termination condition has been met. These methods are heuristic and vulnerable to error, as they are capable of improving weak predictions. Neural network-based methods merge unregulated and supervised learning by training feed-forward classification model with an additional penalty from an auto-encoder or other unregulated data embedding [33].

3.2.4 Reinforcement Learning

The combination of reinforcement learning (RL) and deep learning is Deep Reinforcement learning. This area of research has been able to solve a wide variety of complex decision-making tasks which a computer had previously been out of control. Deep RL thus opens up various new applications in fields such as education, robotics, smart grids, banking, and many more. A learning reinforcement function is to train an agent that communicates with his environment. By doing actions, the agent arrives at different situations known as states. Actions result in incentives that may be both positive and negative [25]. A main feature of RL is that an agent is practicing positive behaviour. That means it incrementally modifies or acquires new habits and skills. Another important aspect of RL is that it uses experience of trial and error (as opposed to, for example, dynamic programming which assumes complete knowledge of the environment a priori). Thus, the RL agent does not need full environmental awareness or control; it just needs to be able to communicate with the environment and gather information. Here, the agent has only one objective, maximizing his total reward over an episode [34].

3.3 Deep Learning Models

Many common deep learning networks including Recursive Neural Network (RvNN), RNN, CNN, and profound generative models are discussed in this section. Since deep learning has evolved quite rapidly, however, several new networks and new architectures emerge every few months, which is out of the reach of this study. We will cover the most applicable deep learning technologies available at the moment in supervised, unsupervised, and reinforcement learning.

Deep learning addresses problems encountered when applying back-propagation type algorithms to deep networks with many layers. A key idea is to learn not only the nonlinear mapping between input and output vectors but also the underlying structure of data (input) vectors. To achieve this goal, unsupervised pre-training is used. This is achieved in practice by using RBMs or autoencoders in each hidden layer as building blocks in forming deep neural networks [63] (Fig. 2).

Supervised Learning	Convential Neural Network (CNN)	AlexNet, 2012
		ZFNet 2013
		VGGNet 2014
		GoogLeNet 2015
		ResNet 2015
		Inception -ResNet 2017 ,
		Xception 2017
		PolyNet 2017 .
Unsupervised Learning	Auto -Encoder	Sparse Autoencoders , 2013
		Denoising Autoencoders
		Contractive Autoencoders
		Variational Autoencoders
	Recurrent neural networks	Stacked RNN
		LSTM
	Deep Generative Network	Deep Belief Network (DBN)
		Deep Boltzmann Machine
		Generative Adversarial Networks
Reinforcement Learning	Q-Learning	

Fig. 2 Deep learning models

3.3.1 Convolutional Neural Network (CNN)

The Convolutionary Neural Network also known as ConvNet or CNN is one of the most widely used supervised models of deep learning. The Convolutionary Neural Network is a profound learning technique consisting of several layer numbers. ConvNets is motivated by the visual biological cortex. The visual cortex has tiny regions of cells that are sensitive to different visual field areas. Different brain neurons respond to varying features. This concept of getting a particular function on other neurons is the basis behind ConvNets.

ConvNets' key concept is to get local features at higher layers from the input (usually an image) and merge them towards more complex features at the lower layers. It is however computationally exorbitant due to its multilayered architecture, and training such networks on a large dataset takes many days. So normally these deep networks are trained on GPUs. Convolutionary neural networks are so effective on visual tasks that almost all conventional methods are performed above. ConvNets has three key features: local receptive field, weight sharing, and (pooling) subsampling [36].

Local Receptive Field: In the traditional neural network, every neuron or hidden unit is connected to every neuron in the previous layer or to every input unit. Convolutional neural networks, however, have a local receptive field architecture, i.e. each hidden unit can only connect to a small area of input called a local receptive field. This is done by raising the filter/weight matrix to less than the input. With the local receptive field, neurons can extract elementary visual features such as edges, corners, and endpoints.

Weight Sharing: Weight sharing applies to all receptive fields in a layer using the same filter/weights. In ConvNet, because the filters are smaller than the input, each filter is applied at each input location, i.e., the same filter is applied to all receptive local fields. ConvNet consists of a series of different layer styles to accomplish the various tasks. A typical convolutionary neural network consists of: Convolutionary layer, Activation function layer (ReLU), Pooling layer, Completely linked layer, and Dropout layer. Such layers are stacked to create a complete architecture for ConvNet. The layers of traditional and activation functions are normally stacked together, followed by an optional layer of pooling. The fully connected layer is the last layer of the network, and the output of the last fully connected layer is the class of the input image scores. In addition to the above mentioned main layers, ConvNet may include optional layers such as batch normalization layers to improve training time and drop-out layers to address the issue of over-fitting.

Subsampling (Pooling): Subsampling decreases the incoming spatial size, thereby reducing network parameters. There are few subsampling techniques available, and max-pooling is the most common sub-sampling technique.

The first functional convolution-based architecture was LeNet which used back-propagation for training the network. LeNet was designed to identify handwritten digits (MNIST), and it was introduced to read vast numbers of handwritten checks in the United States. Unfortunately, the solution didn't get much attention because it didn't scale well into bigger issues. Differentiated CCN architecture with different architecture was proposed since 2012 (Different CNN architecture: AlexNet,2012, ZFNet 2013, VGGNet 2014, GoogLeNet 2015, ResNet 2015, Inception-ResNet 2017, Xception 2017, and PolyNet 2017). As the accuracy of new ConvNets models continued to improve, researchers started to concentrate on how to reduce the size and complexity of the current ConvNet architectures without compromising on precision. The intention was to design a model with very few parameters, while maintaining high precision [2017 SqueezeNet, 2017 ShuffleNet].

3.3.2 Auto-Encoder

An auto-encoder is one of unsupervised learning methods in the deep learning community. It can automatically learn effective and robust features from a large amount of unlabeled data [41]. Autoencoders are simple learning circuits which aim to transform inputs into outputs with the least possible amount of distortion [56]. The main principle of autoencoder follows from the name. "Auto" presents that this method is unsupervised and "encoder" means it learns another representation of data. Specifically, auto-encoder learns a encoded representation by minimizing the loss between the original data and the data decoded from this representation [57]. It consists of two modules: a feedforward encoder module that maps the input to a code vector or hidden representation z in the bottleneck layer and a decoder module that tends to reconstruct the input sample from z. Auto-encoders can be viewed as the supervised learning using MLPs, aiming to obtain the optimized network parameters at the lowest reconstruction error. By adding the penalty or changing the architectures of the encoder and the decoder, one can obtain the regularized autoencoders to cope with different tasks. It is noted that the hidden representations produced by the encoder can be used as the input for other machine learning methods, helping to initialize a network. In addition, autoencoders are often used to compress the input data, which is also known as dimension reduction, through restricting the number of output neurons of the encoder. Since the final output of the decoder is nothing but the approximation of the input, we are only concerned with the feature transformation occurring during encoding, which is believed to extract the essential information from the original representations at the lowest possible reconstruction error. The encoder and decoder are often implemented by the Multi-Layer Perceptron (MLPs), with linear or non-linear functions [42].

Autoencoders were first introduced by Hinton and the PDP group [40] to address the problem of "back-propagation without a teacher", by using the input data as the teacher. Together with Hebbian learning rules, autoencoders provide one of the fundamental paradigms for unsupervised learning and for beginning to address the mystery of how synaptic changes induced by local biochemical events can be coordinated in a self-organized manner to produce global learning and intelligent behavior.

In 1989, Baldi and Hornik [43] investigated the relationship between a one-layer autoencoder and principal component analysis (PCA). They found that the new compressed features learnt by linear autoencoder is similar to principal components. However, the computational complexity of training an autoencoder is much higher than PCA because the major computational complexity of PCA is based on matrix decomposition. This limits the application of linear autoencoders. Later, with the involvement of non-linear activation functions, autoencoder becomes non-linear and is capable of learning more useful features [44] than linear feature extraction methods. Non-linear autoencoders are not advantaged than the other non-linear feature extraction methods as it takes long time to train them.

The recent revival of interest in autoencoders is due to the success of effectively training deep architectures because traditional gradient-based optimization strategies are not effective when hidden layers are stacked multiple times with non-linearities [57]. Bengio [46] achieved success in training a stacked autoencoder on the MNIST dataset with a similar greedy layerwise approach. This training approach overcomes the problematic non-convex optimization which prevents deep network structure. Subsequent studies show that stacked autoencoder model can learn meaningful, abstract features and thus achieve better classification results in high-dimensional data, such as images and texts [57].

More recently, auto-encoders have taken center stage again in the "deep architecture" approach [56] where autoencoders, particularly in the form of Restricted Boltzmann Machines (RBMS), are stacked and trained bottom up in unsupervised fashion, followed by a supervised learning phase to train the top layer and fine-tune the entire architecture. The bottom up phase is agnostic with respect to the final task and thus can obviously be used in transfer learning approaches. These deep architectures have been shown to lead to state-of-the-art results on a number of challenging classification and regression problems.

Auto-encoder has many extensions such as Sparse Autoencoders [45], Denoising Autoencoders [46], Contractive Autoencoders [47] and Variational Autoencoders [48], demonstrate a promising ability to extract meaningful features, especially in image processing and natural language processing. Yet, these methods only consider data reconstruction and ignore to explicitly model its relationship. [49] proposes a Generalized Autoencoder (GAE) which focuses on modifying autoencoder to consider data relationship by adding a weighted relational function. Specifically, it minimizes the weighted distances between reconstructed instances and the original ones [57].

3.3.3 Recurrent Neural Networks

Recurrent neural networks or RNNs for the processing of sequential data are a family of neural networks. Most recurrent networks can handle variable-length sequences as well. Sequences can be of variable size, have complicated time and pattern connections, and have high - dimensional. Many examples of sequential datasets include language, genomes, music, voice, language, handwriting, price changes in stock markets, and even images which can be broken down into a series of patches and viewed as a sequence [60]. To go from multilayer networks to recurrent networks, one of the early ideas found in machine learning and statistical models of parameter sharing across different parts of a model needs to be taken advantage of. Sharing parameters allows the model to be extended and applied to examples of various forms (different lengths, here) and generalized through them. If we had separate parameters for each time index value, we would not be able to generalize to sequence lengths not shown during training, nor share the statistical strength over different sequence lengths and over different time positions. Such sharing is extremely significant if a specific piece of information may occur at multiple locations within the sequence.

For RNNs, the layers have the primary function of introducing memory, not hierarchical processing. New information is inserted in each 'layer' (every network iteration), and this information can be passed on to the network for an infinite number of network updates, effectively providing limitless memory capacity to the RNN. Whereas DNN input is presented only at the bottom layer, and output is produced only at the highest layer, RNNs generally receive input and produce output at each time step. Common RNNs are not Clearly contain various time scales and any temporal hierarchy directly from the input signal It needs to be indirectly encoded in the dynamics of the network [59].

Deep RNN can be described as L-layered, and N-neurons per layer. Suppose we have an input time series of $s(t)$ of dimensionality of Nin, and a target time series of y. Note that the network structure basically offers varying time scales. The bottom layer of the input signal has fading memory. The next layer has a fading memory of the secret state of the bottom layer, and therefore a fading input memory that extends deeper in the past for each additional layer, and so forth [60].

By considering transitions separately, we can find various forms of depth of a Deep-RNN: Deep-input-to-hidden, Deep-to-hidden, and Hidden-to-output. Any such option has a different meaning. The approach of a deep transformation, But it does pose a potential problem. As the addition of a deep transition increases the number of nonlinear steps the gradient has to take when propagated back in time, it may become more difficult to train the model to capture long-term dependencies.

An RNN can be further expanded by stacking several recurrent hidden layers on top of each other in yet another way. We call this model a stacked RNN (sRNN), in order to differentiate it from the other variants proposed. A model like this is aimed at enabling each recurrent level to function at a different time scale. A stacked RNN performs comparatively better and more importantly, the stacking approach will synthesize time information in-layer To remove temporal-spatial hierarchical features rather than simple linear, deep spatial characteristics [51].

LSTMs [52] are basically better versions of RNNs, which can read longer data sequences. However, while an RNN can learn about dependencies, it can only learn about the new knowledge. LSTM can solve this problem, as it can understand the context along with the recent dependence. Therefore, LSTM is a special form of RNN where understanding context can help to be useful LSTM networks are similar to RNNs with one significant difference being that memory cells replace hidden layer updates. This makes them better at defining and revealing long-range data dependencies that are essential for sentence structures. An LSTM neural network consists of both special layers of LSTM that can interpret sequential word data as well as the densely connected ones as described above. When the data passes through the layers of the LSTM, it goes through the layers that are tightly connected.

The Deep LSTM is one of the most important models of profound learning. LSTM prevents the issue of gradient degradation by using the gate modules. Nonetheless, LSTM also requires much more computational time. Deep learning models increase the hidden layers of neural networks instead of The Centers of the Neurons. This idea can successfully avoid the local minimums in and the problem of determining the structure. The Deep LSTM is one of the most important models of profound learning. LSTM prevents the issue of gradient degradation by using the gate modules. Nonetheless, LSTM also requires much more computational time. Deep learning models increase the hidden layers of neural networks instead of the Centers of the Neurons [50].

3.3.4 Deep Generative Network

Unlike the deep neural networks, deep generative models derive sample data from the trained model, allowing for qualitatively verifying what these models have learned, for example by visualizing images the model has learned. Here we present three profound generative models, specifically Deep Belief Network (DBN) and Deep Boltzmann Machine (DBM), which use restricted Boltzmann machines (RBMs) as the basic elements for their development [61]. We also present Generative Adversarial Networks (GAN) as a special form of neural network model where two networks are simultaneously trained, one focusing on image generation and the other focusing on discrimination [54].

RBM [53] is a two-layer generative neural network based on probabilistic binary units that functions in a stochastic manner. One layer is referred to as visible layer and denoted v, while the other layer is referred to as secret layer and denoted by h. Each unit within one layer is connected to all other layer units, and these connections are bi-directional. There is no linkage between units of the same layer. RBM neurons form a two-part graph and are therefore called Restricted Boltzmann Machine as opposed to Boltzmann Machine. Boltzmann machines are generative non-deterministic (or stochastic) Deep Learning models with only two types of nodes hidden and visible nodes. No output nodes do exist.

DBN first conduct greedy and unsupervised layer-wise learning, whereby each layer is represented by an RBM. DBN is equipped to recreate its inputs in an unsupervised manner (without goal labels). DBN first conduct greedy and unsupervised layer-wise learning, whereby each layer is represented by an RBM. DBN is equipped to recreate its inputs in an unsupervised manner (without goal labels). This supervised learning is mostly done using gradient descent after the unattended greedy layered-wise training. Deep neural networking methods have been used in wide-ranging applications such as the creation of predictive feature spaces for object recognition in natural images and medical diagnostics. An significant aspect of this layer-wise learning mechanism is that each additional layer raises a lower limit on the data's log likelihood, as long as the number of features per layer is not reduced. To learn a deep, hierarchical model of the data, this layer-by-layer training can be repeated several times [64].

A DBM is also structured by Hierarchically stacking multiple RBMs. Unlike DBN, however, after stacking RBMs [61] all layers in DBM still form an undirected generative model. There are several reasons why Deep Boltzmann machines are interesting. First, like deep belief networks, DBM's have the ability to learn increasingly complex internal representations, which are considered a promising way to solve artifacts and the issues of speech recognition. Second, high-level representations can be built from a wide supply of unlabeled sensory inputs, and very minimal labeled data can then be used to fine-tune the model just slightly for a particular task at hand. Finally, unlike deep belief networks, the approximate inference procedure can incorporate top-down feedback in addition to an initial bottom-up pass, enabling deep Boltzmann machines to better propagate uncertainty about ambiguous inputs, and thus deal more robustly with them [62].

Generative Adversarial Network (GAN) is a class of deep learning models invented by Ian Goodfellow in 2014 [48]. GAN models have two primary networks known as the generation network and the network is discriminatory. The first neural network is the generator network, which is responsible for generating new fake instances of data that look like training data. The discriminator tries to differentiate between the real data generated by the generator network and the false (artificially generated) data [55]. The GANs goal is to try to trick the discriminator network and the discriminator network is trying to stop being fooled.

Due to its usefulness in counteracting the domain transition, and efficacy in producing new image samples, the adversarial training scheme has gained interest in both academia and industry. This model has achieved state of the art performance in many tasks related to image generation.

There is still a need to train (optimize) the network to communicate the proper feature of transform. To this end, we can explore two different methods of training: one direct, and one indirect. The direct training method is to compare the true and the generated distributions of probability, and to backpropagate the difference (the error) through the network. This is the principle this Generative Matching Networks

(GMNs) rules. The true and generated distributions aren't compared for indirect training. Rather, the generative network is trained by having these two distributions go through a downstream task selected in such a way that the process of generative network optimization regarding the downstream task forces the distribution generated to be similar to actual distribution.

4 Deep Learning for COVID-19: Approaches and Future Directions

Additional efforts have used machine Learning and particularity deep learning to better understand SARS-CoV-2 infection severity and infectivity. Every day there is a new research article tried to understand this disease and help the community to face this pandemic. In the following we try to explore recent research work that utilize the use of deep learning models in providing guide and solutions to deal with SARS-CoV-2. Deep learning-based researches can be classified into three categories: Prediction based, medical image diagnosis based, and finally genome based approaches. Most research works focus on using deep learning models in the analysis of medical images. Figure 3 shows a generic view of using deep learning approaches within COVID-19 research directions.

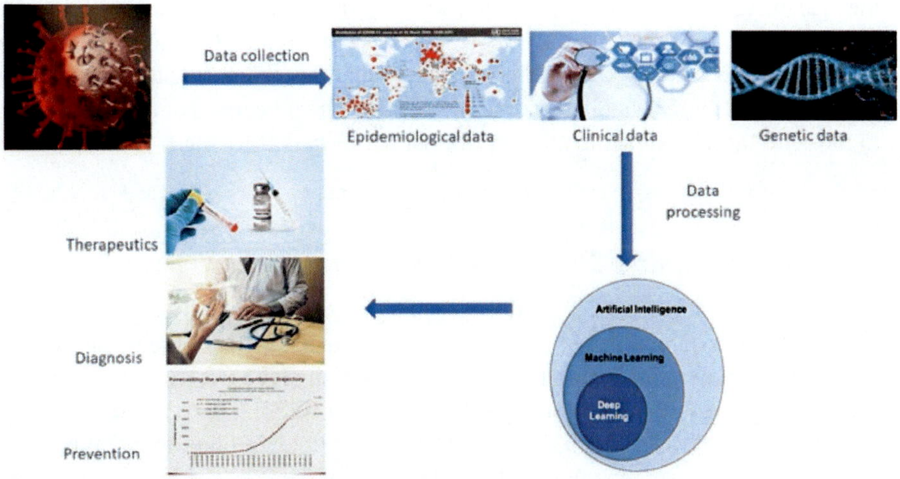

Fig. 3 Deep learning for COVID-19

4.1 Prediction Based Approaches

The prediction will allow the required decisions to be taken on the basis of transmission development, such as raising the lockdown time, carrying out the sanitation procedure, and providing the daily resources [65]. To this end, mathematical models are adopted at the cutting edge machine learning technology, such as supporting vector regression (SVR) and Deep learning using standard deep neural network (DNN) and recurrent neural networks (RNN) with long short-term memory (LSTM) cells, and polynomial regression (PR). Machine learning and deep learning methods are applied to calculate the expected number of confirmed, recovered, and globally death cases.

Authors in [65] uses a regular DNN consisting of a thin input layer with 128 neurons, three hidden thin layers with 256 neurons and an output layer consisting of a single neuron, while the RNN consists of three stacks of LSTM layers with 64 neurons combined with a 10% dropout to prevent overfitting and a single neuron output layer. The mean squared error (MSE) is the most commonly utilized objective function and the root mean square error (RMSE) as a measurement function for the regression models evaluation. The trained models are validated against the training data because of the limited availability of the data and then used to predict the number of actual, recovered and death cases for the next 10 days.

Authors in [71] Proposed the use of COVID-19 incident prediction models in Iran using data from Google Trends. Although the predictions are not very precise, they can be useful in building accurate estimates from more aggregation of data for a base idea. In [71] Long-term memory (LSTM) is used as an artificial neural recurrent network to forecast time series where data is sequential. By storing the past in hidden states they can more reliably forecast the outputs. The objectives of this research was to measure through time the number of positive COVID-19 cases.

4.2 Medical Image Diagnosis Based Approaches

The most common test technique currently used for COVID-19 diagnosis is a real-time reverse transcription-polymerase chain reaction (RT-PCR). Chest radiological imaging such as computed tomography (CT) and X-ray have vital roles in early diagnosis and treatment of this disease. Even if negative findings are obtained, signs can be observed by analyzing radiological images of patients due to the low RT-PCR sensitivity of 60–70%. [67] states that CT is a sensitive method for the detection of COVID-19 pneumonia, and can be considered as an RT-PRC screening tool. CT results are found over a long period of time after symptoms start, and patients typically have normal CT within the first 0–2 days. A researcher at Tongji Hospital in Wuhan, China, has concluded that the CT should be used to identify COVID-19 as the primary screening or diagnostic method. As the rate of newly affected global regions is increasing every day, and the number of new patients in those areas is rising geometrically, providing them with the necessary treatment has become a difficult or sometimes impossible task. Hence, the availability of an Artificial Intelligence (AI) system can be useful in quickly and accurately identifying COVID-19 patients using X-ray and CT images. The X-ray and CT scans were used quite often to recognize or identify the patients with COVID-19, according to various studies presented in the literature. Hence, here we present some deep learning-based systems for detection of COVID19 and localization of infected regions from both X-ray and CT images.

Authors in [66] contains sample classification and segmentation models from two separate identification and localisation modalities. Due to a shortage of publicly available COVID-19 X-ray samples, they built a model with a pneumonia data set for chest X-ray image analysis, and then used a Transfer Learning (TL) method for COVID-19 sample retraining. [66] suggested X-ray and CT imaging methods for the identification and localization of contaminated areas. They proposed the use of the Inception Recurrent Residual Neural Network (IRRCNN) model for the COVID-19 detection task, and the NABLA-N network is used for X-ray and CT image segmentation of infected regions. The proposed method of diagnosis of pneumonia demonstrates approximately 87.26% precision of testing. Both detection and segmentation methods have very good precision for X-ray pictures. There are, however, certain limitations of this study that need to be addressed in the near future. First, they need to train and test the COVID Det X-ray model with more COVID-19 samples. Second, since in the initial implementation the COVID-Det CT model is trained and evaluated on only 300 samples, the model provides very good accuracy in detection.

Authors in [68] described a color probability DL approach for augmenting radiographs Overlay planned to improve pneumonia diagnosis. Unlike popular approaches to the classification of entire images, this method specifically learns pixel-level probabilities of pneumonia through the parenchyma of the lungs. That gives natural transparency and explanation. Of 5 recent COVID-19 epidemiological and case-study studies, a total of 10 frontal chest x-rays from 5 patients treated in China and the United States were collected. DL model is implemented as a U-Net equipped 22 K radiologist-annotated radiographs, generating maps of the risk of pneumonia overlaid on an input radiogram.

Authors in [67] introduced the binary (COVID vs. No-Findings) and multi-class (COVID vs. No-Findings vs. Pneumonia) classification models. This model has achieved 98.08% classification accuracy for binary classes and 87.02% for multi-class instances. In this analysis the DarkNet model was used as a classifier for the real-time object detection method you only look once (YOLO). Seventeen convolutionary layers were implemented, and different filters were introduced on each layer. This model can be used to help radiologists validate their initial screening, and can also be used for immediate screening of patients via cloud. This model is equipped with 125 X-ray images from the chest that are not in a normal shape and have been acquired in haste. Diagnostic tests performed in recovered patients after 5–13 days are found to be positive. This crucial finding shows us that the virus can continue to spread in recovered patients.

For COVID-19 chest X-ray detection which was named COVID-19 Detection Neural Network (COVNet) a DL-based method was proposed. The ReseNet50 used

COVNet as backbone. This model was tested from 3322 patients on 4356 chest CTs and showed 0.96 and 0.95 scores respectively for the Area Under Curve and Community Acquired Pneumonia (CAP) score [14]. The study also revealed CT findings and compared asymptomatic and symptomatic patients for COVID-19 detection and the results showed no substantial differences in age, gender distribution, or comorbidities for symptomatic and asymptomatic cases [72]. A deep neural network with a high architectural diversity and limited long-range connectivity was used by one of the most important and recent models, COVID-Net [73]. This model has been experimenting with two open access data repositories, one for the COVID-19 detection dataset and the other for the pneumonia detection data. The experimental results showed that approximately 83.5% were tested accuracy. In this case, both databases were fused to conduct training and test the proposed method [73]. However, we have trained a model with a pneumonia dataset in our proposed approach, and used Transfer Learning (TL) with a COVID19 dataset to train the same model. Recently a deep learning-based detection of COVID-19 from volumes of chest 3D CT with small labeled samples was proposed. A pretrained UNet model for 3D lung region segmentation was applied for each patient, and then a 3D-CNN model was applied to predict the probability of COVID-19 infections. The model was trained on 499 CT volumes and tested on 131 CT volumes and got 0.959 ROC AUC with sensitivity and specificity respectively 0.907 and 0.911. Overall, this model had an accuracy of about 90% [74]. Even, another one latest study showed very good accuracy for pneumonia detection tasks in which the model grouped together three separate categories: COVID-19, Influenza-A viral pneumonia and safe CT picture cases. The VNet based VNET-IR-RPN is used in pulmonary tuberculosis for region of interest segmentation. This deep learning-based method showed around 86.7% testing accuracy from CT images [75].

Authors in [69] proposed using the GAN network to generate more images to help detect COVID-19 with maximum accuracy from the available x-ray images. The key concept is to compile all possible COVID-19 images that already exist and use the GAN network to create more pictures. In the collected dataset the number of images is 307 images for four different class types. The groups are the bacterial COVID-19, regular pneumonia, and pneumonia virus.

Authors in [7] examined three different scenarios, the first one that was included 4 classes from the dataset while 3 classes were included in the second scenario, and 2 classes in the third scenario. All the scenarios included the class COVID-19 as it was the key target to be identified in this research. The Googlenet was chosen as the key deep transfer model in the first example, as it achieved 80.6% in precision testing. The Alexnet was selected as the main deep transfer model in the second scenario, as it achieved 85.25% in test accuracy while in the third scenario, which included 2 classes (COVID-19 and normal), Googlenet was chosen as the main deep transfer model, as it achieved 100% accuracy in testing and 99.9% accuracy in validation. One open door is to apply the deep models with a larger benchmark dataset for future works. [76] classified features obtained from various CNN (Convolutional Neural Network) models using X-ray images with support vector machine (SVM) classification. Their study states the best performance was provided by ResNet50 model with SVM classifier.

4.3 Genome Based Approaches

Bartoszewicz et al. [70] utilize reverse complement neural networks developed from CNN and LSTM architectures to identify whether a virus has the potential to infect a host body using its viral genome sequence, and apply machine learning interpretation skills techniques to identify the components of the sequence that are most correlated with infectious.

After studying RNA sequences that were available on the GISAID database, the authors concluded that COVID-19 is highly homologous to SARS-CoV-1. They found this homology by translating the RNA sequences into protein sequences and then building a 3D protein model using homology modeling (comparing the proteins and constructing an atomic-resolution of them) [77]. A DFCNN (Deep Fully Convolutional Neural Network) was used to identify and rank the protein-ligand interactions for performing virtual screening quickly since no docking or molecular dynamic simulation is needed. For clarification, a molecular dynamic simulation refers to a computer simulation method used for analyzing the physical movements of atoms and molecules; docking refers to the intentional removal of parts for analysis. With these techniques, the authors were able to identify potential drugs for COVID-19 by performing drug screening against four chemical compound databases. The proposed DFCNN architecture was trained on the protein-ligand dataset from the PDBbind database. The model was able to predict three potential drugs with a score of higher than 0.997 and three more with a score greater than 0.99.

DeepMind [78] is one of the top research companies in the world. On April 2, 2019, they submitted their research to the journal Nature on "Improved protein structure prediction using potentials from deep learning." It was accepted and published in December 2019. On January 15, 2020, they've reviewed this research in a blog post named AlphaFold: Using AI for scientific discovery, which is now helping identify the protein structures that are associated with COVID-19.

This research mainly addresses the problem of protein folding. You can think of proteins like large, complex molecules. Their three-dimensional structure changes as they perform different operations. The authors gave us an example of why identifying the structure of proteins is essential. The proteins are combinations of amino acids that are typically encoded in DNA. But the DNA does not contain information regarding the protein structure. Using traditional techniques, it would take ages to count all possible configurations of a typical protein before reaching the true 3D structure. The 3D structure from a complex protein sequence is predicted using deep learning and research like AlphaFold, we can now much more easily identify the 3D structures of proteins.

The working of Alphafold is divided into two phases. In the first phase standard biology techniques are used, and new protein fragments are created by repeatedly replacing the pieces of existing proteins. These structures are continuously improved with the help of a generative adversarial neural network (GAN). The protein structure from the output of the GAN has two properties: Distances between pairs of amino acids Angles between chemical bonds that connect those amino acids. In the second phase, the distance and the angles are improved by the Gradient Descent algorithm until they achieve the best scores. Below is an image of the entire work-flow of AlphaFold.

5 Conclusion

As observed in China, this COVID-19 growth can be decreased and quenched by raising the number of susceptible individuals from those infected. This can be done by being unsocial through restraint and adopting the lockdown initiative. This article provides a thorough overview of the historical deep learning models and recent state-of-the-art methods recommended for trying to solve COVID-19 pandemic challenges with Deep Learning (DL) technologies. Deep Learning (DL), is a largely developed machine learning class. Deep learning is an effective instrument to counter the current pandemic. If we take this opportunity to collect data, pool our expertise, and integrate our skills, we can save a lot of lives—now and in the future.

References

1. Fowler, P.: A chance to be our best. J. Pharm. Pract. Res. **50**, 122–123 (2020). https://doi.org/10.1002/jppr.1655
2. Xie, X., Muruato, A., Lokugamage, K., LeDuc, J., Menachery, V., Shi, P.: An Infectious cDNA Clone of SARS-CoV-2. Cell Host Microbe. Published (2020). https://doi.org/10.1016/j.chom.2020.04.004
3. Di Gennaro, F., Pizzol, D., Marotta, C., Antunes, M., Racalbuto, V., Veronese, N., Smith, L.: Coronavirus diseases (COVID-19) current status and future perspectives: a narrative review. Int. J. Environ. Res. Public Health **17**, 2690 (2020). https://doi.org/10.3390/ijerph17082690
4. Adnan, M., Suliman, S., Kazmi, A., Bashir, N., Siddiquea, R.: COVID-19 infection: origin, transmission, and characteristics of human coronaviruses. J. Adv. Res. **24**, 91–98 (2020)
5. Dargan, S., Kumar, M., Ayyagari, M.R., et al.: A survey of deep learning and its applications: a new paradigm to machine learning. Arch. Computat. Methods Eng. (2019). https://doi.org/10.1007/s11831-019-09344-w
6. Ongsulee, P.: Artificial intelligence, machine learning and deep learning. In: 2017 15th International Conference on ICT and Knowledge Engineering, Bangkok, pp. 1–6 (2017). https://doi.org/10.1109/ICTKE.2017.8259629
7. Loey, M., Smarandache, F., Eldeen, N., Khalifa, M.: Within the Lack of Chest COVID-19 X-ray Dataset: A Novel Detection Model Based on GAN and Deep Transfer Learning. preprint
8. Bruns, D., Kraguljac, N., Bruns, T.: COVID-19: facts, cultural considerations, and risk of stigmatization. J. Transcultural Nurs., pp. 1–7. https://doi.org/10.1177/1043659620917724
9. Gandhi, R.T., Lynch, J.B., del Rio, C.: Mild or moderate Covid-19. N. Engl. J. Med. (2020). https://doi.org/10.1056/NEJMcp2009249

10. Dai, X.: ABO blood group predisposes to COVID-19 severity and cardiovascular diseases. Eur. J. Prev. Cardiol. The European Society of Cardiology 2020. Article reuse guidelines: sagepub.com/journals-permissions. https://doi.org/10.1177/2047487320922370
11. Bhattacharyya, S., Maulik, U., DuttaQuantum: Inspired computational intelligence. Res. Appl., -33-83 (2017). ISBN: 978-0-12-804409-4. https://doi.org/10.1016/C2015-0-01859-7
12. Palm, G., McCulloch, W., Pitts, W.: A Logical Calculus of the Ideas Immanent in Nervous Activity, pp. 229–230. Springer, Berlin, Heidelberg (1986). ISBN 978-3-642-70911-1
13. Linnainmaa, S.: The Representation of the Cumulative Rounding Error of an Algorithm as a Taylor Expansion of the Local Rounding Errors, (In Finnish). Master's Thesis, Department of Computer Science, University of Helsinki, Helsinki, Finland (1970)
14. Ivakhnenko, A.G.: Polynomial theory of complex systems. IEEE Trans. Syst. Man Cybern. MC-1 **4**, 364–378 (1971)
15. Werbos, P.: Backpropagation through time: what it does and how to do it. Proc. IEEE **78**(10), 1550–1560 (1990). https://doi.org/10.1109/5.58337
16. Olivier, D., Bengio, Y.: Shallow vs. deep sum-product networks. In: Advances in Neural Information Processing Systems, vol. 24, pp. 666–674 (2011)
17. Weng, J., Ahuja, N., Huang, T.: Cresceptron: a self-organizing neural network which grows adaptively. International Joint Conference on Neural Networks (IJCNN) **1**, 576–581 (1992)
18. Williams, T., Li, R.: Wavelet Pooling for Convolutional Neural Networks. Published as a conference paper at ICLR 2018
19. Weng, J., Ahuja, N., Huang, T.S.: Learning recognition and segmentation using the cresceptron. Int. J. Comput. Vis. **25**, 109–143 (1997). https://doi.org/10.1023/A:1007967800668
20. Freund, Y., Haussler, D.: Unsupervised Learning of Distributions on Binary Vectors Using Two Layer Networks. Technical Report UCSC-CRL-94-25. University of California, Santa Cruz (1994)
21. Goodfellow, I.J., Pouget-Abadie, J., Mirza, M., Xu, B., Warde-Farley, D., Ozair, S., Courville, A., Bengio, Y.: Generative Adversarial Nets
22. Segmentation of neuronal structures in EM stacks challenge. In: IEEE International symposium on biomedical imaging. http://tinyurl.com/d2fgh7g (2012)
23. Ciresan, D.C., Giusti, A., Gambardella, L.M., Schmidhuber, J.: Deep neural networks segment neuronal membranes in electron microscopy images. In: Advances in Neural Information Processing Systems (NIPS), pp. 2852–2860 (2012)
24. Traian, T.: NVIDIA Tegra Inside Every Audi 2010 Vehicle. Retrieved 2016-08-03 from http://news.softpedia.com/news/NVIDIATegra-Inside-Every-Audi-2010-Vehicle-131529.shtml
25. Nayyar, Z.: Feature Engineering in Machine Learning (2015). https://doi.org/10.13140/RG.2.1.3564.3367
26. Huang, J., Dong, Q., Gong, S., Zhu, X.: Unsupervised Deep Learning by Neighbourhood Discovery
27. Caron, M., Bojanowski, P., Joulin, A., Douze, M.: Deep clustering for unsupervised learning of visual features. In Proceedings of the European Conference on Computer Vision (ECCV), pp. 1–18 (2018)
28. Dizaji, K.G., Herandi, A., Deng, C., Cai, W., and Huang, H. Deep clustering via joint convolutional autoencoder embedding and relative entropy minimization. In Proceedings of the IEEE International Conference on Computer Vision (ICCV), pp. 5747–5756 (2017)
29. Yang, B., Fu, X., Sidiropoulos, N.D., Hong, M.: Towards k-means-friendly spaces: Simultaneous deep learning and clustering. In: Proceedings of the International Conference on machine learning (ICML), pp. 1–14 (2017)
30. Zhang, R., Isola, P., Efros, A.: Split-brain autoencoders: unsupervised learning by cross-channel prediction. In:The IEEE Conference on Computer Vision and Pattern Recognition (CVPR), pp. 1058–1067 (2017)
31. Wu, Z., Xiong, Y., Stella, X., Lin, Y.: Unsupervised feature learning via non-parametric instance discrimination. In: IEEE Conference on Computer Vision and Pattern Recognition (CVPR), pp. 3733–3742 (2018)

32. Nair, V., Alonso, J., Beltramell, T.: RealMix: Towards Realistic Semi-Supervised Deep Learning Algorithms (2019)
33. Kingma, D., Rezende, D., Mohamed, S., Welling, M.: Semi-supervised learning with deep generative models. Proceedings of Advances in Neural Information Processing Systems **27**, 3581–3589 (2014)
34. Lavet, V., Henderson, P., Islam, R., Bellemare, M., Pineau, J.: An Introduction to Deep Reinforcement Learning (2018). arXiv:1811.12560v2
35. Sabour, S., Frosst, N., Hinton, G.E.: Dynamic routing between capsules. In: 31st Conference on Neural Information Processing Systems (2017)
36. Arif, W.M., Ahmed, B., Saduf, A., Iqbal, K.: Advances in Deep Learning: Basics of Supervised Deep Learning. Springer, Singapore, pp. 13–29 (2020). https://doi.org/10.1007/978-981-13-6794-6-2
37. Wang, H., Hiksha, B.: On the Origin of Deep Learning (2017). arXiv:1702.07800v4 [cs.LG]
38. Sutskever, I., Vinyals, O., Le, Q.V.: Sequence to sequence learning with neural networks. In: Advances in Neural Information Processing Systems (2014)
39. Schmidhuber, J.: Deep learning in neural networks: an overview. Neural Netw. **61**, 85–117 (2015)
40. Rumelhart, D.E., Hinton, G.E., Williams, R.J.: Learning internal representations by error propagation. In: Parallel Distributed Processing. Vol. 1: Foundations. MIT Press, Cambridge, MA (1986)
41. Yuan, F.N., Zhang, L., Shi, L., Xia, J.T., Li, X.: Theories and applications of auto-encoder neural networks: a literature survey. Chin. J. Comput. **42**, 203–230 (2019). https://doi.org/10.11897/SP.J.1016.2019.00203
42. Jiang, P., Maghrebi, M., Crosky, A., Saydam, S.: Unsupervised deep learning for data-driven reliability and risk analysis of engineered systems. In: Handbook of Neural Computation, pp. 417–431 (2017)
43. Baldi, P., Hornik, K.: Neural networks and principal component analysis: Learning from examples without local minima. Neural Netw. **2**(1), 53–58 (1989)
44. Japkowicz, N., Hanson, S.J., Gluck, M.A.: Nonlinear autoassociation is not equivalent to pca. Neural Comput. **12**(3), 531–545 (2000)
45. Deng, J., Zhang, Z., Marchi, E., Schuller, B.: Sparse autoencoder-based feature transfer learning for speech emotion recognition. In: 2013 Humaine Association Conference on Affective Computing and Intelligent Interaction (ACII). IEEE, pp. 511–516 (2013)
46. Vincent, P., Larochelle, H., Bengio, Y., Manzago, P.-A.: Extracting and composing robust features with denoising autoencoders. In: Proceedings of the 25th International Conference on Machine Learning. ACM, pp. 1096–1103 (2008)
47. Rifai, S., Vincent, P., Muller, X., Glorot, X., Bengio, Y.: Contractive auto-encoders: explicit invariance during feature extraction. In: Proceedings of the 28th International Conference on Machine Learning (ICML-11), pp. 833–840 (2008)
48. Goodfellow, I., Pouget-Abadie, J., Mirza, M., Xu, B., Warde-Farley, D., Ozair, S., Courville, A., Bengio, Y.: Generative adversarial nets. In: Advances in Neural Information Processing Systems, pp. 2672–2680 (2014)
49. Wang, W., Huang, Y., Wang, A., Wang, L.: Generalized autoencoder: a neural network framework for dimensionality reduction. In: Proceedings of the IEEE Conference on Computer Vision and Pattern Recognition Workshops, pp. 490–497 (2014)
50. Wen, Y., Lie, X., Gonzalez, J.: Fast training of deep LSTM networks. Springer International Publishing, pp. 3–10 (2019) Isbn: 978-3-030-22796-8
51. Pang, B., Zha, K., Cao, H., Shi, C., Lu, C.: Deep RNN Framework for Visual Sequential Applications, pp. 423–432 (2018)
52. Hochreiter, S., Schmidhuber, J.: Long short-term memory. Neural Comput. **9**(8), 1735–1780 (1997)
53. Sja, F., Christian, I.: An Introduction to Restricted Boltzmann Machines, pp. 14–36 (2012). https://doi.org/10.1007/978-3-642-33275-3-2

54. Goodfellow, I.J., Pouget-Abadie, J., Mirza, M., Xu, B., Warde-Farley, D., Ozair, S., Aaron, C., Bengio, Y.: Generative Adversarial Nets. NIPS preceding (2014)
55. Creswell, A., White, T., Dumoulin, V., Arulkumaran, K., Sengupta, B., Bharath, A.A.: Generative adversarial networks: an overview. IEEE Signal Process. Mag. **35**(1), 53–65 (2018)
56. Baldi, P.: Autoencoders, unsupervised learning, and deep architectures. In: JMLR: Workshop Workshop on Unsupervised and Transfer Learning, pp. 27–37 (2012)
57. Meng, Q., Catchpooley, D., Skillicornz, D., Kennedy, P.J.: Relational Autoencoder for Feature Extraction (2018). arXiv: 1802.03145v1 [cs.LG]
58. Pascanu, R., Gulcehre, C., Cho, K., Bengio, Y.: How to Construct Deep Recurrent Neural Networks (2014). arXiv:1312.6026v5 [cs.NE]
59. Hermans, M., Schrauwen, B.: Training and analyzing deep recurrent neural networks. Advances in Neural Information Processing Systems **26**, 190–198 (2013)
60. Mikami, A.: Long Short-Term Memory Recurrent Neural Network Architectures for Generating Music and Japanese Lyrics. Ph.D, Computer Science Department, Boston College (2016)
61. Suk, H.: Deep Learning for Medical Image Analysis (2017). https://doi.org/10.1016/B978-0-12-810408-8.00002-X
62. Salakhutdinov, R., Hinton, G.: Deep Boltzmann Machines. In: Proceedings of the 12th International Conference on Artificial Intelligence and Statistics (AISTATS), Clearwater Beach, Florida, USA (2009). Volume 5 of JMLR: W-CP 5
63. Khan, A., Zameer, A., Jamal, T., Raza, A.: Deep Belief Networks Based Feature Generation and Regression for Predicting Wind Power (2018). arXiv:1807.11682
64. Karhunen, J., Raiko, T., Cho, K.: Deep learning: a short review. In: Advances in Independent Component Analysis and Learning Machines (2015). http://dx.doi.org/10.1016/B978-0-12-802806-3.00007-5
65. Punn, N., Sonbhadra, S., Agarwal, S.: COVID-19 Epidemic Analysis using Machine Learning and Deep Learning Algorithms. Display the preprint in perpetuity. medRxiv preprint. https://doi.org/10.1101/2020.04.08.20057679. This version posted 11 Apr 2020
66. Alom, M., Rahman, M., Nasrin, M., Taha, M., Asari, K.: COVID-MTNet: COVID-19 Detection with Multi-Task Deep Learning Approaches (2020)
67. Ozturka, T., Talob, M., Yildirimc, E., Baloglud, U., Yildirime, O., Acharyaf, U.: Automated detection of COVID-19 cases using deep neural networks with X-ray images. Comput. Biol. Med. https://doi.org/10.1016/j.compbiomed.2020.103792
68. Hurt, B., Kligerman, S., Hsiao, A.: Deep learning localization of pneumonia: 2019 coronavirus (COVID-19) outbreak. J. Thorac. Imaging **35**(3), W87–W89 (2020)
69. Shadab, S., Alam Khan, T., Afrin Neezi, N., Adilina, S., Shatabda, S.: DeepDBP: Deep Neural Networks for Identification of DNA-binding Proteins. https://doi.org/10.1101/829432
70. Bartoszewicz, J.M., Seidel, A., Renard, B.Y.: Interpretable Detection of Novel Human Viruses from Genome Sequencing Data (2020). https://doi.org/10.1101/2020.01.29.925354
71. Ayyoubzadeh, S.M., Ayyoubzadeh, S.M., Zahedi, H., Ahmadi, M., Kalhori, S.R.: Predicting COVID-19 incidence through analysis of google trends data in iran: data mining and deep learning pilot study. JMIR Public Health Surveill. **6**(2), e18828 (2020). https://doi.org/10.2196/18828
72. Lin, L., Qin, L., Xu, Z., Yin, Y., Wang, X., Kong, B., Bai, J.: Artificial intelligence distinguishes COVID-19 from community acquired pneumonia on chest CT. Radiology, p. 200905 (2020). https://doi.org/10.1148/radiol.2020200905
73. Linda, W., Wong, A.: COVID-Net: A Tailored Deep Convolutional Neural Network Design for Detection of COVID-19 Cases from Chest Radiography Images (2020). arXiv preprint arXiv:2003.09871
74. Chuansheng, Z., Deng, X., Fu, Q., Zhou, Q., Feng, J., Ma, H., Liu, W., Wang, X.: Deep Learning-based Detection for COVID-19 from Chest CT using Weak Label. medRxiv (2020)
75. Xiaowei, X., Jiang, X., Ma, C., Du, P., Li, X., Lv, S., Yu, L.: Deep Learning System to Screen Coronavirus Disease 2019 Pneumonia (2020). arXiv preprint . arXiv:2002.09334
76. Sethy, P.K., Behera, S.K.: Detection of Coronavirus Disease (COVID-19) Based on Deep Features (2020)

77. Zhang, H., Saravanan, K.M., Yang, Y., Hossain, M.T., Li, J., Ren, X., Wei, Y.: Deep Learning Based Drug Screening for Novel Coronavirus 2019-nCov (2020). Preprints 2020, 2020020061. https://doi.org/10.20944/preprints202002.0061.v1
78. Computational predictions of protein structures associated with COVID-19. https://deepmind.com/research/open-source/computational-predictions-of-protein-structures-associated-with-COVID-19

Digital and Emerging Technologies for Monitoring and Controlling COVID-19

Monitoring COVID-19 Disease Using Big Data and Artificial Intelligence-Driven Tools

Sally M. Elghamrawy, Ashraf Darwish, and Aboul Ella Hassanien

Abstract With the daily huge growth in the number of confirmed COVID-19 cases, COVID-19 extremely threatens public health, countries' economic, social life, and the international relations around the world. The accurate diagnosis based on a huge amount of data has become a serious issue that effect the disease control, especially in the widespread countries. To monitor COVID-19, big data analytics tools and Artificial Intelligence (AI) techniques play a significant role in many aspects. The integration between both technologies will help healthcare workers for early and accurately diagnose COVID-19 cases. In addition, the strategic planning for crisis management is supported by aggregation of big data to be use in the epidemiologic directions. Moreover, AI and big data driven tools presents visualization for COVID-19 outbreak information that help in detecting risk allocation and regional transmissions. In this chapter, a review of recent works related to COVID-19 containment using AI and big data techniques is introduced, showing their main findings and limitations to make it easy for researchers to investigate new techniques that will help in COVID-19 pandemic.

Keywords COVID-19 · Coronavirus · Artificial intelligence · Big data · Big data analytics tools · Deep learning

S. M. Elghamrawy (✉)
Computer Engineering Department, MISR Higher Institute for Engineering and Technology, Mansoura, Egypt
e-mail: sally_elghamrawy@ieee.org

A. Darwish
Faculty of Science, Helwan University, Helwan, Egypt
e-mail: ashraf.darwish.eg@ieee.org

A. E. Hassanien
Faculty of Computers and Artificial Intelligence, Cairo University, Giza, Egypt
e-mail: aboitcairo@cu.edu.eg

S. M. Elghamrawy · A. Darwish · A. E. Hassanien
Scientific Research Group in Egypt, Giza, Egypt

A. E. Hassanien and A. Darwish (eds.), *Digital Transformation and Emerging Technologies for Fighting COVID-19 Pandemic: Innovative Approaches*, Studies in Systems, Decision and Control 322, https://doi.org/10.1007/978-3-030-63307-3_10

1 Introduction

COVID-19 cases had been spread widely in different countries with increasing in the death rates, the rate of information evolution has become extremely fast and it became a critical issue to handle this huge amount of data every day. In May 29th 2020, the number of coronavirus cases was 5,955,507 and the deaths was 363,245 in the world [1] and the rate is rapidly increasing. Big data analytics tools must handle and analyse the massive amount of big medical data quickly due to the fact that time is very significant issue in healthcare applications that effect the containment of disease.

Advanced data analytics tools are needed to extract knowledge from patients' data to develop original methods to early diagnose cases with COVID-19. Big data analytics is the process of analysing COVID-19 huge Data collected from different sources (Chest CT scans and X-Ray images, RT-PCR data, CBC, sensors, social networks and medical reports). COVID-19 big data Analytics is used to extract hidden patters, valued perceptions relationships from the confirmed and suspected cases. Healthcare workers then take these results and conclude some finding that help in CONID-19 pandemic containment. The main tasks of COVID-19 big data Analytics tools are shown in Fig. 1.

This chapter is structured as follows: The recent works for monitoring COVID-19 based on big data are presented in Sect. 2. Section 3 highlight the AI driven

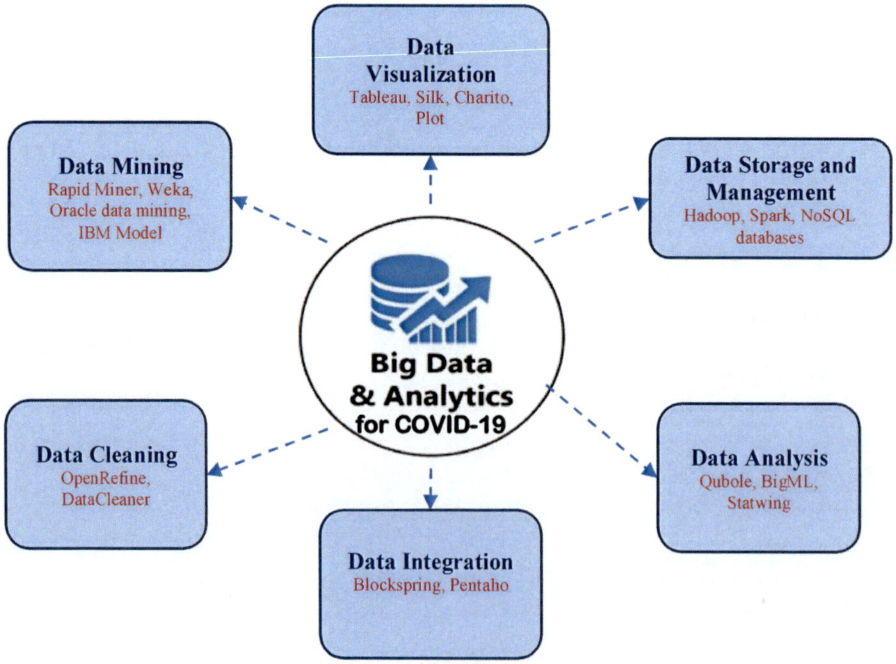

Fig. 1 The main tasks of COVID-19 big data analytics tools

tools against fighting COVID-19. Section 4 highlights a case study about the role of Big Data and Geographic Information System technologies against COVID-19 in Germany. Section 5 presents the problems and challenges. Section 6 concludes this chapter.

2 Recent Studies in Monitoring COVID-19 Using Big Data

Recent studies used big data analysis tools to decrease the chance of spreading COVID-19 by monitoring COVID-19 disease and detecting the possible areas of infection. It is thus vital to utilize big data and intelligent analytics tools for monitoring COVID-19 disease to improve the community health. Many researchers exploit bid data and intelligent tools to monitor COVID-19 disease in different ways, as shown in Fig. 2.

A recent study in [2] has presented an approach for tracking passengers' data traffic from a big-data source. This tracking will show the probability of forecasting COVID-19 and prevent the further spread of novel COVID-19. They collect tracking data from the official flight companies and digital big data sources extracted from data from citizens' mobile phones from the WeChat app, in order to provide information for risk management.

Authors in. [3], at Wuhan city, use big data storage tool to store and estimate number of trippers from Wuhan by using three different sources: the global flight bookings, the daily number of foreign and the national traveller volumes from and to Wuhan city. These data are then used to predict the forecasting of the epidemic based on the number of cases transferred from Wuhan to other cities.

Coronavirus (Covid-19 outbreak has proved the need for proactive containment. All countries must rapidly utilize their resources to save their people lives and their economic stability. The authors in [4] stated that Taiwan used big data analytic

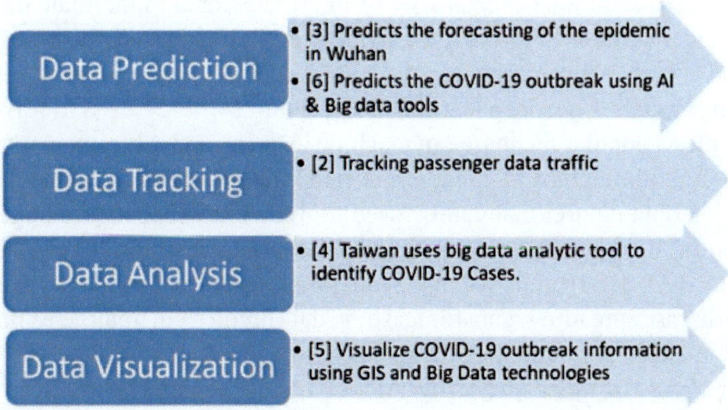

Fig. 2 Recent research areas for monitoring COVID-19 using big data

tools to present different approached to identify and contain COVID-19 cases early to protect the public health. They stated that they use big data storage advantage to apply integration between national health insurance database and immigration database. This huge database and its tools are used to dynamically generate alerts of COVID-19 confirmed cases and severe cases during the medical examination of the patient, thus identify and diagnosis Taiwan's cases and take rapid action.

Authors in [5] used the geographic information systems (GIS) to aggregate different big data sources and quickly create visualization for COVID-19 outbreak information. They stated that the tools used helped in predicting the risk allocation and regional transmissions.

An integration between big data and AI tools is a critical issue for the discovering and preventing the spread of COVID-19. The authors in [6] investigated the importance of AI techniques with big data tools. They stated that using the 3V's of big data helped in creating different datasets that are used by many researchers and models. They reviewed a number of models that used AI and big data tools to contain the diseases like GoogleFlue [7] and BlueDot [8] that predicted the COVID-19 outbreak and sent alerts.

3 COVID-19 Artificial Intelligence-Driven Tools

The global threats and epidemics caused by the outbreak novel coronavirus COVID-19, recognised in late 2019, lead to researchers" attention to find solutions for its containment. Besides the medical procedures and treatments, different Artificial Intelligence AI tools based on machine learning algorithms (ML) are used for diagnosing, data analysis, decision making and predicting processes in health care field. In this context, AI-based tools help in detecting COVID-19 outbreaks and also anticipate the nature of their proliferation throughout the world. However, the main concept of AI-based tools is the need for adequate training data. Often, the traditional ML techniques needs a set of interpreted data to facilitate the process of training the classifiers. This process is identified as supervised learning. However, AI researchers do not prefer to wait for complete data sets for training, verifying, and testing the models. The Deep learning DL techniques can be used to solve this issue. DL becomes a popular technique in machine learning and AI framework for big data analysis. It is a branch of ML that has been applied successful in many applications with its big data nature, that can be used to mine knowledge from huge amount of data sets. In this context, this paper categorizes the capabilities of AI-driven tools to monitor COVID-19 into six main categories, as shown in Fig. 3, namely, Early alerts to disease spreading, data forecasting, data prediction, treatments management, outbreak management and data diagnosis.

Early alerts to disease spreading: AI tools can expect infectious disease outbreaks, as the BlueDot [8] model that used AI tools for predicting the COVID-19 outbreak and sent alerts on December 31 before the World Health Organization did. ***Data Forecasting***: Forecasting COVID-19 data help in the visualization of COVID-19

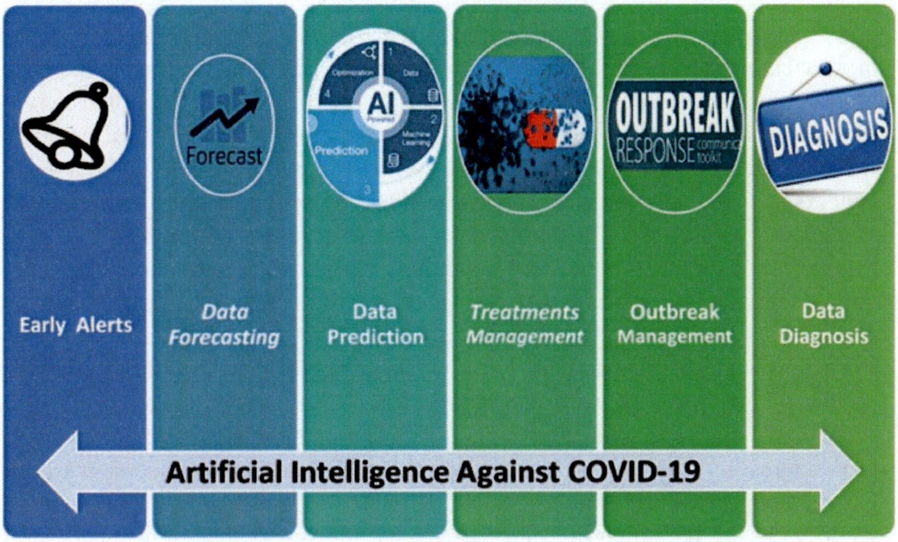

Fig. 3 Artificial intelligence against COVID-19's outbreak

pandemic which in turns help many decision makers to have a global overview of the current situation. In addition, the data collected by AI tools for predicting the forecasting prevent the further spread of novel coronavirus (2019-nCoV). ***Data Prediction***: Neural network and Machine learning techniques are used for training models, using data from the COVID-19 pandemic, to predict the spreading of diseases. In addition, it can be used to predict the infected areas and the patient response to treatment.

Treatments Management: Before COVID-19 pandemic, AI tools were widely used for predicting and devolving new treatments for many diseases. For COVID-19, healthcare workers stated that using AI tools save their efforts in developing and testing new treatments. Moreover, AI tools accelerate the process of finding (predicting) the appropriate treatment [9]. Number of authors [24, 25] devolved an AI-driven model that uses a deep learning system to predict treatments related to COVID-19. ***Outbreak Management***: AI driven tools are used to manage the pandemic by scanning community area for citizens with suspected or confirmed cases, then the collected data is used to present a measurement for the ideal social distance. China used these tools at airports and train stations across for recognizing citizens with *high temperature and detect the person with no mask. This tool will manage virus spreading.* ***Data Diagnosis***: Using AI tools for diagnosis will contribute in accurately and early detecting of COVID-19 using either text data or images, which will save lives, limit the spread of the disease and help healthcare workers to provide the appropriate treatment to patients. Many efforts had been made to utilize these AI capabilities for monitoring COVID-19 in number of ways.

Authors in [10] discuss the significance of the AI-driven tools and their applicable models for training and testing. To gain greater trust during the decision-making process, several types of data are required to be used instead of depending on one type of data. Hence the use of digital and interactive data has been explored in the context of active learning. Furthermore, despite the prevalence of COVID-19 (worldwide), AI-based methods are expected to act as population-wide training/test paradigms. Additionally, the Authors provide an Active Learning (AL) model schema for time series data. Comments from experts are used in AL in conjunction with the decisions of each category of data. Because the DL is data-dependent, a separate DL is used for different data types. Multimedia and multimodal data are used to make the final decision possible. In addition, Marcello and Vayena in [10] discussed how countries that use digital technology to counter the ongoing COVID-19 pandemic and demonstrate how widespread data collection will help inhibit the COVID-19 pandemic, but should not ignore privacy and public trust. Global best practices for upholding standards for data collection and fair handling should be established. Some concerns were addressed in this paper, for example (1) Detailed data management practices like data collection and data processing, (2) data processing level, data quality and security checks, data integrity flaws, data breaches, incomplete or unsuccessful de-identification and data set prejudices, (3) data privacy and data protection challenges. All the issues addressed are imperative for exploiting big data and smart analytics and making good use of them for public health.

BoXul [11] Uses various resources to compile and curate individual- databases from state, provincial, and local health records, as well as additional online information. All data is geocoded and includes major symptoms and dates (start, date of acceptance and confirmation) and travel history where appropriate. It is crucial to build accurate, actual, and reliable data on emerging disease outbreaks, which can help generate clear evidence to support which support decision-making in public health. The static data set was uploaded to fgshare, which contains a fixed version of the data record when submitted, which will be modified which downloadable from https://github.com/beoutbreakprepared/nCoV2019.

In a study [12], Li Tan aimed to review and analyze the time cycles of complete blood counts of treated and death cases in order to gain the main signs of virus development. The authors analysed and reported several study results namely (1) Prediction of forecast in COVID-19 cases based on blood lymphocyte percentage (LYM%), (2) Established time of LYM percentage model from treated COVID-19 cases, (3) Validated of the TLM in disease classification in COVID-19 cases, (4) LYM% indicates the severity degree of the infection of COVID-19 and (5) analysis of possible reasons for Lymphpenia in patient.

The authors in [13] discuss the possible use of four integrated emerging technologies, namely the Internet of Things, big data analytics, AI and Blockchain, to advance two conventional public health approaches to tackle COVID-19: (1) surveillance, identification and prevention of COVID-19; and (2) Affect mitigation on health care indirectly linked to COVID-19.

Coronavirus' first public dataset on Twitter has been published by Emily et al. [14] describing a multilingual Twitter dataset about COVID-19. The authors of this paper collected dataset continuously since January 22, 2020. A collected Twitter dataset is accessible to the researchers at (https://github.com/echen102/COVID-19-TweetIDs).

The authors of this paper thought that the data set can be used to study the dynamics of online conversation in the context of a planetary outbreak with unprecedented dimensions and impacts. This dataset can also help track false information about scientific Coronaviruses and unverified rumors, or enable you to understand fear and panic. Ultimately, the tweets dataset will enable enlightened solutions and describe targeted policy interventions to combat this global crisis. Given the confirmation of the transmission of COVID-19 from person to person and the ship's space and population density are relatively limited, it is crucial to assess the transportability of COVID-19 and to predict the potential magnitude of the epidemic on board the cruise ship Diamond Princess in the future.

The latest research on predicting and estimating the reproductive number of novel coronavirus (COVID-19) reported by Sheng in [15] at the primary stage of disease outbreaks on board the ship and expected new regular cases over the next ten days. All data were taken from the official website (Japan's Ministry of Health, Labor, and Welfare, 2020) which reported a COVID-19 case in Japan.

To assess COVID-19's portability, the authors applied an early R package to detect (R0) in the primary stage of the outbreak. Serial separator distribution mandatory for R0 estimate, and there was inadequate information about mass states to estimate serial separation. Therefore, researchers hypothesized that the sequence interval for COVID-19 on the ship was equal to the time interval for COVID-19 in Wuhan, China. Then, authors installed the value of the serial separator with the gamma distribution. The "get_R" function was used to estimate the maximum probability (ML) to obtain an R0 distribution. Also, they derive other statistics to estimate the R0 distribution using the boot strategy with 1000 reconfiguration to obtain a wide range of possible R0 values. In the graph format, they show these R0 values and measure the average and quadratic range of those R0 values. The "Projections" R package was used to model rational epidemiological pathways and to forecast future everyday occurrences. The forecasting and simulation were created by the appropriateness of current daily occurrence data, distribution of sequential intervals and estimation of R0 in a model based on the assumption that daily incidence is almost subject to the daily infection-determined Poisson distribution.

Recently, different studies [9, 16–18] implemented AI- algorithms to diagnose COVID-19 using CT and X-ray Chest images. The authors used deep learning techniques to distinguish COVID-19 disease from other viral pneumonia to help healthcare workers to provide the appropriate treatment to patients.

McCall [19] studied the influence of AI on the coronavirus outbreak and stated that the choice of using AI application for monitoring and controlling the impact of COVID-19 is much better than the choice of closing borders and trades.

From another perspective, AI tools are not only used from diagnosis COVID-19, but they also help in finding the appropriate treatment. Richardson et al. [20] predict that medicines associated with AP2-Associated Protein Kinase 1 (AAK1) disrupting these proteins prevent viral entry into the lung alveolar epithelial.

4 Case Study: The Role of Big Data and Geographic Information System Technologies to Curb the COVID-19 Outbreak in Germany

The prevention capacity of the potential coronavirus outbreak has a notable correlation with possessing a sustainable real-time detection for the daily infected cases. Many nations search for trusted options that able to diminish the disease spread peak, in which the number of daily fatalities and infected cases approach their highest value, and reduce the continuation of the peak span. These phenomena could be achievable by having a real-time identification system that can help to analyze and allocate not only the potential cases but also regions where the infection rate might be extraordinary. Germany was one of those countries which had this real-time detection and tracing for cases by using sort of technologies such as Big data and Geographic Information System (GIS) [21]. The contribution of these technologies boosts Germans to dominate the outbreak of the pandemic in the shortest time.

The German government agencies have mapped confirmed and suspected cases, fatalities, and recovered cases to identify which regions of the country the COVID-19 pandemic infections have occurred [22] besides providing aggregation, interpretation, and investigation for immense data become achievable using Big data technology which assists to handle a bunch of data efficiency in a reasonable time. This allows them to dominate the spreading of the virus in addition to collect and compile all information relevant to the pandemic, and they ultimately eradicate the virus in a short-term span as shown in Fig. 4 [23]. The infection rate was increased dramat-

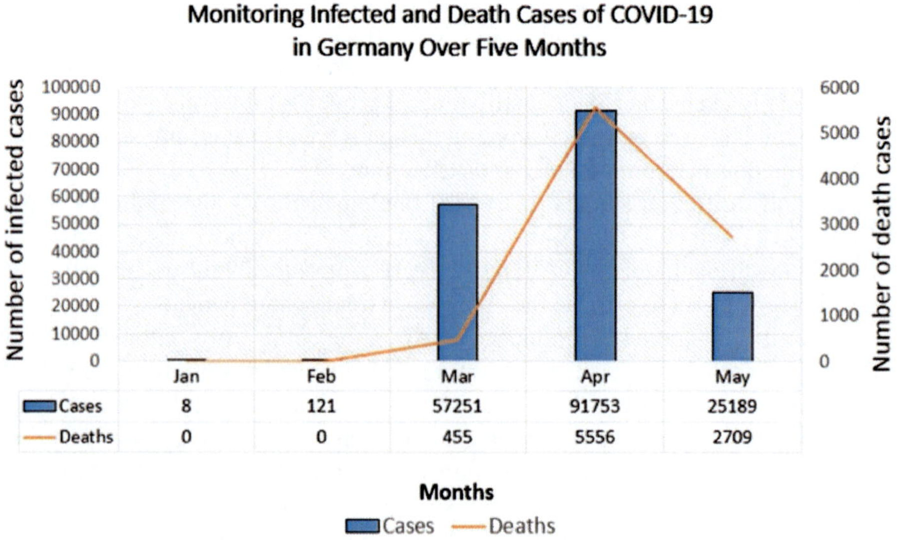

	Jan	Feb	Mar	Apr	May
Cases	8	121	57251	91753	25189
Deaths	0	0	455	5556	2709

Fig. 4 Representation for the number of infected cases and fatalities over five months in Germany from January to May in 2020

ically by the beginning of March when the infected cases were more than 50,000, whereas the fatalities raised to exceed five hundred cases.

In April, disease spread in German reached its peak when the mortality rate was 81.8% and the infection rate was around 62% comparing to the cases in March. However, the determination of virus propagation supported the government agencies to control the situation and drop down the number of both confirmed and fatality cases in the next month, in which the number of infected cases decreased for about 66,000 cases and death cases around 3000 cases.

To sum up, the observation of the pandemic cases and isolating them as well as suspected regions where the virus spread have remarkable outcomes and advantages that could be achieved by integrating some technologies such as Big Data, and GIS in the procedures of retailing the virus outbreak.

5 Problems, Challenges and Future Trends

Although the efforts from the whole world to understand COVID-19, there many issues remain unsolved and cause problems and challenges. In the next, we explore such problems and challenges and present to the researchers and academics some future research points from the AI point of view.

1. The COVID-19 has become a global challenge to the health-care sector worldwide. Moreover, the data and information about COVID-19 is not enough for research and studying. Therefore, data centres and knowledge base systems can play an important role to provide sufficient knowledge about this virus as the vaccination and therapy are currently under development and evaluation.
2. There are some statistical methods have been used in studying the variance of the virus by using a model for a completely randomized design; however, more advanced AI and machine learning techniques can be utilized to deal with the big data problems and challenges. Therefore, some important factors such as the pre-existing diseases, patient's age and severity of the infection.
3. Social media reports are representing a challenging for healthcare workers and it generates huge amount of data. The tweets of people on social media such as Twitter and Facebook can be used to analyse the response to the virus as positive or negative by using the sentiment analysis for their tweets about the response to the virus.
4. Big data technology and GIS have a major task in combating COVID-19 in different areas such as aggregation of multi-source big data, prediction of regional transmission, rapid visualization of epidemic information, spatial monitoring of COVID-19, Identification of spatial risk distribution and control level selection, and social-emotional instruction and avoidance of panic. The key problem is to find new strategies for adapting traditional methods and improving accuracy and

speed to provide more accurate information for fast social management. In addition, GIS use suffers from data processing difficulties and data heterogeneity. This includes cooperation from academic institutions, businesses and governments to jointly promote the formulation of applicable policies and strategies. New policies and approaches for a technological knowledge-acquisition method focused on broad spatial data cantered on social operations remain a new challenge.

5. Choosing the prediction model of COVIID-19 is considered a major problem. In this context, epidemiological and biological information should be represented by this model. In this case, such model needs parameters estimation to be compared with a simple model.
6. AI technology can help protect staff in this COVID-19 condition. The importance of AI comes in the form of Robotics to reduce the burden on clinicians in the outbreak of COVID-19 during the treatment and other mission of the patients. However, not all world countries have the Robotics resources.
7. Using social media information, mobile payment apps and other means and data on travel around the world, machine learning techniques can be used to predict the infected locations of COVID-19. However, using such models is still not enough and need more efforts from the researchers to assist the decision makers and governments.

6 Conclusion

The outbreak of COVID-19 pandemic has infected many confirmed cases and thousands of deaths according to WHO statistics. At the moment of writing this paper, the number is still increasing since COVID-19 is easily spreadable via respiratory droplets. In this context, many efforts had been made to contain this disease by the early detection of COVID 19 to help in limiting its spreading. Thus, many researchers have been developed some tools using AI techniques and big data analytic tools for early prediction and diagnosis of COVID 19 cases. These models had been a great help to clinicians and decision-makers for making the recommendations for treatments. Finally, the value of integrating AI and big data tools strongly comes into play by providing clinicians and decision-makers with helpful information on COVID-19 disease, which will reduce their efforts.

References

1. WHO: Coronavirus disease 2019 (COVID-19) Situation Report—66 (2020). https://www.who.int/docs/default-source/coronaviruse/situation-reports/20200330-sitrep-70-covid-19.pdf?sfvrsn=7e0fe3f8_2. Accessed 31 Mar 2020
2. Wu, J.T., Leung, K., Leung, G.M.: Nowcasting and forecasting the potential domestic and international spread of the 2019-nCoV outbreak originating in Wuhan, China: a modelling study. Lancet **395**(10225), 689–697 (2020)

3. Zhao, X., Liu, X., Li, X.: Tracking the spread of novel coronavirus (2019-nCoV) based on big data. medRxiv (2020)
4. Wang, C.J., Ng, C.Y., Brook, R.H.: Response to COVID-19 in Taiwan: big data analytics, new technology, and proactive testing. Jama. **323**(14), 1341–1342 (2020)
5. Zhou, C., Su, F., Pei, T., Zhang, A., Du, Y., Luo, B., Cao, Z., Wang, J., Yuan, W., Zhu, Y., Song, C.: COVID-19: challenges to GIS with big data. Geogr. Sustain. (2020)
6. Long, J.B., Ehrenfeld, J.M.: The Role of Augmented Intelligence (AI) in Detecting and Preventing the Spread of Novel Coronavirus (2020)
7. Lazer, D., Kennedy, R.: What We Can Learn from the Epic Failure of Google Flu Trends: WIRED (2020). https://www.wired.com/2015/10/canlearn-epic-failure-google-flu-trends/. Published 2015. Accessed 31 Jan 2015
8. Niller, E.: An AI Epidemiologist Sent the First Warnings of the Wuhan Virus: WIRED (2020). https://www.wired.com/story/aiepidemiologist-wuhan-public-health-warnings/. Published 2020. Accessed 31 Jan
9. Khalifa, N.E.M., Taha, M.H.N., Hassanien, A.E., Elghamrawy, S.: Detection of Coronavirus (COVID-19) Associated Pneumonia Based on Generative Adversarial Networks and a Fine-Tuned Deep Transfer Learning Model Using Chest X-ray Dataset (2020). arXiv preprint arXiv: 2004.01184
10. Marcello Lenca, M., Vayena, E.: On the responsible use of digital data to tackle the COVID-19 pandemic. Nat. Med. (2020). https://doi.org/10.1038/s41591-020-0832-5
11. BoXu1: Epidemiological data from the COVID-19 outbreak, real-time case information. Sci. Repost (2020)
12. Tan, L., et al.: Lymphopenia predicts disease severity of COVID-19: a descriptive and predictive study. Signal Transduct. Targeted Ther. **5**(1), 1–3 (2020)
13. Ting, D.S.W., Carin, L., Dzau, V., et al.: Digital technology and COVID-19. Nat. Med. (2020). https://doi.org/10.1038/s41591-020-0824-5
14. Chen, E., Lerman, K., Ferrara, E.: COVID-19: The First Public Coronavirus Twitter Dataset (2020). arXiv:2003.07372 [cs.SI]
15. Zhang, S., Diao, M., Yu, W., Pei, L., Lin, Z., Chen, D.: Estimation of the reproductive number of novel coronavirus (COVID-19) and the probable outbreak size on the Diamond Princess cruise ship: a data-driven analysis. Int. J. Infect. Dis. **1**(93), 201–204 (2020)
16. Li, L., Qin, L., Xu, Z., Yin, Y., Wang, X., Kong, B., Bai, J., Lu, Y., Fang, Z., Song, Q., Cao, K.: Artificial Intelligence Distinguishes COVID-19 from Community Acquired Pneumonia on Chest CT. Radiology, p. 200905 (2020)
17. Fischer, A.M., Varga-Szemes, A., van Assen, M., Griffith, L.P., Sahbaee, P., Sperl, J.I., Nance, J.W., Schoepf, U.J.: Comparison of artificial intelligence–based fully automatic chest CT emphysema quantification to pulmonary function testing. Am. J. Roentgenol., pp. 1–7 (2020)
18. ELGhamrawy, S.M.: Diagnosis and Prediction Model for COVID19 Patients Response to Treatment based on Convolutional Neural Networks and Whale Optimization Algorithm Using CT Images. medRxiv. 1 Jan 2020
19. McCall, B.: COVID-19 and artificial intelligence: protecting health-care workers and curbing the spread. Lancet Digit. Health **2**(4), e166–e167 (2020)
20. Richardson, P., Griffin, I., Tucker, C., Smith, D., Oechsle, O., Phelan, A., Stebbing, J.: Baricitinib as potential treatment for 2019-nCoV acute respiratory disease. Lancet **395**(10223), e30–e31 (2020)
21. Zhou, C., Su, F., Pei, T., Zhang, A., Du, Y., Luo, B., Cao, Z., Wang, J., Yuan, W., Zhu, Y., Song, C., Chen, J., Xu, J., Li, F., Ma, T., Jiang, L., Yan, F., Yi, J., Hu, Y., Liao, Y., Xiao, H.: COVID-19: challenges to GIS with big data. Geogr. Sustain. **1**(1), 77–87 (2020)
22. Böhmer, M.M., Buchholz, U., Corman, V.M., Hoch, M., Katz, K., Marosevic, D.V., Böhm, S., Woudenberg, T., Ackermann, N., Konrad, R., Eberle, U.: Investigation of a COVID-19 outbreak in Germany resulting from a single travel-associated primary case: a case series (2020)
23. Novel Corona Virus 2019 Dataset: Kaggle, [Online]. Available: https://www.kaggle.com/sudalairajkumar/novel-corona-virus-2019-dataset?select=COVID19_line_list_data.csv. Accessed 2020

24. Elghamrawy, S.: An H_2O's deep learning-inspired model based on big data analytics for Coronavirus Lisease (COVID-19) diagnosis. In Big data analytics and artificial intelligence against COVID-19: Innovation Vision and Approach (pp. 263–279). Springer, Cham. (2020)
25. Abdel-Hamid, N.B., ElGhamrawy, S., El Desouky, A. and Arafat, H.: A dynamic spark-based classification framework for imbalanced big data. J. Grid Comput. **16**(4), 607–626 (2018)

Drones Combat COVID-19 Epidemic: Innovating and Monitoring Approach

Adarsh Kumar, Mohamed Elsersy, Ashraf Darwsih, and Aboul Ella Hassanien

Abstract With the daily rapid growth in the number of newly confirmed and suspected Coronavirus cases, Coronavirus extremely threatens public health, countries' economic, social life, and international relations around the world. In the fight against Coronavirus, Unmanned Aerial Vehicles (UAV) or drones can play a significant role in many aspects to limit the spread of this pandemic. Also, the strategic planning of many governments such as in China for controlling this crisis is supported by the use of drones for the Coronavirus outbreak. This chapter explores the possibilities and opportunities of UAV, also called drones in fighting Coronavirus. Drones are introduced, showing their main findings to make it easy for researchers to investigate new techniques that will help the healthcare sector worker and reduce the spread of Coronavirus pandemic. The chapter also presents some problems and challenges that can help healthcare sectors and curbing the Coronavirus spread.

Keywords COVID-19 · Coronavirus · Unmanned aerial vehicles · UAV · Drones

A. Kumar (✉)
Department of Systemics, School of Computer Science, University of Petroleum and Energy Studies, Dehradun, India
e-mail: adarsh.kumar@ddn.upes.ac.in

M. Elsersy
Higher Colleges of Technology, Al Ain City, Abu Dhabi, United Arab Emirates
URL: http://www.egyptscience.net

A. Darwsih
Faculty of Science, Helwan University, Cairo, Egypt
URL: http://www.egyptscience.net

A. E. Hassanien
Faculty of Computers and Artificial Intelligence, Cairo University, Cairo, Egypt

M. Elsersy · A. Darwsih
Scientific Research Group in Egypt, Cairo, Egypt

A. E. Hassanien and A. Darwish (eds.), *Digital Transformation and Emerging Technologies for Fighting COVID-19 Pandemic: Innovative Approaches*, Studies in Systems, Decision and Control 322, https://doi.org/10.1007/978-3-030-63307-3_11

1 Introduction

The concept of UAVs generally denoted to as drones, is huge, instead. Considering the broad range of existing configurations this is understandable [1]. Any aerial car that is not autonomously or remotely controlled by an on-board human operator for a flight shall be called a UAV in this exercise [2]. UAV vary in scale from large military drones with a wingspan of approximately two hundred feet to inch-wide commercially available micro-drones. Their flight range stages around the operator are limited to three feet to specialist marine drones that can fly over 17,000 miles without landing. A generic version of their average flight height, which can be anything from a few feet to an elevation of 65,000 ft, may also be available [3, 4]. Most commercial drones are adopting a similar style these days. The main configuration consists of a microcontroller serving as a flight controller, usually with 4 or up to 8 vehicles and propellers, a radio antenna, a portable speed controller, and a battery mounted on soft plastic or metal frame [5]. In the same way, gyroscopes and other sensors are mounted to improve the mid-air stability of the drone and a global positioning system (GPS) can be used for navigation. Full hobby drones are often fitted with an aerial camera and a gimbal for extra video stabilization. Also, new sensors may be included, but the faster performance and weight shift may be included [6–8].

With the daily rapid growth in the number of newly confirmed and suspected COVID-19 cases, COVID-19 extremely threatens all the world countries. Infected cases data of COVID-19 can be detected using several medical means. These means of COVID-19 medical diagnosis include clinical characteristics and radiologist's diagnosis. Clinical characteristics involve human temperature monitoring in which drones play an important role to detect human temperature of people on the streets and isolated areas.

The rest of this chapter is organized as follows. Section 2 explores the impact of drones on societies in the era of COVID-19. Section 3 presents the proposed monitoring framework. In Sect. 4, some problems and challenges are presented. Section 5 concludes this chapter.

2 The Societal Impact of Drones in the Age of COVID-19

The emergence of the latest and evolving technology has a tremendous physical, social, and personal impact. They most usually impact exercise, how we do things, how we perform duties, how we attain dreams. Whilst learning new movement skills and opportunities. Drones can carry several sensors, transmitters, and imaging devices. When drones continue to proliferate, they will affect sectors ranging from recreation to agriculture, from construction to transport. Several associations of organizational and industrial standards were established to either model drones and or to help integrate them with the present infrastructure. They also developed new and innovative market-centered services and carrier structures, fostering this era's common expectation and related business models.

1. Protection, security from harm and health, freedom from fear of harm are fundamental human rights that most nation's constitutions should safeguard and protect. Today's use of drones in civil airspace has raised questions about the challenging terms of these fundamental rights. These issues are answered mostly to both the program and the user. Worries about the middle generation around the battery's lifestyles increase the power, airworthiness, and durability of the aircraft. The key issue of commercial drones operating around the public domain is that minor mistakes could lead to collisions that place public safety and well-being at risk. If they crash into public structures such as power poles or drift into airports and other secure airspaces, they may lead to life-threatening, hazardous circumstances.
2. The statistics carried on drones running within the purchaser area through cameras and sensors can be even more precious to attackers and easier to target out of the doors of the army and business climate. When a private UAV is hacked, the owner may find it very difficult to identify the information leak and ensure the security of the information in addition to ownership claims. Three. Drones now face complicated coverage and legal responsibility troubles concerning ensuring their commercial use. Those issues get up from the complexity of different factors which include the distinct sorts of injuries, methods for air manage, and unsure private of the legal guidelines. Coverage groups have yet to extend UAV insurance policies, so little or no suggestion of precedent can follow. Personal damage and privacy infringement are the most serious issues, given that many UAVs are flying over populated areas. On the way to choosing a personal injury case, insurance providers will have the right of access to a wide variety of details. They want the place of departure and landing, the course of the flight, the purpose of the flight, the altitude, etc. In the same way, monitoring agencies want to consider the potential of drones to capture a large number of privacy infringement data that can either be stored onboard or sent to a remote server. It's also very difficult to explain the purpose behind the collection and use of records.

COVID-19, a global pandemic that has killed more than 164,000 people and injured more than two million people worldwide, makes us rethink how governments, businesses, and societies around the world operate with minimal or no physical contact. Today, frontline warriors and heroes in the country are doctors, medical staff, local police, private security guards, and college rejection. Technologies like artificial intelligence, big data, geographic information systems, mapping, site technology, and standalone machines play an increasing role in responding to the COVID-19 epidemic. However, in this war against this invisible enemy, drones play a major role by helping the authorities and people in various ways to prevent further spread of coronavirus. The main ways in which drones are used to inspire others about ways in which new technologies can be used to combat the spread of COVID-19 are shown in Fig. 1, for examples. Drones can offer many advantages including:

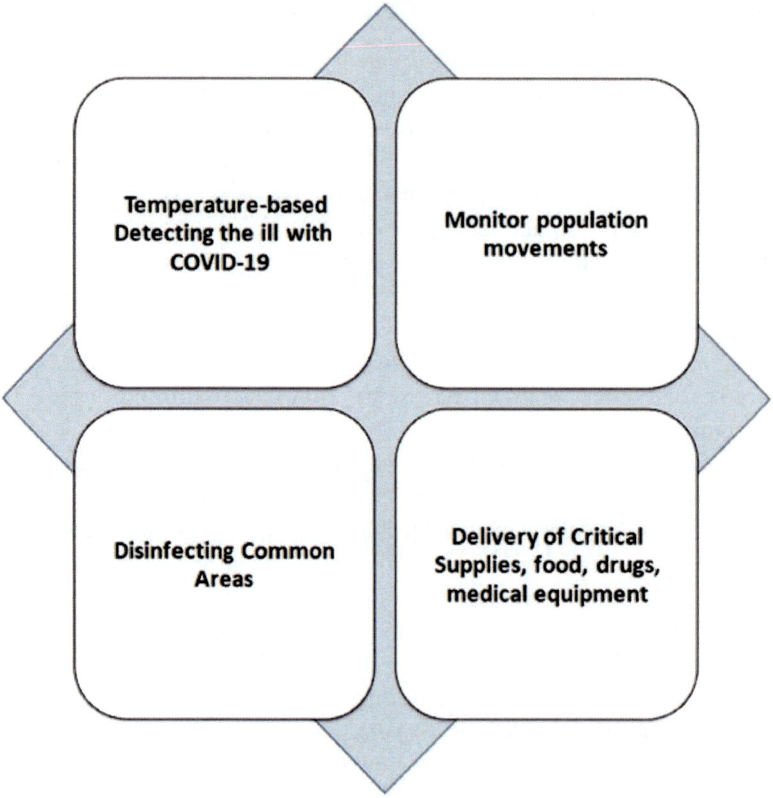

Fig. 1 Drones ways to combat COVID-19

- Temperature-based Detecting the ill with COVID-19
- Monitor population movements
- Disinfecting Common Areas
- Delivery of critical supplies, food, drugs, medical equipment.

An indispensable activity to stop the epidemic is detecting the temperature and distinguishing the healthy from the unhealthy in order to isolate the latter. To do this, until now a person needs to participate, in addition to the patient, in order to check patient temperature, or take a test to detect COVID-19. Drones are used to read the temperatures of people leaving or during the quarantine period, and use drones equipped with a thermal scanner to search for temperature anomalies in people on the ground [9].

To monitor crowds, drones can be installed and very useful for broadcasting important information, especially in areas that lack open channels of communication. For example, the police authority in many countries in Europ such as Spian have been used a dr one equipped with a loudspeaker to inform people of the established guidelines regarding the emergency that was imposed [9].

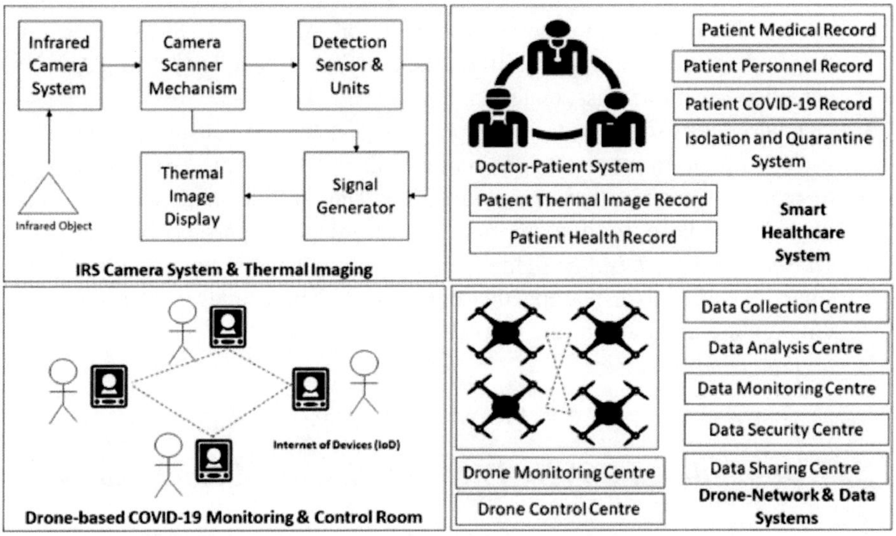

Fig. 2 Integration and interoperation of various sub-systems in drone-based COVID-19 operation

One of the primary tasks during this race against coronavirus is clearing up large, hard-to-reach areas or places. Laying off human resources in some extreme locations, or minimizing human intervention is some of the benefits of using agricultural drones for these tasks []. In addition, drones can offer advantages to monitor the movement people during the epidemic.

Figure 2 shows the integration and interoperation of various sub-systems in Drone-based COVID-19 operation that include patient personal and medical records, its monitoring and medication, and thermal image collection. This operation helps the audience to know more about COVID-19 nature of spreading and chain. Thus, create a general awareness for safety measures to tackle influenza viruses such a COVID-19. Using the drone, COVID-19 chain can be broken and day-to-day necessities can be covered. The details of drone-based integration and interoperation of various sub-systems are explained as follows.

2.1 IRS Camera System and Thermal Imaging

1. *Infrared Camera System*: In this sub-system, a camera lens collects the object image and forwards it to the scanner for image-based properties identification. In this identification process, the collected images are considered as raw images of a video for further processing.
2. *Camera Scanner Mechanism*: In this sub-system, the camera scans the image for initial processing. During this initial processing, the camera identifies the thermal properties in the images. If the image exhibits clear thermal properties then the

extracted data is sent to the signal generator. Whereas, if the scanner fails to detect clear thermal properties then the collected image is sent to a specialized detection sensor and associated units.

3. *Detection Sensor & Units*: The specialized detection sensors and units apply various filters for extracting the required data. These sensors use various types of materials such as silicon, germanium, zinc sulfide, zinc selenide, calcium fluoride, sapphire, sodium chloride, etc. for image properties extraction.
4. *Thermal Image Display*: This sub-system considers the mobile, PDA, or smart devices for displaying thermal images. These images help measure the temperature and identify the COVID-19 affected area.
5. *Signal Generator*.

2.2 Drone-Based COVID-19 Monitoring and Control Room

This system is used to monitor and control COVID-19 activities remotely. Here, a set of smart devices is interconnected for sharing the necessary information between different parties. Thereafter, it is used for various other purposes such as making preparations for purchasing the medical and other types of equipment, and medicines. COVID-19 pandemics and hotspots can be monitored remotely using this system. Further, statistics collected from drones can be shared in the public domain for creating awareness and protections. This system helps the government, and non-profit NGOs to extend their help for patients living in COVID-19 affected areas and hotspots.

2.3 Drone-Network and Data Systems

1. **Drone Monitoring Centre**: This sub-system monitors drone activities. The drone device performance, movements, services, and operations are observed in this sub-system.
2. **Drone Control entre**: In this sub-system, controlling the party of instructing the monitoring party to move the drone in the area of COVID-19 patients and start the necessary operations.
3. **Data Collection Centre**: This sub-system focuses on the types of data to be collected from COVID-19 affected areas and prepared a well-defined template. This template and the data is forwarded for necessary processing and decision making.

   (a) **Data Analysis Centre**: In this sub-system, analysis is performed over collected data. Here, if data is collected from sensor devices then patient medical data is prepared for present and future condition movements. There is an option to collect thermal images for data analysis as well.

(b) **Data Monitoring Centre**: In this center, various statistics are displayed for decision making. These statistics include drone networks QoS measurements, Type and area of drone's services, COVID-19 patient statistics, etc.
(c) **Data Security Centre**: In this sub-system, data security is ensured at storage, processing, and transmission stages. The security is ensured through cryptography primitives and protocols applied at different levels.
(d) **Data Sharing Centre**: In this sub-system, data is shared with trusted system parties. Here, many parties need data for analysis and decision making. For example, data monitoring centers, security centers, drone monitoring centers, etc. are system parties that need the data for their jobs. Thus, whom to share what data is decided in this sub-system.

2.4 Smart Healthcare System

1. **Patient Thermal Image Record**: This is the most important record for medico and non-medico persons associated with COVID-19 operations. The thermal images reflect the temperature or density-based movements that need sanitization regularly.
2. **Patient Health Record**: The patient's health record is important to consider when a person is found to be affected by COVID-19 or similar viruses. Thus, medication needs patient history that is possible through maintain historical records. This is made feasible through the patient health record sub-system.
3. **Patient Medical Record**: In this sub-system, the patient's medical records are maintained. The medical records re different from health records in terms of specialized data. In medical records, disease-specific records are maintained. Whereas, health record means the general medial practices and routine followed to maintain the health.
4. **Patient Personnel Record**: Like health and medical records, patient's personnel records are important as well. These records are necessary to have personal identification and authentic payment sources for treatment.
5. **Patient COVID-19 Record**: This record is recently found to be necessary for finding the chain of airborne diseases like COVID-19. Individual patient's symptoms, treatments, and cure are considered as single- clinical trials. Thus, it is helpful in the identification of the right treatment when data is big and it is properly analyzed [10].
6. **Isolation and Quarantine System**: This system takes care of activities required when patients get affected or it is found to be COVID-19 positive case. Thus, processes and cares are defined, monitored, and practiced in this sub-system. Based on the transmission behavior of COVID-19 in humans, authors in [11] develop three quarantine models for this epidemic taking into account the compartments including exposed population, migrant population, home insulation, infectious population, quarantine population in hospitals and recovered

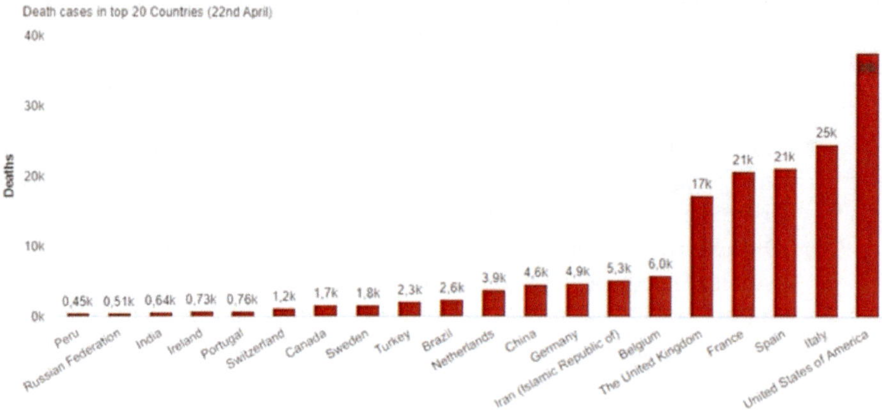

Fig. 3 COVID-19 death data analysis

population. Local and global asymptotic stability was demonstrated for all three models. Extensive numerical simulations are performed to determine analytical results with appropriate examples.

Figure 3 shows the death data analysis steps applied for collecting the real-time statistics. A similar analysis could be performed for observing the variations in COVID-19 cases and analyze the different causes of this virus and its variations. In a drone-based system, similar statistics could be analyzed for thermal images, necessities of sanitization, medication delivered and needs, etc.

3 The COVID-19 Proposed Monitoring Framework Based on Drones

In the current situation of spreading of COVID-19 in 2020, drones will play an important role in supporting the health sector, where drones can be placed in airports for the detection and diagnosis of passengers after providing them with the necessary sensors such as measuring passenger's temperature as in China in 2020 [12, 13]. They used some codes at travel checkpoints, including hanging from drones at highway tollbooths. Drivers are required to scan them before their cars are allowed to enter cities, a process that can track the location of people by their Chinese resident identity card number. Figure 4 describes the proposed multi-layered framework which consists of 4 layers for drone-based COVID-19 monitoring.

Layer-1: Wearable sensor network: This layer considers the wearable devices that can be used for measuring a patient's medical records. This data is important for patient's profiling and storing medical histories for long term treatments.

Layer-2: Drone network: The drone-based network is very popular to collect the data from a sensor-based IoT network constructed at layer-1. Here, a single or set

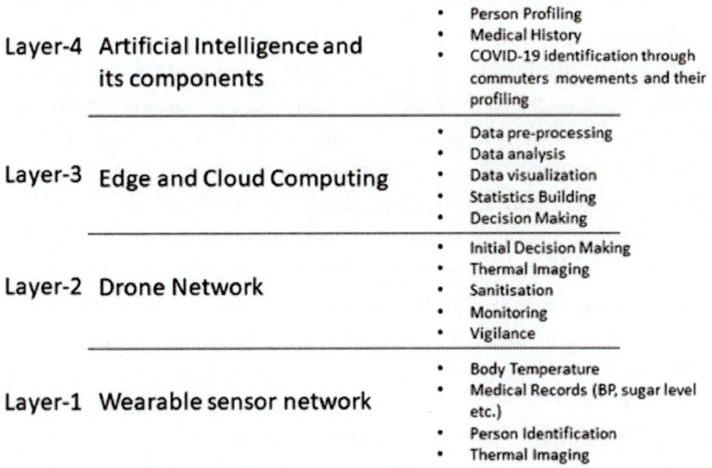

Fig. 4 A proposed multi-layered framework for drone-based COVID-19 monitoring and industry 4.0 processes

of drones communicates freely over a geographical area for collecting the patient's medical records continuously. In the healthcare system, finding a patient's medical records for fixing the type of treatment and surgical operation is very important. Presently, this record is not available easily, securely, and timely. A drone-based system is an automated process of continuously collecting the records and it can be stored over an edge or cloud servers for long term usage. A drone network can collect the patient's thermal images that can be used for analysis and COVID-19 case identification [14]. Thereafter, drones can be used for monitoring, vigilance, alert systems, directing the instructions, and sanitization. The major challenge of a drone network is to avoid collision and overlapping of area coverage. The movement of drones can be controlled through artificial intelligence and neural networks.

Layer-3: Edge and Cloud Computing: This computing has multiple roles. In the COVID-19 pandemic situation, this computing helps build an initial patient profiling that can be used for subsequent COVID-19 treatments including medication and sanitization. Thus, the data collected from the drone network is stored here for data analysis and it can be further used for analysis and initial level decision making. This layer is more useful in the identification of similarities and differences in cases arises from multiple COVID-19 hotspots after data analysis. Further, this computing can direct the drones' movement by self-computing analysis or decision received from layer-4.

Layer-4: Artificial Intelligence and Its Components: This layer is more useful in person or patient profiling. Here, data collected after analysis from edge or cloud computing can be linked to building profiling. For example, a commuter based profiling can help in identification of zero-COVID-19 case. This can further lead to the identification of causes, symptoms, and variations timely. Subsequently, this profiling will be helpful not just for COVID-19 cases but for all other contagious diseases.

Figure 4 shows a case study where a set of commuters movements are observed. In this simulation, commuters are allowed to move freely and randomly in any direction. The simulation considers the spreading of COVID-19 cases when an affected person comes closer to other people with a distance shorter than 0.5 m. This simulation is executed with an assumption that people do not have the right to shake a hand or any other type of physical contact. Further, drone-movements are shown in Fig. 5.

These drones have a fixed zone for sanitization. Figure 5 shows the ground-view of people's movement and Fig. 6 shows the integration of drones for COVID-19 monitoring, sanitization, and thermal images. The thermal images generated by simulation is shown in Figs. 6 and 7. These images are exchanged with edge and cloud computing servers for further analysis. This analysis can be used for help in the identification of COVID-19 cases, causes, the chain of spreading, and other medical records [15]. Thus, it will be useful in present scenarios.

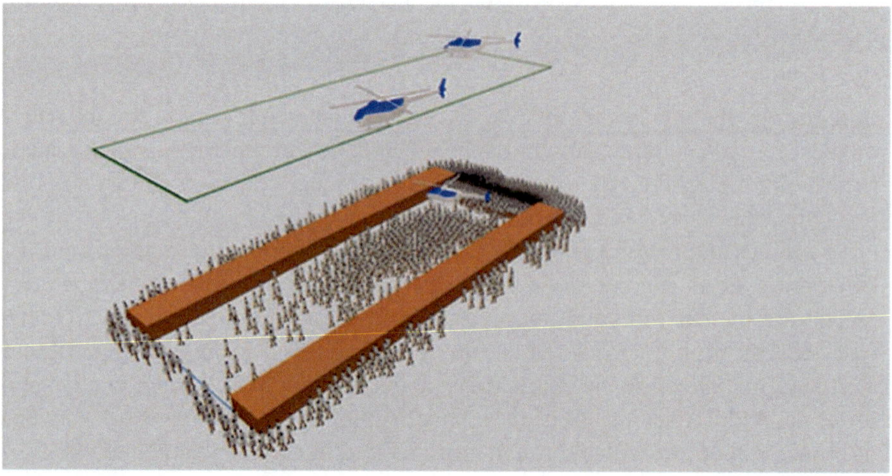

Fig. 5 Drone-based COVID-19 monitoring

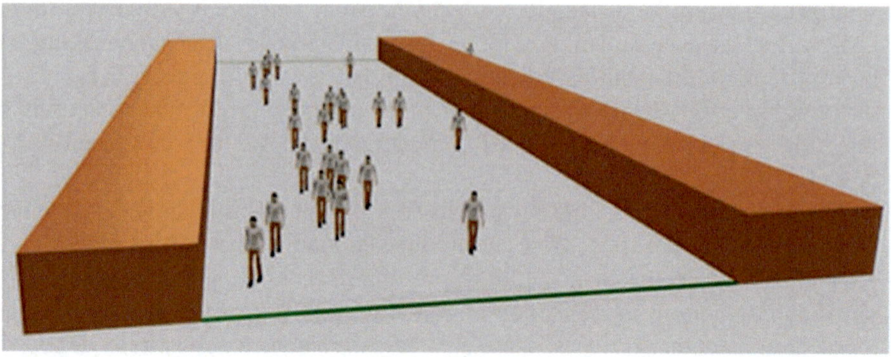

Fig. 6 The area considered for pedestrian movement and COVID-19 observations

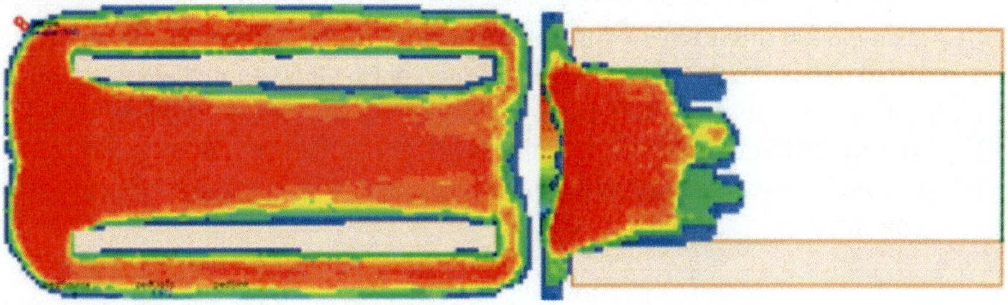

Fig. 7 Simulated thermal images used for data analytics and COVID-19 case predictions

It can also be placed in the main streets to follow the movement of citizens, such as the train station and bus stops. Some governments can rely entirely on the drones to identify the injured and limit their movements to implement social isolation for them during specific periods to limit the spread of the COVID-19 epidemic.

4 Problems and Challenges

In the next, the problem, challenges, and future trends are presented to represent research problems for researchers and academics.

5 Security Issues

Damages from malicious users are more intense inside the case of drones data compared to the network and protection needs to be one of the core issues that any architecture. Drones should cope with attacks on drones data that can probably cause taking control of these drones or crashing them. Examples of assaults are shown as follows:

(i) Physical Challenges. Drones data and images are subject to a large number of physical challenges that could hinder their service, and likewise prevent their missionary achievement. Their presence presupposes an external threat, which should be considered. Intrusion (animals, humans), environmental factors (wind, temperature) affect the operation of UAV and can result in injury. Examples of these difficulties include complex obstacles or the existence of civic structures, trees, electric cables, and houses. Those risks can be avoided by using generation to identify neighboring visitors to the air site and techniques to prevent collisions.
(ii) GPS Spoofing Attacks The basic principle in GPS spoofing is the transmission of false GPS signs to the drone's controlled computer, usually more powerful

than the satellite's real GPS sign. Consequently, the sufferer could use the fabricated indicators as opposed to the original ones. GPS spoofed warnings provide the drone with particular incorrect proximity and let it adjust its course for you to track the drone. GPS enables drones navigation, and they can be easily spoofed due to non-encryption of the warnings, which instantly affects operator commands.

6 Scheduling Drones Charging Problem

Popular methods for recharging and replacing batteries limit drones' full use in most applications. A peer-to-peer network of drones and charging stations is an incredibly promising choice for enabling drones to be used in multiple programs with the support of their flight time. Charging stations are small, thus requiring an appropriate, accurate and most cost-effective set of rules for scheduling to serve the most desirable drones Consequently, the use of drones with COVID-19 along with a restricted battery, barriers in phrases of carrying heavy masses, short flight time and more than one drones cooperative timing problems are various drawbacks.

7 Big Data Governance

As with large records of COVID-19 patients, the demanding situations of drones facts encompass storage, switch, analysis, seek, sharing, visualization, and information privacy. Similarly, drones data consisting of geospatial imagery, motion pictures, binaries, and many others. Therefore, we need to extend the current polices of drones records governance and processing along with archiving the source of statistics for later processing [16].

8 Trajectory Prediction

Researchers want to construct a trajectory prediction version of the use of time series forecasting to expect the complete flight path of an incomplete drone trajectory as a result of crashes inclusive of hitting a sudden tree/wall or through any method during monitoring the people to detect the infected people of COVIID-19 [17, 18].

9 Routing and Direction Planning

A fundamental question approximately how we enforce routing of drones to scan the infected areas or borders of COVID-19 is whether or not we reserve the whole direction before the beginning of the experience. Optimizing the course of each drone must be taken into consideration. Within the future, swarm intelligence can be applied to this vicinity to the most excellent path computation.

10 Conclusions and Future Perspective

Due to possible epidemics and potential global risks, the outbreak of the latest COVID-19, detected in late 2019, needs particular attention. In this context, drones have gained many advanced capabilities that managed it to improve the performance of the most sophisticated missions such as innovative monitoring of COVID-19 people, infected areas, and borders of countries. As well as, applications where a rapid response is a necessity to resume action in COVID-19. Recent studies will depend on using drones to decrease the chance of spreading COVID-19 by monitoring COVID-19 disease and detecting the possible areas and borders of infection. COVID-19 outbreak has proved the need for proactive containment which can mainly depend on the recently emerging technologies such as drones. Also, the related and previous studies showed that coronavirus could survive on inanimate objects and surfaces however; the presence of coronavirus in the environment has not been discussed. Therefore, in the future, some emerging technologies such as drones can be deployed to deal with this problem.

References

1. Newcome, L.R.: Unmanned aviation: a brief history of unmanned aerial vehicles. AIAA (2004). https://doi.org/10.2514/4.868894
2. Carrillo-Larco, R.M., Moscoso-Porras, M., Type-Rondan, A., Ruiz-Alejos, A., Bernabe-Ortiz, A.: The use of unmanned aerial vehicles for health purposes: a systematic review of experimental studies. Glob. Health Epidemiol. Genom. **e13** (2018). Published 27 Jun 27 2018. https://doi.org/10.1017/gheg.2018.11
3. O'Brien, S.: Drone Startups Swoop Up Millions, 2015, January 7. Retrieved 14 Oct 2015 from http://money.cnn.com/2015/01/07/technology/ghostdrone/
4. Aggarwal, S., Kumar, N.: Path planning techniques for unmanned aerial vehicles: a review, solutions, and challenges. Comput. Commun. **149**, 270–299 (2019)
5. Cummings, M.L., Bruni, S., Mercier, S., Mitchell, P.J.: Automation Architecture for Single Operator, Multiple UAV Command and Control. Massachusetts Institute of Technology Cambridge (2007)
6. Jain, U., Rogers, M., Matson, E.T.: Drone forensic framework: sensor and data identification and verification. In: IEEE Sensors Applications Symposium (SAS), pp. 1–6 (2017). https://doi.org/10.1109/sas.2017.7894059

7. Sciancalepore, S., Ibrahim, O.A., Oligeri, Di Pietro, R.: PiNcH: an effective, efficient, and robust solution to drone detection via network traffic analysis. Comput. Netw. **168**, 107044 (2020)
8. Darwish, A., Hassanien, A.E.: Wearable and implantable wireless sensor network solutions for healthcare monitoring. Int. J. Sens. **11**(6), 5561–5595 (2011). https://doi.org/10.3390/s110605561
9. Chamola, V., Hassija, V., Gupta, V., Guizani, M.: A comprehensive review of the COVID-19 pandemic and the role of IoT, drones, AI, blockchain, and 5G in managing its impact. IEEE Access **8**, 90225–90265 (2020)
10. Yaacoub, J.-P., Noura, H., Salman, O., Chehab, A.: Security analysis of drones systems: attacks, limitations, and recommendations. Internet Things **11** (2020, September). (Article 100218)
11. Mishra, B.K., Keshri, A.K., Rao, Y.S., Mishra, B.K., Mahatoe, B., Ayeshaf, S., Rukhaiyyar, B.P., Saini, D.K., Singhi, A.K.: COVID-19 created chaos across the globe: three novel quarantine epidemic models. Chaos Solitons Fractals **138**, 109928 (2020, September)
12. Tsai, C.-W., Lai, C.-F., Chao, H.-C., Vasilakos, A.V.: Big data analytics. In: Big Data Technologies and Applications. Springer, pp. 13–52 (2016)
13. Kumar, A., Gupta, P.K., Srivastava, A.: A review of modern technologies for tackling COVID-19 pandemic. Diabetes Metab Syndr. **14**(4), 569–573 (2020, July–August). https://doi.org/10.1016/j.dsx.2020.05.008
14. Harvey, M.C., Rowland, J.V., Luketina, K.M.: The drone with the thermal infrared camera provides high-resolution georeferenced imagery of the Waikite geothermal area, New Zealand. J. Volcanol. Geoth. Res. **325**, 61–69 (2016)
15. Lee, E.J., Shin, S.Y., Ko, B.C., Chang, C.: Early sinkhole detection using a drone-based thermal camera and image processing. Infrared Phys. Technol. **78**, 223–232 (2016)
16. Sujath, R., Chatterjee, J.M. Hassanien, A.E.: A machine learning forecasting model for COVID-19 pandemic in India. Stoch Environ. Res. Risk Assess (2020). https://doi.org/10.1007/s00477-020-01827-8
17. Choudhary, G., Sharma, V., You, I.: Sustainable and secure trajectories for the military Internet of Drones (IoD) through an efficient medium access control (MAC) protocol. Comput. Electr. Eng. **74**, 59–73 (2019)
18. Hassanien, A.E., Salam, A., Darwish, A.: Artificial intelligence approach to predict the COVID-19 patient's recovery. EasyChair Prepr. (2020, April 22)

3D Printing Supports COVID-19 Pandemic Control

Mourad R. Mouhamed, Ashraf Darwish, and Aboul Ella Hassanien

Abstract At the end of December last year a new type of coronavirus has appeared in Wuhan, China, with new properties the researchers named it COVID-19. In February, the world health organization considers it a world pandemic; it had spread in most world countries. This virus attacks the respiratory system, which makes failure in the system's function. The effect of this crisis touched all the fields life, where all countries applied quarantine and roadblock that makes a real shortage in most of the ple needs. Besides biological scientists' efforts, the computer scientists proposed many ideas to fight this epidemic using emergent technologies. This chapter is covering 3D printing principals the latest efforts against COVID-19 as one of the emergent technologies. 3D printing technology helps to flatten the curve of the outbreak of the virus by reducing the effect of shortage in the supply chain of medical parts and all personal protective equipment (PPE) (i.e. face masks and goggles), where it provides the extensive customization capability.

Keywords Coronavirus · COVID-19. 3D printing · Additive manufacturing · Drug repurposing

1 Introduction

Throughout history, humans have suffered from the spread of many epidemics, starting with the plague of Antonia in 165, which lasted for fifteen years, and claimed the lives of five million people through the Spanish flu pandemic at the beginning of the twentieth century, which caused the elimination of nearly fifty million people.

M. R. Mouhamed (✉) · A. Darwish
Faculty of Science, Helwan University, Cairo, Egypt
e-mail: mouradraafat@yahoo.com

A. E. Hassanien
Faculty of Computers and Artificial Intelligence, Cairo University, Cairo, Egypt

M. R. Mouhamed · A. Darwish · A. E. Hassanien
Scientific Research Group in Egypt (SRGE), Cairo, Egypt

A. E. Hassanien and A. Darwish (eds.), *Digital Transformation and Emerging Technologies for Fighting COVID-19 Pandemic: Innovative Approaches*, Studies in Systems, Decision and Control 322, https://doi.org/10.1007/978-3-030-63307-3_12

Today the world is facing a new threat: the new Crohna epidemic (COVID 19), which started the outbreak for the first time in Wuhan, Hubei, China, in December 2019 and was recognized by the World Health Organization as a pandemic on March 11, 2020. China began experiencing an epidemic on the thirteenth day. From February 2020, the number of injuries exceeded fifteen thousand injured in one day. Chinese governments face the necessary measures to combat this virus. The steps allow the entire Hubei province, with a population of 56 million, to be quarantined and a field hospital built in ten days to treat patients. But since then, it appears that China has taken control of the disease, while your numbers worldwide have increased by about 13 times in two weeks. When the World Health Organization declared it a global pandemic, WHO President Tidros Gebresos called on "countries taking urgent and decisive action." Indeed, it had an effective outcome of the virus blockade in China, which made it to reduce the number of daily infections on March 19. On the other hand, the virus has spread to European countries, so these countries became a new hotspot, according to WHO reports. Coronaviruses (order Nidovirales, family Coronaviridae, subfamily Coronavirinae) are viruses that surrounded by a positive sense, single-stranded RNA genome [1]. Among sizes of genome ranging length between 26 and 32 (KB), Coronaviruses have large genomes for RNA viruses. Coronavirus is one of the major pathogens that primarily targets the human respiratory system [2]. Past coronaviruses outbreak involves the Middle East respiratory syndrome (MERS)-CoV and severe acute respiratory syndrome (SARS)-CoV that are enormous public health threat [3]. During the writing of these words, the number of infected people reached more than three million with 250 thousand death quarter of them in the United States of America [4]. Figure 1 shows the cumulative number of cases and death per day.

With the current COVID-19 rapid outbreak, the great number of people that need a diagnosis, treatment, and isolation overburdened global health systems. Weaknesses are apparent beyond the board, from staffing, critical medical-grade devices, and facilities for reliable and fast testing to hospital beds availability. 3D printing companies and design engineers proposed many efforts to fight this epidemic, one of these efforts presented, some of designers and companies helped the population by present their works and put them free on their websites. In [5], Andre et al. summarize community-driven proposals based on Free and Open Source scientific

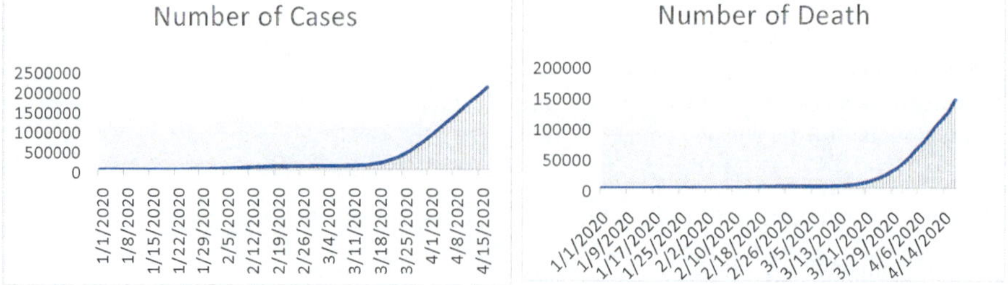

Fig. 1 Cumulative number of cases and death due to COVID-19

and medical Hardware (FOSH) currently being deployed and developed to sustain access to personal protective equipment (PPE), patient treatment and diagnostics. The authors focused on three crucial topics, firstly patient treatment, focusing on ventilators, secondly, personal protective equipment (PPE) like visors and masks, then focused on tools of diagnosis, specially test-kits and scientific equipment. The medical equipment immense majority is densely controlled by a few firms of specialty medical that sell little amounts because, through 'normal' times, a hospital of medium-sized only requires a handful. These firms have historically aggressively defended their intellectual copyrights [6, 7] to the human life's detriment (as a new example, imagine that a manufacturer to sue a maker for 3D printing life-saving valves in Italy for violation of patent) [8]. Additionally, entities of non-practicing continue to try to actively stop medical treatments from being used, even during the current COVID-19 epidemic [9]. Placing aside the patenting absurdity and then prevent others from utilizing evident inventions in regular times [10], in the pandemic days where millions of lives are in danger, it is intuitively clear that this greed and selfish type is no longer acceptable. One of the best open sources is manual and low-cost respirator, which is developed and presented by researchers at MIT, this respirator could be used in the treatment of COVID-19 patients that are in the critical case. The MIT Emergency Ventilator Project's ventilator, called as "E-Vent," transforms readily available materials such as bag valve masks to help people with problems of breathing into available respirators [11].

The rest of this chapter is organized as follows. Section 2 describes the basics and background. Section 3 presents 3D printing applications in the medical area. Section 4 discuss the integration of 3D printing with COVID-19 dosage forms and control systems. Section 5 presents problems and challenges. Section 6 concludes this chapter.

2 Basics and Background

2.1 3D Printing Technology

3D printing is an additive manufacturing (AM) method for building complex geometries plus a wide range of structures from the digital 3D model. The 3D printing process consists of successive printing materials layers that are formed on top of each other. Charles Hull developed this technology in 1986 in a process known as stereolithography (SLA), which was followed by subsequent developments such as fused deposition modeling (FDM), powder bed fusion, contour crafting (CC) and inkjet printing [12]. The main parts (see Fig. 2) of this technology are Methods, materials, software, and hardware, these parts will be discussed in detail.

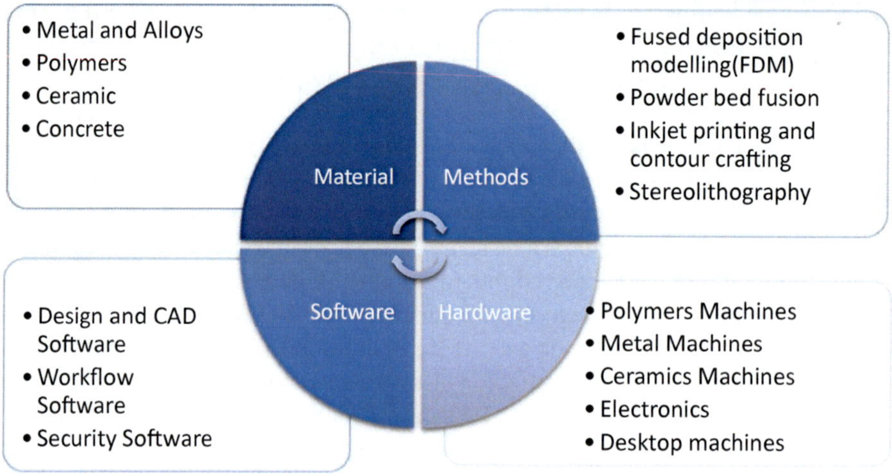

Fig. 2 The main parts of 3D printing technology

2.2 Methods

Different printing methods have been used to manufacture polymer composites. Among them, some methods are well-established, such as inkjet 3D printing, fused deposition model (FDM), and stereolithography. In contrast, other methods are still used by only small researchers' groups. Each method has its benefits and weaknesses in fabricating composite products. The fabrication method selection depends on the starting materials, resolution, and speed of processing, costs, and final results performance [13]. Each method has its pros and cons also the suitable material and application; we will discuss two of them in detail.

1. ***Fused Deposition Modeling (FDM)***

Fused deposition modeling (FDM) printers are the most usually used printers for building polymer composites. Thermoplastics such as polycarbonate (PC), Polylactic Acid (PLA), and Acrylonitrile butadiene styrene (ABS), are widely used due to their low melting temperature. FDM printers operate by controlled extrusion of thermoplastic filaments.

The main advantages of the FDM method are the low cost, high speed, and the printing process simplicity; on the other hand, the shortage in a suitable material that is using in the printing process considered a drawback of this method. Also, the complete removal of the support structure used during printing may be difficult [13]. This method efficiency depends on many parameters such as layer thickness, raster width, printing orientation, air gap, and raster angle [14]. This method usually used indefinite applications like Rapid prototyping, Toys, and Advanced composite parts.

2. ***Inkjet head 3D printing***

The technology of powder-liquid 3D printing was advanced at the Massachusetts Institute of Technology (MIT) in 1993 as a technology of fast prototyping [15]. Powder processing is the bases of this technology, as shown in Fig. 4. at the beginning, the powder is spreading on the platform of build and then selectively combined into a patterned layer by depositing a binder of liquid during inkjet printhead, that is eligible to move in the direction of X + Y. After a desired form of the 2D pattern, the lowers of the platform and the powder next layer is spread. This process is reiterated, and finally, the unbounded powder must be taken away to get firm products.

This method has many advantages like quick printing, room temperature processing environment, and printing tremendous structures. Nevertheless, it has a drawback like Maintaining workability, Coarse resolution, Lack of adhesion between layers, and Layer-by-layer finish. The main parameters of this method are the particle size distribution of ceramics, the ink viscosity, as and content of the solid, well as printing speed, the rate of extrusion, and the size of the nozzle. The contour crafting is the same method for inkjet printing, but it is being used to building massive structures. This method has the extruding capability of concrete paste or soil by utilizing larger nozzles and high pressure. Contour crafting has been prototyped to be used for construction on the moon [16].

3. ***Stereolithography (SLA)***

Stereolithography utilizes photopolymers that can be remedy by a laser of ultraviolet (UV). It is controlled in a required path to shoot in the resin reservoir, and the photocurable resin will polymerize into a 2D patterned layer. After the curing of each layer, the platform downs and another uncured resin layer is ready to be patterned. On the other side, it is expensive, relatively slow, and the printing materials range is minimal. Likewise, the reaction kinetics and the complexity of the curing process. The exposure and light source energy is the leading factors controlling each layer thickness [17]. SLA can be effectively used for the complex nanocomposites' fabrication [18].

2.3 Material

The structure, applications, and hardware of each method make a variety of the materials that have been used. This subsection presented a brief discussion about three basics materials that are used in biomedical applications like (metal and alloys, polymers, and ceramics) that used in previous methods.

1. ***Metals and Alloys***

The additive manufacturing of metals is showing outstanding perspectives on growth. The companies' number of selling AM systems went from 49 in 2014 to 97 in 2016, around 49% connected with AM of metals. It is used in many applications such as (defense, biomedical, automotive industries, and aerospace industry) [19]. It does not ease the traditional manufacturing to produce a complex geometry; the AM of metals support to resolve this issue. The 3D printing process of metals depends on the metal melting utilizing a laser electron beam or any energy source, after the melting process the metal is dropped layer by layer and then convert into the solid-state. The most commonly used methods for metals printing are direct energy deposition (DED) and powder bed fusion (PBF). Still, there are many developed methods recently, like direct metal writing [20], binder jetting [21], friction stir welding [22], cold spraying [23], and diode-based processes [24]. These methods can realize higher speed and accuracy. Many materials of metals such as aluminum alloys, stainless, titanium alloys, and nickel-based alloys can be fabricated utilizing the PBF method [25]. The advantages of this type of material are Multifunctional optimization, Mass-customisation, Reduced material waste, Fewer assembly components Possibility to repair damaged or worn metal parts on another hand there are many challenges appeared like Limited selection of alloys Dimensional inaccuracy and poor surface finish Post-processing may be required (machining, heat treatment or chemical etching)

2. ***Polymers***

One of the biggest popular materials in 3D printing and AM manufacturing is polymer because it is easy to adapt to a different process of 3D printing. Photopolymer resins can polymerize when activated by UV light in stereolithography 3D printing. Polymers for additive manufacturing are found in the form of thermoplastic filaments, reactive monomers, resin, or powder. Further developments using new resins have improved strength and temperature resistance. Thermoplastic polymers (e.g. acrylonitrile-butadiene-styrene copolymers (ABS) [26], polycarbonate (PC) [27] and polylactic acid (PLA) [28]) could be processed by various 3D printing methods. The capability of employing 3D printing of polymers and composites has been explored for several years in many industrial applications, such as aerospace, architectural, toy fabrication, and medical fields. Some of the benefits of fabricating composites using 3D printing include the ability to customize geometry with high accuracy. The pros of this type of material, that Fast prototyping, Cost-effective, Complex structures, Mass-customization. On the other hand, many challenges faced this type like (Weak mechanical properties, Limited selection of polymers and reinforcements, anisotropic mechanical properties (especially in fibering forced composites).

3. ***Ceramics***

The biomaterials play an important role in the field of AM manufacturing, the material that is used in this material is advanced ceramics. The main methods of 3D printing ceramics are inkjet (suspension), powder bed fusion, paste extrusion, and stereolithography. Inkjet is deemed to be the main method of making dense ceramic samples that may not need post-treatment [29]. A stable suspension with controlled rheology that easily flows does not clog at the nozzle and has an effective drying process is required for 3D inkjet printing [30]. Breaking and thinning of the printed filament is also deemed to be an important aspect of inkjet printing of ceramic suspensions and the viscoelastic behavior of inks is the controlling factor [31]. The main advantages of this material are Controlling porosity of lattices, Printing complex structures and scaffolds for the human body, organs, Reduced fabrication time, and better control of composition and microstructure. However, many challenges appeared during the uses of this material like Limited selection of 3D-printable ceramics Dimensional inaccuracy and poor surface finish, and Post-processing (e.g. sintering) may be required.

2.4 3D Printing Applications

Nowadays AM or 3D printing technology plays a great role to help people in their daily purpose and their crisis like the recent one COVID-19. In the following, A recent Wohlers' report [19] forecasts that the additive manufacturing industry will grow from $6.1 billion in 2016 to $21 billion by the year 2020. The biomedical market represents 11% of the total AM market share today and is going to be one of the drivers for AM evolution and growth. 3D printing technology can be used to print 3D skin [32], drug and pharmaceutical research [33], bone and cartilage [34], replacement tissues [34], organ [35], printing for cancer research [36] and last models for visualization, education, and communication.

The advantages of 3D printing technology in the context of health application raise in many examples like generating a 3d printing skin that can be used in cosmetics and test pharmaceutical. Where there is no need to use animal skin to test the chemical product, also it can be used in medical instruments like respiratory and ventilator.

2.5 3D Printing of Medical Devices

The process of printing a physical object by 3D printer happened in five steps see Fig. 3. The first one is 3D digital model design must be prespecified sizes or matched to a patient's anatomy. The second step is converting the digital design to a buildable file that is sent to the printer. This file will often slice the design into layers, include

Fig. 3 THE process of printing 3D medical device

additional support material to aid printing, and tell the printer where to build the device on the platform of the printer to prepare it for printing. The third step is to build or print the device by considering the quality of material used in the printing process. The post-processing is the fourth step, where includes cleaning to remove residual debris, controlled cooling (also called annealing), and/or additional steps such as drilling, cutting, polishing, and sterilization. The last step is the process of validation and verification by checking the functional features and geometrical features to be sure that the product is within defined specifications, as long as specified processing parameters are monitored and controlled. The results from validation are submitted to the FDA [37] to show that the device meets regulatory requirements and is reasonably safe and effective for its intended use. Each device or type of device has its own set of tests that may be based on FDA Guidance documents, international standards, or internal process controls.

2.6 3D Printing Fights COVID-19

In this subsection, a comprehensive study about the effort that has been presented and still is presenting to fight COVID-19 will be discussed. Due to the quick spread of this pandemic, a great shortage in hospital supplements, protective medical equipment, and also all supply chain processes happened. 3D printing companies and researchers present many models that can help to decrease this shortage [38]. Many designers present open-source models for all peoples, these models like ventilators, face mask, face shield, swap test, and respirator valves. Many examples of these efforts will be discussed as follow.

1. ***Face mask***

HP's Multi Jet Fusion 3D printing technology is producing a newly designed face mask that can be made on HP's MultiJet Fusion 540 but is more suitable for the 4200 and 5200. The masks are 3D printed in PA-12. The Czech Institute of Informatics, Robotics, and Cybernetics at the Czech Technical University in Prague (CIIRU CTU) have developed a new 3D printable respirator called the CIIRC RP95. The CIIRC RP95-3D has been successfully tested and certified to EN 140:1999 as an FFP3 safety half mask. The mask is ready for external filter P3 R (manufactured by SIGMA Lutín). CIIRC estimates that one HP 3D printer can produce between 50 and 70 masks daily [39]. The Chilean/US-based company Copper 3D developed a range of antibacterial 3D printing materials that have been proven to eliminate 99.99% of fungi, viruses, and bacteria, as well as a large number of microorganisms. Developed for prosthetics, surgical equipment, and wound dressings, these materials are one example of how 3D printing can be used to limit the spread of COVID-19. Copper 3D also recently developed a solution to combat the shortage of N95 masks called NanoHack, with an open-source design.

2. ***Ventilators***

The FDA recently authorized an Emergency Use Authorization (EUA) for ventilators, ventilator tubing connectors, and ventilator accessories, which could include items such as 3D-printed tubing connectors for multiplexing ventilator use.

Joshua in [40] presented a study to determine if the designs are an open-source both in spirit (license) as well as practical details (e.g. possessing accessible design source files, bill of materials, assembly instructions, wiring diagrams, firmware, and software as well as operation and calibration instructions). Next, the existing Internet and gray literature are reviewed for open-source ventilator projects and designs. One Giant Lab. They have congregated to help in the COVID-19 pandemic by developing open source solutions and of most relevance to this study, on a project specifically on the development of open-source ventilators. Their documentation and information are freely available. Although just starting, as of 17 March 2020, they have over 2500 registered volunteers and over 9000 on their Slack team.

Another example was presented by Volkswagen, that has announced a task force that will assess and adapt its car-making capacity and manufacturing facilities to the production of hospital ventilators and medical devices. Volkswagen has over 125 industrial 3D printers and is assessing how it can use 3D printing to produce in-demand medical equipment to tackle the COVID-19 pandemic.

In the UK, Health Secretary Matt Hancock encouraged manufacturers to work together to urgently provide more ventilators to the NHS. In response, Ricoh 3D announced it would assist in their production. HP Inc also pledged to mobilize its production to deliver critical medical equipment to combat the pandemic and has already supplied hospitals near its R&D centers with over 1000 3D printed parts.

Medical experts across the country have said that, with the spread of the new coronavirus, there may not be enough ventilation devices to roam around. The New York Times reported that hospitals in the U.S. have approximately 160,000–200,000 ventilators, while the Critical Care Medicine Association has estimated that more than one million coronary patients in the U.S. may need ventilation devices. Therefor Professor Dan Abbott, who teaches architectural and engineering design at Southern Maine Community College in South Portland, used a 3D printer to create a starfish-shaped connector that allows up to four patients to use the same ventilator.

3. ***Face shield***

3D Hubs is leveraging its global network to support funding and coordinating the efforts of 3D printing communities to fight the epidemic. to raise accessibility to this critical equipment, 3D Hubs has released the COVID-19 Manufacturing Fund, gathering these projects with both the production funds and means they want to quickly fabricate vital equipment like ventilator parts and protective masks. Many efforts have been presented like, managed over 20 thousand face shield requests, partnered with the University of Colorado Denver and Make4Covid. (CU Denver) produce up to 10 thousand protective face shields for hospitals in Denver, also Prusa produced key components for the face shield RC2, already being used 'in the field'.

4. ***Respiratory valves***

The CEO of Isinnova, a Brescia-based engineering company and his fellow, Alessandro Romaioli, at first, they visited the hospital directly to examine the valves and went about quickly creating a prototype. Then they tested it successfully on a patient, Isinnova teamed up with Lonati (local manufacturing company) to produce the valves by colossal quantity. SLS 3D printer form Lonati, beside Issinova's 3D printers, were put to run, and the two companies collaborate to produce one hundred respirator valves in one day.

5. ***Quarantine room building***

By increasing the number of infected people that are needed to quarantine, Winsun, a company of architectural 3D printing established in China, has sent to Xianning Central Hospital in the Hubei Province fifteen 3D printed quarantine rooms. The hospital is out of Wuhan's area, the virus's epicenter, thus it felt dangerous when COVID-19 had just started spreading. A shortage of hospital beds rapidly became an insistent problem for staff as the patients' number grows exponentially during the spread first few weeks.

6. ***Test swabs***

3D printed COVID-19 test swab is another device that is offered by Formlabs, a Boston-based 3D printing company. Every printer can produce three hundred test swabs at a time, which makes be able to produce 75,000–150,000 swabs every day. This development will quickly supply hospitals with access to big amount of these main COVID-19 test components.

3 Integration of 3D Printing with COVID-19 Dosage Forms

Various health and manufacturing disruptions are expected due to 3D printing. For more than 20 years, 3D printing applications have been limited to prototypes. In recent years, many 3D printed elements have been introduced to the healthcare and manufacturing market such as medical devices (prostheses), dental instruments (crowns and bridges), prostheses (kidneys and heart) for research purposes, implants, and many others. A large number of researches has been done to produce several dosage forms and many are still under development. Different dosage forms of COVID-19 can be developed such as IR tablet, Bilayer tablet Capsule, Polypill. SEDDS (Self-emulsifying drug delivery system), and Implants as shown in Fig. 4. The global supply chain for these vital products has been disrupted by widespread closures and reduced travel. Now, 3D printing is proving more flexible and adaptable manufacturing methods. Unfortunately, they are also less suitable for producing large numbers of items, and there are unanswered questions about safety and quality control [41]. The ongoing COVID-19 crisis highlighted the importance of a robust drug supply

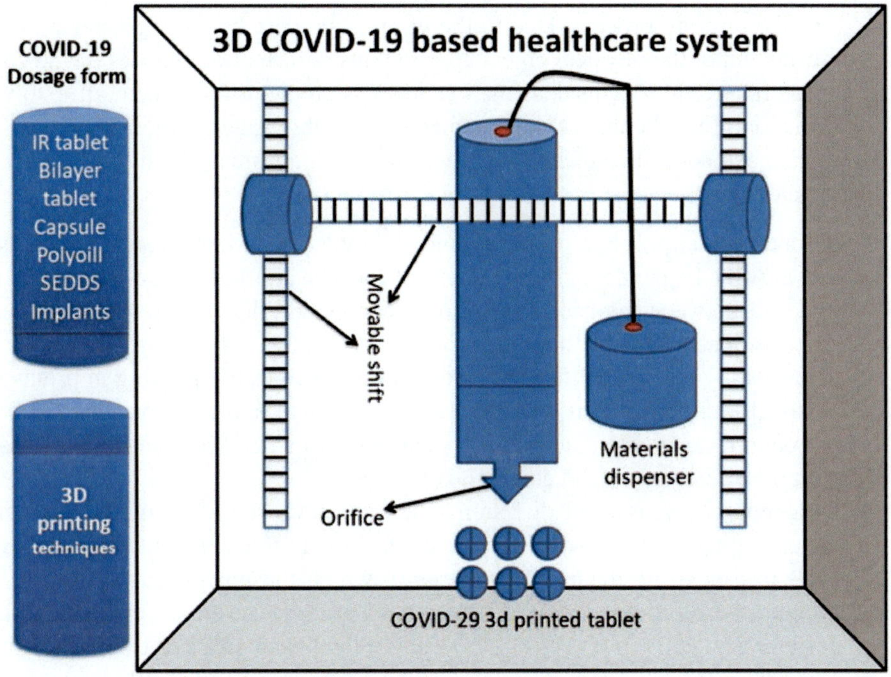

Fig. 4 3D COVID-19 dosage form system Adapted from [43]

chain that can be intensively and flexibly intensified to produce life-saving drug treatments. Oral solid dosage forms (OSDF) 3D printing could be a vital part of the emergency drug production response to support vulnerable patients in rural and other isolated locations. In the context of the current epidemic, the relevance of different 3D printing technologies will depend on the physicochemical properties, unit dose strength, and BCS classification of reused drug compounds currently being tested for COVID-19. Furthermore, the deployment strategy should focus on simplifying dosage forms and formulas, reducing the size and complexity of printing systems, and real-time quality assurance via Process Analysis Technologies (PAT).

4 Problems and Challenges

Besides the great advantages that have been offered by additive manufacturing or 3D printing technology, there are many drawbacks and challenges. 3D printing technology helps to build a complex structure, also printing free design and customization. On the other hand, the disadvantages like a high cost for many numbers of items from one configuration, the shortage of applications in a large structure, the anisotropic of mechanical properties, and limitation of material. Also, many methods like SLS and SLA need high energy resource in the process of a laser beam that increases the cost of the printing process. Due to the printing process by adding layer by layer, a deformation of the physical object appears; this deformation reduces the mechanical performance [42]. This deformation depends on the printing method, so to void deformation; it must enhance the parameters of methods, as mentioned before.

In the next, several challenges are identified with a primary focus on COVID-19 pandemic

- A fundamental problem of management complexity in healthcare systems as in the COVID-19 pandemic that can affect most of the healthcare systems is the trade-off between the need to control expenses and deliver effective healthcare services. Therefore, we have to explore possibilities offered by technology by investigating the criticalities and potentialities such as 3D printing in producing personalized medications for COVID-19 as the nature of the virus changes from one country to another. 3D printing technologies can provide success in terms of both the effectiveness and cost of healthcare.
- Drug repurposing is a new direction to design a new drug for COVID-19 by using the used drugs with SARS and other epidemics. In this context, a key challenge leads to a possible model of synthesis in which the potential role of drug 3D printing in drug compounds is effective. A 3D printing can empower compounding designing and effective drug for COVID-19 for sustainable healthcare systems.
- Nowadays, the fundamental role of technology with COVID-19 is indeed and widely recognized. Next-generation genome sequencing sophisticated modeling techniques, miniaturized biosensors, remote-monitoring devices, etc. are a few challenges of the massive investments in smart health technologies based on 3D printing technology. To contribute to filling this gap, there is a necessity to use 3D printing technology to overcome these challenges.
- With the advantages of 3D printing such as high reproducibility, complex standardized drug manufacturing processes, and the ability to produce dosage forms characterized by innovative and unique dosage forms, personalized drug dosing for COVID-19, and complex drug-release profiles, undoubtedly, the challenge to conventional drug production is imminent.
- The use of 3D printing technology with COVID-19 presents opportunities for cyber-attacks to impact the physical world. This is because 3D printing technology needs internet connectivity to function and is often connected to internal networks. This allows for useful features, such as remote diagnosis troubleshooting, which also opens up the potential for a cyber-attack that compromises the systems remotely.
- 3D design and 3D printing can play a significant role in helping us get through the COVID-19 outbreak. By enabling anyone to manufacture about any design, 3D printing can produce useful gadgets on-demand like air filter holders and door openers as well as face shields. Manufacturers and designers and supply lines all over the world are currently under a lot of pressure to provide enough protective gear for both healthcare professionals and the general public. 3D printed gadgets can contribute to this need and help home users to better protect themselves against COVID-19.

- The application of 3D printing technology for COVID-19 is affected by several cost factors, such as cost of materials for drug delivery, utility, and technological maintenance. Also, the implementation of 3D printing is associated with various forms of related investment, including hardware, software, and system integration.

5 Conclusion

This chapter presents the latest efforts of additive manufacturing or 3D printing technology to fight the outbreak of new epidemic COVID-19. The 3D printing technology based on four components like printing method, printing material, the software of the printer, and the hardware; in this chapter, we discussed two crucial methods (FDA and SLA) that are commonly used in the health care application. Also, materials like Polymers, metal alloys, and ceramics are discussed in detail by giving the pros and cons of each material. The important application of 3D printing like healthcare was discussed briefly by presenting the 3D printed devices that can help in this pandemic, like supporting the hospitals by the protective equipment such as face mask, face shield, and also some parts like respirator valves. But on the other hand, many problems and challenges are facing this technology and especially medical devices. Future work will focus on these challenges and how to overcome these problems and challenges in the future.

References

1. Schoeman, D., Fielding, B.C.: Coronavirus envelope protein—current knowledge. Virol. J. (2019). https://doi.org/10.1186/s12985-019-1182-0
2. Rothan, H.A., Byrareddy, S.N.: The epidemiology and pathogenesis of coronavirus disease (COVID-19) outbreak. J. Autoimmun. (2020). https://doi.org/10.1016/j.jaut.2020.102433
3. Carty, M.F., DiNicolantonio, J.J.: Nutraceuticals have the potential for boosting the type 1 interferon response to RNA viruses including influenza and coronavirus. Prog. Cardiovasc. Dis. (2020). Available online https://doi.org/10.1016/j.pcad.202.02.007
4. https://www.worldometers.info/coronavirus/. Accessed 2020
5. Andre, M.C., Jennifer, C.M., Lucia, L.P., Tom, B.: Leveraging open hardware to alleviate the burden of COVID-19 on global health systems. Preprints (2020). https://doi.org/10.20944/preprints202003.0362.v1)
6. Pagano, U.: The crisis of intellectual monopoly capitalism. Camb. J. Econ. **38**(6), 1409–1429 (2014)
7. Peters, J.: Medical company threatens to sue volunteers that 3D-printed valves for life-saving coronavirus treatments. The Verge (2020). https://www.theverge.com/2020/3/17/21184308/coronavirus-italy-medical-3d-print-valves-treatments. Accessed 2020
8. Masnick, M.: SoftBank Owned Patent Troll, Using Monkey Selfie Law Firm, Sues To Block Covid-19 Testing, Using Theranos Patents. TechDirt (2020). https://www.techdirt.com/articles/20200316/14584244111/softbank-owned-patent-troll-using-monkey-selfie-law-firm-sues-to-block-covid-19-testing-using-theranos-patents.shtml. Accessed 2020
9. Kinsella, N.S.: Against intellectual property. J. Libertarian. Stud. **15**(2) (2001)
10. https://www.inputmag.com/tech/mit-is-developing-open-source-$100-ventilators-in-response-to-covid-19-shortages. Accessed 2020

11. Boldrin, M., Levine, D.K.: Against intellectual monopoly. Cambridge University Press, Cambridge, vol 8 (2008). https://doi.org/10.1017/CBO9780511510854
12. Ngo, T.D., Kashani, A., Imbalzano, G., Nguyen, K.T.Q., Hui, D.: Additive manufacturing (3D printing): a review of materials, methods, applications, and challenges. Compos. Part B Eng. **143**, 172–196 (2018)
13. Wang, X., Jiang, M., Zhou, Z., Gou, J., Hui, D.: 3D printing of polymer matrix composites: a review and perspective. Compos. B **110**, 442–458 (2017)
14. Sood, A.K., Ohdar, R., Mahapatra, S.: Parametric appraisal of mechanical property of fused deposition modeling processed parts. Mater. Des. **31**(1), 287–295 (2010)
15. Sachs, E.M., J.S. Haggerty, M.J. Cima, Williams, P.A.: Three-dimensional printing techniques. Google Patents, 1993
16. Khoshnevis, B.: Automated construction by contour crafting—related robotics and information technologies. Autom. ConStr. **13**(1), 5–19 (2004)
17. Melchels, F.P., Feijen, J., Grijpma, D.W.: A review on stereolithography and its applications in biomedical engineering. Biomaterials **31**(24), 6121–6130 (2010)
18. Manapat, J.Z., Chen, Q., Ye, P., Advincula, R.C.: 3D printing of polymer nanocomposites via stereolithography. Macromol. Mater Eng. **302**(9), 1600553 (2017)
19. Wohlers, T.: 3D printing and additive manufacturing state of the industry Annual Worldwide Progress Report Wohlers Report (2017)
20. Chen, W., Thornley, L., Coe, H.G., Tonneslan, S.J., Vericella, J.J., Zhu, C., Duoss, E.B., Hunt, R.M., Wight, M.J., Apelian, D.: Direct metal writing: controlling the rheology through microstructure. Appl. Phys. Lett. **110**(9), 094104 (2017)
21. Bai, Y., Williams, C.B.: An exploration of binder jetting of copper. Rapid Prototyp. J. **21**(2), 177–185 (2015)
22. Sharma, A., Bandari, V., Ito, K., Kohama, K., Ramji, M., BV H.S.: A new process for design and manufacture of tailor-made functionally graded composites through friction stir additive manufacturing. J. Manuf. Process **26**, 122–130 (2017)
23. Sova, A., Grigoriev, S., Okunkova, A., Smurov, I.: Potential of cold gas dynamic spray asadditive manufacturing technology. Int. J. Adv. Manuf. Technol. **69**(9–12), 2269–2278 (2013)
24. Matthews, M.J., Guss, G., Drachenberg, D.R., Demuth, J.A., Heebner, J.E., Duoss, E.B., Kuntz, J.D., Spadaccini, C.M.: Diode-based additive manufacturing of metals using an optically-addressable light valve. Optic Express **25**(10), 11788–17800 (2017)
25. Herzog, D., Seyda, V., Wycisk, E., Emmelmann, C.: Additive manufacturing of metals. Acta Mater. **117**, 371–392 (2016)
26. Postiglione, G., Natale, G., Griffini, G., Levi, M., Turri, S.: Conductive 3D microstructures by direct 3D printing of polymer/carbon nanotube nanocomposites via liquid deposition modeling. Compos. Appl. Sci. Manuf. **76**, 110–114 (2015)
27. Yang, J.-U., Cho, J.H., Yoo, M.J.: Selective metallization on copper aluminate composite via laser direct structuring technology. Compos. B Eng. **110**, 361–370 (2017)
28. Zhuang, Y., Song, W., Ning, G., Sun, X., Sun, Z., Xu, G., Zhang, B., Chen, Y., Tao, S.: 3D–printing of materials with anisotropic heat distribution using conductive polylactic acid composites. Mater. Des. **126**, 135–140 (2017)
29. Travitzky, N., Bonet, A., Dermeik, B., Fey, T., Filbert-Demut, I., Schlier, L., Schlordt, T., Greil, P.: Additive manufacturing of ceramic-based materials. Adv. Eng. Mater. **16**(6), 729–754 (2014)
30. Derby, B.: Additive manufacture of ceramics components by inkjet printing. Engineering **1**(1), 113–123 (2015)
31. Bienia, M., Lejeune, M., Chambon, M., Baco-Carles, V., Dossou-Yovo, C., Noguera, R., Rossignol, F.: Inkjet printing of ceramic colloidal suspensions: filament growth and breakup. Chem. Eng. Sci. **149**, 1–13 (2016)
32. Pai, D.: 3D-Printing skin is real: here's what you need to know. Allure News (2017). [Online]. Available https://www.allure.com/story/3d-printing-skin. Accessed 2020
33. Norman, J., Madurawe, R.D., Moore, C.M.V., Khan, M.A., Khairuzzaman, A.: A new chapter in pharmaceutical manufacturing: 3D-printed drug products. Adv Drug Del Rev **108**, 39–50 (2018)

34. Liu, Y., Hamid, Q., Snyder, J., Wang, C., Sun, W.: Evaluating fabrication feasibility and biomedical application potential of in situ 3D printing technology. Rapid Prototyp. J. **22**(6), 947–955 (2016)
35. CL. V.: Medical application for 3d printing: current and projected uses. Med. Devices **39**(10), 1–8 (2014)
36. Knowlton, S., Onal, S., Yu, C.H., Zhao, J.J., Tasoglu, S.: Bioprinting for cancer research. Trends Biotechnol. **133**(9), 504–513 (2015)
37. https://www.fda.gov/medical-devices/3d-printing-medical-devices/process-3d-printing-medical-devices. Accessed 2020
38. https://3dprintingindustry.com/news/3d-printing-community-responds-to-covid-19-and-coronavirus-resources-169143/. Accessed 2020
39. https://www.ciirc.cvut.cz/covid-2/. Accessed 2020
40. Pearce, J.M.: A review of open source ventilators for COVID-19 and future pandemics [version 1; peer review: 2 approved, 1 approved with reservations]. F1000Research **9**, 218 (2020). https://doi.org/10.12688/f1000research.22942.1
41. Hsiao, W.-K., Lorber, B., Paudel, A.: Can 3D printing of oral drugs help fight the current COVID-19 pandemic (and similar crisis in the future)? Expert Opin. Drug Deliv. (2020). https://doi.org/10.1080/17425247.2020.1772229
42. Parandoush, P., Lin, D.: A review on additive manufacturing of polymer-fiber composites. Compos. Struct. **182**, 36–53 (2017)
43. Pravin, S., Sudhir, A.: Integration of 3D printing with dosage forms: a new perspective for modern healthcare. Biomed. Pharmacother. **107**, 146–154 (2018)

The Role of Social Robotics to Combat COVID-19 Pandemic

Lamia Nabil Mahdy, Kadry Ali Ezzat, Ashraf Darwish, and Aboul Ella Hassanien

Abstract As the COVID-19 pandemic grows, the shortening of clinical hardware is expanding. A key bit of hardware getting out of sight has been ventilators. The contrast among the organic market is significant to be dealt with ordinary creation strategies, particularly under social removing measures set up. The examination investigates the method of reasoning of human-robot groups to increase creation utilizing preferences of both the simplicity of coordination and keeping up social removing. This chapter highlights the role of social robotic in fighting COVID-19. Also, it presents the requirements of social robotics.

Keywords Coronavirus · COVID-19 · Robots · Social robotics · Social distancing

L. N. Mahdy · K. A. Ezzat (✉)
Biomedical Engineering Department, Higher Technological Institute, Cairo, Egypt
e-mail: kadry_ezat@hotmail.com
URL: http://www.egyptscience.net

L. N. Mahdy
URL: http://www.egyptscience.net

A. Darwish
Faculty of Science, Helwan University, Cairo, Egypt
URL: http://www.egyptscience.net

A. E. Hassanien
Faculty of Computer and Artificial Intelligence, Cairo University, Cairo, Egypt
URL: http://www.egyptscience.net

L. N. Mahdy · K. A. Ezzat · A. Darwish
Scientific Research Group in Egypt (SRGE), Cairo, Egypt

A. E. Hassanien and A. Darwish (eds.), *Digital Transformation and Emerging Technologies for Fighting COVID-19 Pandemic: Innovative Approaches*, Studies in Systems, Decision and Control 322, https://doi.org/10.1007/978-3-030-63307-3_13

1 Introduction

The present reality is confronting COVID-19 pandemic that, since its first case showed up in Wuhan, China in Dec 2019 until 24th of April 2020, has just influenced 2,733,503 individuals (affirmed cases) with 192,208 confirmed passing in 213 nations, zones and regions as indicated in Fig. 1 [1]. The pandemic is further extending and has just prompted immense interruption in pretty much every part of the day by day life. One of the most effective deficiency is of clinical supplies. The most mind-boggling item among the shortage facing hardware is a clinical ventilator that is utilized for counterfeit breathing when a patient, in extreme instances of COVID-19, gets incapable to inhale normally [2]. Notwithstanding, one of the greatest world-wide difficulties is that the number of ventilators accessible in emergency clinics is significantly not exactly required [3] for adequately managing the pandemic because of a far-reaching lack of clinical ventilators and defensive apparatuses [4].

Because of the remarkable conditions and ventilator deficiencies, the administrations have been placing makers into war balance to deliver extra ventilators to satisfy the need [5]. The organizations have been intending to help their creations, for example, Hamilton and Getinge, contrasting with their present creation of 15,000 and 10,000 units, are meaning to deliver 21,000 and 16,000 ventilators separately in 2020 [6]. Non-clinical gadget fabricates, then again, are taking quick measures to repurpose their creation lines to handle the remarkable interest as a social obligation or on calls of the legislatures, for example, Passage and General Electric (GE) have planned to create 50,000 machines in 100 days [7]. Corporate goliaths, Siemens AG and Airbus SE have likewise responded positively to address the interest of ventilators. Dyson has structured another ventilator in 10 days and has plans to create 15,000 ventilators [8]. Be that as it may, the groundbreaking interest is hard to

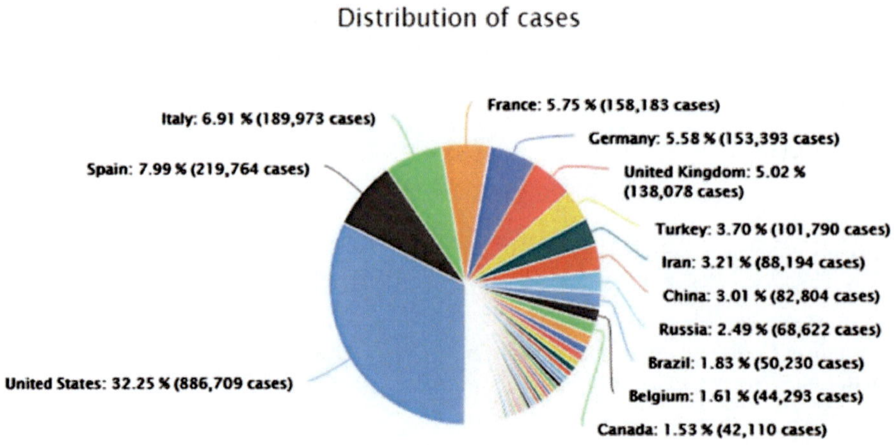

Fig. 1 Countries cases distribution

accomplish as the assembling frameworks of today, focused on large scale manufacturing, are structured with computerization arrangements not adaptable enough for huge scope reconfigurations [9, 10]. Broadly useful computerization arrangements are absent [11]. Moreover, there is a high human association in numerous procedures, for example, gathering [12, 13] while get together is by a long shot the predominant movement in ventilator creation. It maintains another test in the hour of a pandemic (especially on account of COVID-19) that holding fast to isolate rules requires least individuals to go out to work; and (the individuals who do) need to keep up social removing rules.

Mechanization (for example mechanical technology) is a powerful method to supplant human exertion, soothing people from physical undertakings and lift creation volume [14]. Be that as it may, regular robotization is unbendable making it extremely troublesome or practically difficult to repurpose the creation lines on an exceptionally a surprising bit of news, and frequently not ready to exist together with humans. More up to date equipment in the class of modern mechanical technology is a community-oriented robot or cobot [15]. Robots are more secure for people, simpler to incorporate, and are appropriate for mechanization of physical errands while coinciding with people [16, 17]. Other than the high capability of robotic computerization and cases accessible today of utilizing robots for single tedious errands (for example machine tending), their application in the get together is restricted [18].

The rest of this chapter is organized as follows. Section 2 presents the theoretical basics and background. Section 3 shows the role of social robotics in fighting COVID-19. Section 4 highlight the requirements of social robotics. Section 5 concludes this chapter.

2 Basics and Background

2.1 The COVID-19 Pandemic

The characterizing worldwide wellbeing emergency within recent memory is the novel coronavirus COVID-19, and without a doubt, the best test people have seen since World War II [19]. The infection is spreading like a wave in each mainland and nation except Antarctica. It took more than a quarter of a year for the novel coronavirus to influence the initial 10,000 individuals while it took just 12 days to come to the following 100,000 [20]. The quantity of passing starting at now is 33,106 concluding death pace of 4.8%. Thus, the malady was given the status of a pandemic on March 11, 2020, by the World Health Organization (WHO) [21]. The epic coronavirus, in the wake of getting into a human body, assaults the lower respiratory framework, and the casualty may encounter fever, dry and tenacious hack, myalgia, and brevity of breath. While the hazardous circumstance may bring about the type of pneumonia and intense respiratory misery disorder (ARDS). The WHO evaluates that some 80% of the influenced individuals recuperate without requiring any hospitalization [22] be that as it may, genuine breathing troubles are seen in one in every six COVID-19 casualties [23]. In these serious cases, the infection harms the lungs, accordingly, the insusceptible arrangement of the body responds by extending the veins and making the liquid go into the lungs. It makes it hard to breathe, because of which, the oxygen level drops, and can prompt a casualty. The truly sick individuals should be treated in Intensive Care Units (ICUs), where clinical ventilators are required to offer fake breathing help.

WHO's the assertion that the worldwide spread of coronavirus sickness 2019 (COVID-19) is a pandemic has contributed significantly to clearing up disarray in the wording in the expert writing and the media. Conversations on when the wide topographical spread of a malady turns into a pandemic will in general repeat when the world is faced with a developing irresistible sickness. The discussion around the wording utilized for COVID-19 brings up two significant issues. The inquiry is the reason there was hesitance to call the COVID-19 episode a pandemic, and the subsequent inquiry is whether the phrasing is of any useful significance. In practically all great reading material, a scourge turns into a pandemic when there is far-reaching geological appropriation of the malady. For certain weeks, the COVID-19 scourge, which had spread to more than 100 nations, appeared to fit the old style meaning of a pandemic. One could sensibly solicit whether the utilization from the term pandemic would change any of the activities important to control the spread of the infection. There are a few circumstances wherein it could be useful to utilize all around characterized wording to control the spread of an irresistible illness. The assets for controlling a pandemic are both unique, generously bigger, and for the most part considerably more expansive than for a confined flare-up or scourge [24]. Along these lines, the terms utilized for the various circumstances could be confined by the important control gauges. Maybe exceptional to pandemics, these remember significant universal coordination and cooperation for giving a guide to influenced nations, selecting the important assets for advancing examination on meds and antibodies, and creating complex hazard correspondence. Specifically, travel limitations become a significant issue and, even though these are guided by the International Health Regulations, nations have the choice to embrace singularly their hindrances to worldwide travel. This was the situation for COVID-19 [25].

2.2 *Reconfiguring Factories at the Peak of the Pandemic*

Outfitting the reaction endeavors, the administrations over the world are framing procedures together with the makers to reconfigure their manufacturing plants and to increase the creation to fulfill the foreseen need of ventilators. Assembling reconfiguration alludes to rapidly adjust to new errands and prerequisites to expand volume or to deliver an alternate item [26]. It very well may be the capacity to add machines to the existing operational framework rapidly to react quickly and cost successfully to surprising interest flood [27, 28]. Koren [29] characterizes responsiveness of assembling frameworks as to quickly and cost successfully creating new items on existing frameworks if confronting market changes, item changes or if (a piece of) a framework falls flat. The assembling frameworks of today are intended for large scale manufacturing with least item inconstancy at the serious expense. Critical research has been delivered in the previous decades to construct reconfigurable or responsive assembling frameworks [30–32] however their acknowledgment in the industry is without a doubt restricted. Subsequently, traditional assembling frameworks, in any event, while showing a level of adaptability, are regularly difficult to up-scale or down-scale to respond if the interest changes. The intricacies to acknowledge responsive assembling frameworks have been the high starting expense and increment in frameworks' unpredictability. There is an absence of universally useful mechanization arrangements that can be coordinated and reconfigured advantageously and rapidly to support robotization and increment creation volume. Correspondingly, in the hour of current pandemic, just the main organizations furnished with massive specialized and HR are (reluctantly) meaning to address the need of fluctuating requests, while they face difficulties of production network interruptions, a huge number of staff working from their homes, and social separating measures all around however especially where they work in industrial facilities. Mechanization expands creation volume, improves item quality, and calms people from redundant assignments [33]. Mechanical mechanization of physical assignments is incredibly determined by robots in the present assembling. Be that as it may, robots remain too unyielding and are unfeasible when need to exist together with people [34]. It takes a long time to incorporate a robot in a creation framework; with practically zero chance of any alterations without impressive endeavors. Besides, a generous creation increase amid the COVID-19 pandemic is required when people are approached to keep up social separating or ideally remain at home. Just, the least conceivable staffing is permitted at creation floors or mechanical production systems while keeping up a separation of least two meters from one another (as of now exhorted by certain administrations) planned for limiting the spread of COVID19.

COVID-19 considers cytokine storm disorders and immunosuppression As of March 12, 2020, coronavirus ailment 2019 (COVID-19) has been affirmed in 125,048 individuals around the world, conveying mortality of roughly 37%, contrasted and a death pace of under 1% from flu. There is a critical requirement for viable treatment. The current spotlight has been on the advancement of novel therapeutics, including antivirals and antibodies. Collecting proof recommends that a subgroup of patients with extreme COVID-19 may have a cytokine storm disorder. We suggest ID and treatment of hyper inflammation utilizing existing, affirmed treatments with demonstrated security profiles to deliver the prompt need to lessen the rising mortality. The current administration of COVID-19 is steady, and respiratory disappointment from intense respiratory trouble disorder (ARDS) is the main source of mortality. 2 Secondary hemophagocytic lymphohistiocytosis (sHLH) is an under-perceived, hyperinflammatory disorder portrayed by a fulminant and deadly hypercytokinemia with multiorgan disappointment.

3 The Role of Social Robotics in Fighting COVID-19

3.1 Robots

A social robot is a self-governing robot that interfaces and speaks with people or different self-sufficient physical operators by keeping social practices and rules appended to its job. Robots have been utilized in older consideration (assistive living, memory games), Physical recuperation, recovery and preparing, stroke, wounds, education, care of people with intellectual and social handicaps (dementia, mental imbalance).

As of late, there has been an outstanding enthusiasm for sending robots as associates, alluding to them as collective robots (otherwise known as robots) [35–37]. Robots have developed as more up to date equipment in the class of modern mechanical autonomy. They don't require shut off regions for their activity, rather they can coordinate or team up with people and offer their remaining burden to perform undertakings requiring a blend of their best and supplementing abilities. Besides, they are simpler to program (or repurpose) and are versatile [38]. Robots have been effectively incorporated into a few modern applications in the previous years from easy to complex undertakings [39] yet their maximum capacity in labor escalated assignments, for example, in gathering, has stayed undiscovered.

The key thought with robots is the incomplete computerization of a portion of the errands (that does't require human abilities) of a procedure such that robots will in general help or engage the kindred human. Robots are planned for expanding the level of computerization in manual work, in this way opening new applications or areas for robotization. Since the gathering is among the most work serious components of an assembling esteem chain, the mechanization of a small amount of it can conceivably have enormous advantages [40].

For each of these areas, there are extensive developments, as well as opportunities, to be explored in robotics. In the case of clinical care, areas of specific importance include disease prevention, diagnosis and screening, and patient care and disease management.

New ages of robots, from full scale to microscale, could be created to explore high-hazard regions and constantly work to disinfect all high-contact surfaces. For finding and screening, versatile robots for temperature estimation in open territories and ports of section speak to a reasonable utilization of developing innovations. Robotized camera frameworks are normally used to screen various individuals at the same time in huge regions. Consolidating these warm sensors and vision calculations onto self-governing or remotely worked robots could expand the productivity and inclusion of screening. These portable robots could likewise be utilized to over and again screen temperatures of in-/outpatients in different territories of the emergency clinics with information connected to medical clinic data frameworks. By systems administration existing security frameworks with facial acknowledgment programming, it is conceivable to backtrack contacts of contaminated people to alarm other people who may be in danger of disease. It is significant, notwithstanding, to acquaint suitable principles with deference protection. For introductory indicative testing for COVID-19, most nations suggest gathering and testing nasopharyngeal and oropharyngeal swabs. This includes test assortment, dealing with, move, and testing. During a significant episode, a key test is an absence of qualified staff to swab patients and procedure test tests. Computerized or robot-helped nasopharyngeal and oropharyngeal swabbing may accelerate the procedure, lessen the danger of disease, and let loose staff for different assignments. A few people don't create side effects of the infection or harbor the infection right now of testing. In these cases, a blood test to check for immune response appearance could be critical and used to recognize quiet diseases [41]. Computerizing the way toward drawing blood for research center tests could likewise ease clinical staff from an assignment with a high danger of presentation. Scientists are considering mechanical frameworks dependent on ultrasound imaging recognizable proof of fringe lower arm veins for computerized venepuncture. Mechanized multiplex constant tests would permit quick in vitro subjective identification and separation of pathogens. Self-governing automatons or ground vehicles might be utilized for test move just as the conveyance of prescriptions to contaminated patients when development is ill-advised [42]. COVID-19 could be an impetus for creating automated frameworks that can be quickly sent with remote access by specialists and basic specialist organizations without the need of venturing out to bleeding edges. The broad isolate of patients may likewise mean delayed seclusion of people from the social association, which may negatively affect psychological well-being. To address this issue, social robots could be conveyed to give proceeded with social connections and adherence to treatment systems unafraid of spreading illness.

Be that as it may, this is a difficult region of advancement since social associations require assembling and keeping up complex models of individuals, including their insight, convictions, feelings, just as the specific circumstance and condition of the cooperation. Teleoperation is likewise a full-grown innovation that can be utilized for both telemedicine and working from home. As of late, schools, colleges, and organizations in China have received online courses and communications. As 5G data transmission and 4–8 K video become generally accessible, COVID-19 may stamp the tipping purpose of how future associations work. As opposed to dropping huge universal presentations and meetings, new types of social occasion—online instead of in-person participation—may increment. Remote participants may get familiar with utilizing mechanical symbols and controls. In the end, numerous gatherings might be accessible using top-notch low-inactivity augmented reality, with the participants' virtual robot symbols completely versatile and inundated in the meeting setting. These modalities would lessen sickness disease rates and carbon impression all the while. Verifiably, robots have been created to take on dull, messy, and hazardous occupations. Their first wide-spread arrangement was in quite a while, likewise battling irresistible ailments includes a situation that is unacceptable for human specialists yet is appropriate to robots. The encounters with the Ebola flare-up recognized an expansive range of utilization cases, yet subsidizing for a multidisciplinary look into, in association with offices and industry, to meet these utilization cases stays restricted. Presently, the effect of COVID-19 may drive further research to apply autonomy to address the dangers of irresistible ailments. Be that as it may, without supported research endeavors robots will, by and by, not be prepared for the following episode. By encouraging a combination of building and irresistible ailment experts with devoted financing we can be prepared when (not if) the following pandemic shows up [43].

3.2 Social Distancing

The COVID-19, acclimated with a history of pandemics [44, 45], has brought severe social separating measures. Notwithstanding their universal outskirt terminations and suspension of air traffic, the legislatures are utilizing national gatekeepers to limit the open development. The legislatures around the world have requested that as far as possible their social contacts however much as could reasonably be expected. On the off chance that somebody gets side effects of COVID-19, the entire family unit is committed to self-isolate for 14 days [46]. Moreover, individuals in high hazard classification (for example more established than 70 years, pregnant ladies, and individuals with a history of extreme sicknesses for example extreme asthma, lung maladies, and hypertension, and so forth.) are required to self-disconnect [47]. With a higher consistency to the procedure, the pandemic can almost certainly purchase additional time by hindering the spread lastly be turned away with different measures set up, especially antibodies. Social removing is focused for the time until a strain-explicit antibody is created and appropriated which may take a while or even years. The execution of social separating procedures for so long is testing. It carries end to the financial movement, mass joblessness, and lack of products. Industrial facilities, other than makers of clinical supplies, are approached to stay operational with the least conceivable staffing.

Proportionally, keeping up social separating systems in industrial facility territories is likewise testing. Most assembling organizations are operational while keeping up a separation between administrators during and after work. In this situation, robots, machine partners of people, are inalienably ready to help the social removing techniques. It is proposed that the work undertakings are planned such that the immediate communication between people is made restricted, rather association is made among robot and the human where conceivable. The bolts in the given figure show potential commitment among human(s) and robot(s). The commitment between human to human is separated as a social removing measure. Indeed, even with this model of HRC commitment, the move of a pathogen is conceivable from the surface of a work-in-process that is directed from a contaminated human to a robot and afterward to another human. To de-chance this, wearing hand gloves must be guaranteed for every human administrator (Fig. 2).

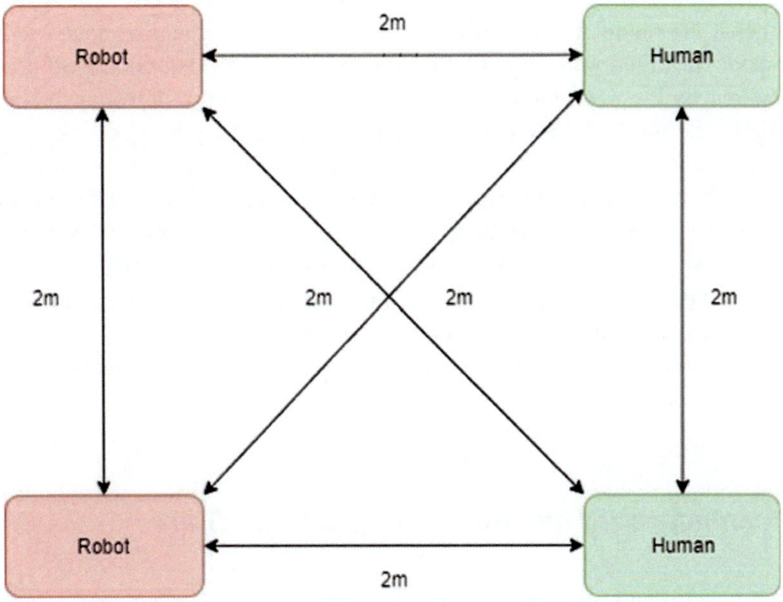

Fig. 2 Social distancing measures adapted with human to robot interaction only and maintaining a distance of 2 m between the two operators

3.3 Enablers for Reconfiguring Factories Through Robots

Makers hold a past filled with confronting cruel market variances notwithstanding financial crunches frequently expecting them to upscale or downscale their ability or to make items they are not knowledgeable about. The strain to adjust to the interest changes gets escalated for specific items and their customary (and firmly related) makers. The exploration introduced in this article has proposed HRC as a potential answer for adaptable and reconfigurable computerization for errands, for example, get together.

4 Social Robotics Requirements to Fight COVID-19

4.1 Modularization

A particular creation framework establishes general engineering both in the mechanical structure of its components (modules) just as in the data stream. The all-inclusive engineering can empower a snappy and financial consideration or rejection of the components of a creation framework for limit or ability modification [48]. Modularization possibly accomplished both in the item being created and in its assembling framework [49, 50]. A reconfigurable HRC gets together cell may address request changes if the modularization is coordinated in its equipment (robots, robot-tables, transports, installations) and programming (robot programs). On the off chance that the interest builds, extra modules might be added to the current framework with little exertion. The methodology is undifferentiated from the LEGO development brick modular approach is a perfect answer for reconfiguring capacity particularly when joining robotization (for example robots) and human abilities. Off the rack robots with shifting properties of mechanical design, degrees of opportunity, value, payload, and security restrictions might be utilized to achieve various sorts of undertakings. Thus, other equipment if having a widespread and particular plan can be included or evacuated when required. As indicated by the gathering multifaceted nature and work errands conveyance among humans and robots, modules can be included framing a get-together cell.

4.2 Standardized and Rapid to Deploy Hardware

Broadly useful and prepared to convey computerization arrangements can make it helpful to upscale robotization level and lift creation volume. It is the ideation of having a creation framework or subsystems that rush to actualize. The processing plant in a container idea describes normalized secluded creation units that are adaptable, brisk to convey, and versatile [51]. The idea of normalized and prepared to convey equipment is accessible in military battles and wellbeing administration units however is elusive in assembling scene. Some equipment arrangements are getting accessible easily of versatility, and reconfigure capacity. Such equipment might be used for creating reconfigurable creation frameworks.

4.3 Software Templates

The most well-known undertakings for robot computerization are palletizing, pick and spot, machine tending, screwing, and so forth. The robots are customized in a machine language to play out any of the appointed undertakings. Nonetheless, an enormous piece of the robot program can be utilized as a layout to reconstruct the robot whenever experienced changes for example in item plan or format (change in pick and spot areas). Prepared to utilize standard program formats can bolster the end clients for quicker joining and reconfiguration. A case of this is Easy Templates created by the Danish Technological Institute (DTI) [52] for UR robots. These are anything but difficult to utilize pre-created formats for most regular automated errands.

4.4 *Digital Twins Digital Twin (DT)*

DT is an information associated portrayal of structure and components of a framework. In the introduced work, a DT can be an expansion of the reproduction models past the structure period of the HRC framework. DT can bolster dynamic in configuration, arranging, and activity of assembling frameworks. Additionally, the convenience of advanced twins to address a few parts of robots has been examined in the writing. Reproduction based DTs offer a protected space to get bits of knowledge into the operational conduct of an unpredictable framework, making improvements and approval of the arrangements before carrying arrangements into this present reality.

5 Conclusion

The scale and the seriousness of the pandemic call mainstream researchers to control national and universal endeavors to battle the illness. Assembling division is feeling the squeeze and is attempting to successfully react to the earth-shattering however ideally transient need of items, for example, a clinical ventilator. The adaptability of most assembling frameworks seen today to adjust to item varieties is obliged inside a restricted limit. The robots with a blend of present-day fabricating advances can be valuable for creation increase in crisis circumstances. As in the hour of crisis request climb for specific items, makers are feeling the squeeze to react rapidly while conveying remarkable outcomes, the measures adjusted are frequently flighty. Subsequently, the adjusted procedures set new headings for the assembling part. The unpredictable methodologies may give new understandings and frequently become another type of creating things. The learnings from utilizing robots during a pandemic, to supplant human endeavors, may conceivably be a stage forward towards human-robot groups in factories of the future.

References

1. World Health Organization, others: Coronavirus disease 2019 (COVID-19): situation report, 70 (2020)
2. Chatburn, R.L.: Understanding mechanical ventilators. Expert Rev. Respir. Med. **4**, 809–819 (2010). https://doi.org/10.1586/ers.10.66
3. Mechanical Ventilators Market Size, Share & Trends Analysis Report. https://www.grandviewresearch.com/industry-analysis/mechanical-ventilators-market. Accessed 31 Mar 2020
4. Formula 1 Helps Boris Johnson Edge Closer to 30,000 Ventilator Target. https://www.bloomberg.com/news/articles/2020-03-24/formula-1-helps-johnson-edge-closer-to-30-000-ventilator-target. Accessed 29 Mar 2020
5. Ebrahim, S.H., Ahmed, Q.A., Gozzer, E., Schlagenhauf, P., Memish, Z.A.: Covid-19 and community mitigation strategies in a pandemic. BMJ. **368**, m1066 (2020). https://doi.org/10.1136/bmj.m1066

6. Ventilators Market Industry Updates—See How Industry Players are preparing against Covid-19 depression. https://www.marketwatch.com/press-release/ventilators-market-industry-updates–see-how-industryplayers-are-preparing-against-covid-19-depression-medtronic-bd-getinge-ab-fisher-paykel-healthcarelimited-dragerwerk-ag-co-kgaa-smiths-group-hamilton-medical-general-electric-company-air-liquide-2020-03-30?mod=mw_quote_news&tesla=y. Accessed 8 Apr 2020
7. Ford, GE plan to produce 50,000 ventilators in 100 days. https://www.cnbc.com/2020/03/30/ford-ge-plan-toproduce-50000-ventilators-in-100-days.html. Accessed 31 Mar 2020
8. James Dyson designed a new ventilator in 10 days. https://edition.cnn.com/2020/03/26/tech/dysonventilators-coronavirus/index.html. Accessed 15 Apr 2020
9. Edmondson, N.F., Redford, A.H.: Generic flexible assembly system design. Assembly Autom. **22**(2),139–152 (2002). https://doi.org/10.1108/01445150210423189
10. ElMaraghy, H., ElMaraghy, W.: Smart adaptable assembly systems. Procedia CIRP **44**, 4–13 (2016)
11. Joergensen, S.N., Hvilshøj, M., Madsen, O.: Designing modular manufacturing systems using mass customization theories and methods. Int. J. Mass Customisation **4**, 171–194 (2012)
12. Parasuraman, R., Wickens, C.D.: Humans: Still vital after all these years of automation. Hum. Factors J. Hum. Factors Ergon. Soc. **50**, 511 (2008)
13. Bahns, E.: It Began with the Pulmotor: The History of Mechanical Ventilation, pp. 10–55. Drägerwerk, Lübeck, Germany, (2015)
14. Krüger, J., Lien, T.K., Verl, A.: Cooperation of humans and machines in assembly lines. CIRP Ann. Manuf. Technol. **58**, 628–646 (2009)
15. Garcia, M.A.R., Rojas, R., Gualtieri, L., et al.: A human-in-the-loop cyber-physical system for collaborative assembly in smart manufacturing. Procedia CIRP **81**, 600–605 (2019)
16. Faccio, M., Bottin, M., Rosati, G.: Collaborative and traditional robotic assembly: a comparison model. Int. J. Adv. Manuf. Technol. **102**, 1355–1372 (2019)
17. Wang, L., Gao, R., Váncza, J., et al.: Symbiotic human-robot collaborative assembly. CIRP Ann. Manuf. Technol. **68**, 701–726 (2019)
18. Malik, A.A., Bilberg, A.: Complexity-based task allocation in human-robot collaborative assembly. Ind. Robot Int. J. Robot. Res. Appl. **46**, 471–480 (2019)
19. COVID-19 pandemic: Humanity needs leadership and solidarity to defeat the coronavirus. https://www.undp.org/content/undp/en/home/coronavirus.html. (2020, April)
20. World Health Organization, others: Coronavirus disease 2019 (COVID-19): situation report, 59 (2020)
21. Landry, M.D., Geddes, L., Moseman, A.P., et al.: Early Reflection on the global impact of COVID19, and Implications for physiotherapy. Physiotherapy (2020)
22. Coronavirus: What are ventilators and why are they important? https://www.bbc.com/news/health-52036948. Accessed 31 Mar 2020
23. Ventilators explained: Key device in fight against coronavirus. https://www.aljazeera.com/news/2020/03/ventilators-explained-key-device-fight-coronavirus200329065155185.html. Accessed 31 Mar 2020
24. Vicentini, F.: Collaborative robotics: a survey. J. Mech. Des. 1–29 (2020)
25. Malik, A.A., Andersen, M.V., Bilberg, A.: Advances in machine vision for flexible feeding of assembly parts. Procedia Manuf. **38**, 1228–1235 (2019)
26. Masood, T., Weston, R.: An integrated modeling approach in support of next-generation reconfigurable manufacturing systems. Int. J. Comput. Aided Eng. Technol. **3**, 372–398 (2011)
27. Masood, T., Weston, R., Rahimifard, A.: A computer-integrated unified modelling approach to responsive manufacturing. Int. J. Ind. Syst. Eng. **5**, 287–312 (2010)
28. Koren, Y., Gu, X., Guo, W.: Reconfigurable manufacturing systems: principles, design, and future trends. Front. Mech. Eng. (2017)
29. Koren, Y., Shpitalni, M.: Design of reconfigurable manufacturing systems. J. Manuf. Syst. **29**, 130–141 (2010)
30. Rashid, A., Masood, T., Erkoyuncu, J.A., et al.: Enterprise systems' life cycle in pursuit of resilient smart factory for emerging aircraft industry: a synthesis of Critical Success Factors'(CSFs), theory, knowledge gaps, and implications. Enterp. Inf. Syst. **12**, 96–136 (2018)

31. Haq, I., Masood, T., Ahmad, B., et al.: Product to process lifecycle management in assembly automation systems. In: 7th CIRP international conference on digital enterprise technology, pp. 476–486 (2011)
32. Perzylo, A., Rickert, M., Kahl, B., et al.: SMErobotics: smart robots for flexible manufacturing. IEEE Robot. Autom. Mag. **26**, 78–90 (2019)
33. Nof, S.Y.: Springer Handbook of Automation. Springer Science & Business Media (2009)
34. Malik, A.A., Masood, T., Bilberg, A.: Virtual reality in manufacturing: immersive and collaborative artificialreality in design of human-robot workspace. Int. J. Comput. Integr. Manuf. 1–16 (2019)
35. Ososky, S., Schuster, D., Phillips, E., Jentsch, F.G.: Building appropriate trust in human-robot teams. In: AAAI Spring Symposium: Trust and Autonomous Systems (2013)
36. Schwartz, T., et al.: Hybrid teams of humans, robots, and virtual agents in a production setting. 12th International Conference on Intelligent Environments (IE), London, 2016, pp. 234–237 (2016). https://doi.org/10.1109/IE.2016.53
37. Correia, F., Petisca, S., Alves-Oliveira, P., et al.: "I Choose… YOU!" membership preferences in human-robot teams. Auton. Robots **43**, 359–373 (2019)
38. Malik, A.A., Bilberg, A.: Framework to implement collaborative robots in manual assembly: a lean automation approach. In: B. Katalinic (ed.) Proceedings of the 28th DAAAM international symposium. DAAAM International, pp 1726–9679 (2017)
39. Villani, V., Pini, F., Leali, F., Secchi, C.: Survey on human-robot collaboration in industrial settings: safety, intuitive interfaces and applications. Mechatronics **55**, 248–266 (2018)
40. Bilberg, A., Malik, A.A.: Digital twin driven human-robot collaborative assembly. CIRP Ann. Manuf. Technol. **68**, 499–502 (2019)
41. Rhodes, A., Ferdinande, P., Flaatten, H., et al.: The variability of critical care bed numbers in Europe. Intensive Care Med. **38**, 1647–1653 (2012)
42. Shakoory, B., Carcillo, J.A., Chatham, W.W., et al.: Interleukin-1 receptor blockade is associated with reduced mortality in sepsis patients with features of macrophage activation syndrome: reanalysis of a prior phase iii trial. Crit. Care Med. **44**, 275–281 (2016)
43. Russell, C.D., Millar, J.E., Baillie, J.K.: Clinical evidence does not support corticosteroid treatment for 2019-nCoV lung injury. Lancet **395**, 473–475 (2020)
44. Glass, R.J., Glass, L.M., Beyeler, W.E., Min, H.J.: Targeted social distancing designs for pandemic influenza. Emerg. Infect. Dis. **12**, 1671 (2006)
45. Fenichel, E.P.: Economic considerations for social distancing and behavioral based policies during an epidemic. J. Health Econ. **32**, 440–451 (2013)
46. Mahase, E.: Covid-19: UK starts social distancing after new model points to 260 000 potential deaths (2020)
47. Thomson, G.: COVID-19: Social distancing, ACE 2 receptors, protease inhibitors and beyond? Int J Clin Pract. **74**(7):e13503 (2020). https://doi.org/10.1111/ijcp.13503
48. Salonitis, K.: Modular design for increasing assembly automation. CIRP Ann. Manuf. Techno. (2014)
49. Shaik, A.M., Rao, V.V.S.K., Rao, C.S.: Development of modular manufacturing systems—a review. Int J Adv Manuf Technol **76**, 789–802 (2015). https://doi.org/10.1007/s00170-014-6289-2
50. Heilala, J., Voho, P.: Modular reconfigurable flexible final assembly systems. Assembly Autom. **21**, 20–30 (2001)
51. Jackson, M., Zaman, A.: Factory-in-a box: mobile production capacity on demand. Int. J. Mod. Eng. **8**, 12–26 (2007)
52. (2020) Danish Technological Institute. https://www.dti.dk/

The 4th Industrial Revolution in Coronavirus Pandemic Era

Mohamed Nagy, Hagar M. Abbad, Ashraf Darwish, and Aboul Ella Hassanien

Abstract The global prevalence of coronavirus disease 2019 (COVID-19) requires a remarkable avenue to endure and restrain it; Although the most advanced and sophisticated healthcare systems around the world could not stand against this pandemic, the synthesis of the fourth industrial revolution manifests its potential to eradicate this virus. This chapter discusses how multiple advanced technologies involve diverse perspectives of fighting the catastrophe, starting from reduction of the spreading of the virus, automated surveillance for infected cases, contribution to retaining the communication as well as social safety during the lockdown, and evolving healthcare medical equipment to the process of developing a vaccine. It also has a vital role in keeping most nations' institutions run remotely, such as education systems, besides the declination of the expected economic losses by running businesses online. Moreover, introducing the essential role of these technologies to monitor the propagation of COVID-19 globally that permits taking precautionary measures earlier and evaluating the current situation of each country individually. Eventually, the influence of these privileges of this revolution and how it has convinced other nations the importance of accelerating and boosting those advanced technologies to defeat the current situation by considering China as a realistic illustration of the efficiency.

Keywords COVID-19 · 4th industrial revolution · Surveillance systems · Artificial intelligence · Cloud computing · Internet of things · Smart healthcare systems

M. Nagy · H. M. Abbad · A. Darwish
Faculty of Science, Helwan University, Cairo, Egypt
e-mail: hagarmosaad@science.helwan.edu.eg

A. Darwish
e-mail: ashraf.darwish.eg@ieee.org

A. E. Hassanien
Faculty of Computers and Artificial Intelligence, Cairo University, Cairo, Egypt
e-mail: aboitcairo@gmail.com

M. Nagy (✉) · A. Darwish · A. E. Hassanien
Scientific Research Group in Egypt (SRGE), Cairo, Egypt
e-mail: mohamednagy@science.helwan.edu.eg

A. E. Hassanien and A. Darwish (eds.), *Digital Transformation and Emerging Technologies for Fighting COVID-19 Pandemic: Innovative Approaches*, Studies in Systems, Decision and Control 322, https://doi.org/10.1007/978-3-030-63307-3_14

1 Introduction

Shorten casualties and endure fewer expenses become the main aim of all the governments all over the world after the spread of Coronavirus. In December 2019, the first infected cases of Coronavirus disease 2019 (COVID-19) [1] have been identified in China and has since broken out globally as a pandemic. Most sophisticated healthcare systems could not survive from this viral pandemic. Moreover, the international economy has been collapsed, because many countries have been suffered from the lockdown which leads to halt numerous services afraid of spreading out the pandemic further. The core of this crisis is that the estimated average of reproduction number (R0) or the basic reproduction number of COVID-19 is between 2.2 and 2.7 which is much higher than the season of flu [2]. Accordingly, the dramatic raising of infected and fatality cases force the attention of nations to find a method that has ability to decline the harms more than developing a vaccine which is a long term solution. One of the reliable, effective, and short term resolutions is the digital revolution which has a remarkable role to prevent the spreading of the pandemic.

The digital revolution or the fourth industrial revolution is a sort of advanced technologies that are used to develop smart systems have ability to defeat healthcare crises. These technologies have a massive computing capacity to analyze and manage tremendous data. Sensors could be attached to machines in order to enhance their observing of the environment and data transmission, these machines might be supported by Artificial Intelligence (AI) to act rationally and perform human tasks efficiently and quickly. These technologies contribute to the production line to improve the product quality which could be exemplified by 3D printer technology. Overall, this digital revolution should have more attention and contribution to fight healthcare pandemics such as COVID-19 [3].

2 The Fourth Industrial Revolution

Digital technology revolution merits and strength have been revealed and become observable after the COVID-19 pandemic outbreak. It is also called the fourth industrial revolution which aim to implement the advanced technology to solve obstacles, this revolution efficiency for tackling diseases is approved. It could be exemplified by China which contained COVID-19 and diminished the rate of mortality and infected cases by assigning advanced technology to dominate almost all aspects of life. Furthermore, the World Health Organization (WHO) has been used this technology to track not only the cases over the globe but also tracking historical data of patients in order to get acknowledge patient zero [4] which is essential to identify the behavior of this virus.

2.1 Phases of Combating Coronavirus

The fourth industrial revolution involves a bunch of advanced technology such as AI, Cloud computing, Big Data, Virtual Reality, Internet of Things (IoT), and robots which are combined as an integral part to suppress the pandemic. Apparatuses supported by this technologies may include hardware sensors with wireless connectivity to monitor specific production line [5] and makes their own decisions, and consequentially reduce human actions especially in medicine. These technologies assist to overcome COVID-19 outbreak in various perspectives such as a diagnose of the viruses as well as their symptoms or releasing telemedicine [6]. In addition to all these, the social distance definitions have serious support to decline the infection rate and interpret numerous data to expose abnormal patterns of the virus genetic to develop a vaccine.

2.1.1 Observing Virus Mutation Directs to a Vaccine

The rapidity of obtaining a vaccine depends on the perception of virus behaviors and mutations. The COVID-19 virus has many more forms or mutated versions [7] with particular properties which could be completely distinct, and might also improve its regular functions such as spreading rate or aggressive behavior that has been happened in Italy [8] as shown in Table 1 [9]. To be exemplified, some research papers represent how various mutated versions of COVID-19 has been spread out in particular countries. Geneticist experiments to identify the spreading behavior and symptoms are tough and takes a long period, however, Artificial Intelligence (AI) and Cloud Computing technologies have an immense influence on enhancing these procedures.

AI machines can identify abnormal patterns in data and classify them into logical groups with high accuracy and speed comparing to humans. Besides, Cloud Comput-

Table 1 COVID-19 mutation: clustering the identified three mutation types of COVID-19 up to its mutation, origin, spreading location

The three strains of coronavirus			
Type	Type A	Type B	Type C
Mutation	Closest to coronavirus found in bats and pangolins	Variation of coronavirus most common in Wuhan	Type C is the 'daughter' of type B
Origin	Considered the 'root of the outbreak'	Derived from type 'A' via two mutations	One mutation different to parent variation
Location	Two sub-clusters-one linked to Wuhan and one common line in America and Australia	Mutates slowly in China but rapidly outside China	Spread to Europe via Singapore

ing allows handling billions of gathered data packages in milliseconds which related to virus, this two advanced technologies increase the ability of identification unusual mutations that supports Geneticists to define symptoms and developing beta-vaccines much faster than before, although, the process of developing a vaccine is considered to take decades.

Indirect assist to acquire vital information relevant to vaccine development which is catching patient-zero [4] does not exclude digital technologies. Introducing patient-zero would be a straight advantage to identify the genetic mutation [10] which occurs to the COVID-19 virus to transform from a wild body to human, this information could be used to break out the nature of this transformation. These tracking procedures to reach to the patient zero needs collecting a huge information which could be done by interactive devices have the Internet of Things (IoT) technology with AI and Cloud Computing, all these technologies has ability to work together in order to make an intense analysis and collecting information automatically.

Surveillance communicated systems are a smooth manner to track historical data of patients, their movements, the period of exposing some symptoms on the case, and the flexibility to trace those who have interacted with infected cases.

To sum up, the fourth industrial revolution has a recognizable and immense part to modify and optimize the process of developing the virus vaccine in both straight-forward and obscure manners.

2.1.2 Introducing Telemedicine Definition and Medical Equipment

Many services have been presented remotely to descend the infected cases mainly for health staff who are more likely to be infected by the virus. For that reason, nations establish a telemedicine system [6] in which healthcare service could be done remotely, the system not only includes remotely healthcare between medical practitioners and patients but also provides transmission of medical tools or regular requirements between medical practitioners. While the worldwide were attempting to improve their medical systems or finding a cure, China was one of the earliest countries that applied a sequence of technologies such as AI, Internet of Things (IoT), and Cloud Computing to keep a limited interaction between medical staff using robots to deliver therapeutic tools, robotics-based treatment to pursue some pharmaceutical requirements to patients, and thermal surveillance robots to monitor the health condition of health staff in addition to outsiders. The wireless connection between them allows data transmission and analysis over internet cloud services that could quickly identify any abnormality regarding the health condition of patients or medical staff and preventing the rate of infection by the pandemic virus COVID-19.

The fourth industrial revolution is involved in the production line of pharmaceutical equipment due to the frequent demand during the COVID-19 outbreak. The unusual spreading of this virus over the population, some health equipment has been used by both medical staff and individuals which diminish health supplies. Whereas an efficient and rapid method becomes obliged for health tools like customized face masks and gloves that can be done using 3D printer technology. Customized face shields [11] have been created and optimized swiftly by 3D printer technology in order to save healthcare providers in advance. They interact directly with infected cases which raises the mortality rate in healthcare providers in the majority of countries where COVID-19 invades.

The integration of Virtual Reality technology in medical training [12] provides a more dynamic, interactive, and realistic environment that enriches the learning style to be more comprehensible, it has also handled the individual awareness about the protection of COVID-19 pandemic. As the novel pandemic has particular treatment methods, holding intensive training about these methods including investigating predicted circumstances is essential, and utilizing it using Virtual Reality technology has been established to be more easier to simulate. Besides, simplifying precautionary measures of the pandemic to individuals has been recommended as an alternative way that would increase the awareness of individuals or even pupils effectively. Even the quality of this technology is observable, but its features are not be employed entirely yet.

Overall, the fourth industrial revolution has important and vital involvement in global healthcare which includes providing remote healthcare, medical supplies exchange between therapeutic practitioners without human interaction, robotics-based treatment for some medical procedures [13], thermal surveillance robots [14] to recognize suspected cases of newcomer strangers to hospitals, participating in medical production for healthcare supplies, and eventually utilizing healthcare environment to be simulated by Virtual Reality technology. These engagements and more have a significant outcome of preventing and control on the pandemic.

2.2 Economic Rescue from Pandemic Consequences

The social and economic impact of COVID-19 invade every region of the world, the reduction of prospective crisis becomes the aim of the majority of realms. Procedures and instructions have been authorized by the World Health Organization to be followed by nations to ensure the safety of public health, these procedures include curfew implementation and shut down airlines globally as well as schools which dramatically injure the global economy [15] and socialism. Technologies are a sole way to lessen these damages by keeping the existence of several businesses by applying remote services, these services have been supported by Cloud Computing technology in which virtual video calls and meetings are accessible by online platforms such as Zoom, Skype, and other platforms.

Collaboration atmosphere becomes achievable remotely by platforms that offer sharing information and resources between users, and handle any expected conflict which makes it similar as working in the company itself. As a result, Education systems use this dynamic environment to do their lectures through online streams and meetings where the lecturer may monitor and communicate to students smoothly and sharing resources and materials as expected.

The fourth industrial revolution aids people to perform their daily life work while the lockdown [16] including their social relationships and practice their daily habits indoors. Online video streaming platforms allow them to communicate with relatives, friends, and colleagues which decline the isolation feeling, furthermore, advanced digital devices provide daily services online such as delivering food from fancy restaurants, do online shopping, and other needs they expected to own. Indeed, Virtual Reality technology has a great impact during the lockdown as supporting individuals to have a virtual interactive environment which is similar to the expected one to be found outdoors.

Network Connection from local quarantines to the worldwide and the World Health Organization has been introduced by advanced technologies that admit the transition of the information about the pandemic to the globe instantly. Although, the harmful impression of the COVID-19 pandemic, researchers claim that the potential control on the pandemic growths starkly with the rhythm of the industrial revolution development, because of the speed of exchanging the pandemic information and new reports which provides a quick and straightforward tracking for the abnormal spreading of the virus.

Briefly, the fourth industrial revolution consolidates the social relationships under the lockdown, in addition, boosts the economy to not fall down due to the consequences of the COVID-19 pandemic by keeping education systems and workforce do their regular daily tasks remotely in a cooperation manner as used to be. Its advantages go beyond the lockdown by surviving social relationships and association between individuals besides doing their ordinary functions online or even having a visual interactive image of the outdoor places using Virtual Reality technology. These technologies proved its successful mechanism to minimize losses of all life aspects, it might overcome pandemics damages entirely If the governments implemented this technology accurately.

3 International Endeavors of Tracing and Restraining the Coronavirus Pandemic

The nations International cooperation compile the spreading direction, death, and infected cases of the pandemic as an integrated structure that can thoroughly managed by using advanced technologies. Processing and analysis interpreted data of the pandemic is a vital concern for the countries and World Health Organization as well, because It is a foundation step to contain the disease. An enormous amount of data has been accumulated from every region around the world that requires proper and secure storage. Big data, cloud computing, and AI technologies assist to manage and interpret this storage effectively as well as applying statistical and numerical methods would be achievable on the massive data of the pandemic by these technologies. Consequentially, different metrics have been considered globally by the World Health Organization in order to observe the invading rate of the COVID-19 pandemic at each country, these metrics consist of the recovered, death, closed, and critical serious cases (Fig. 1).

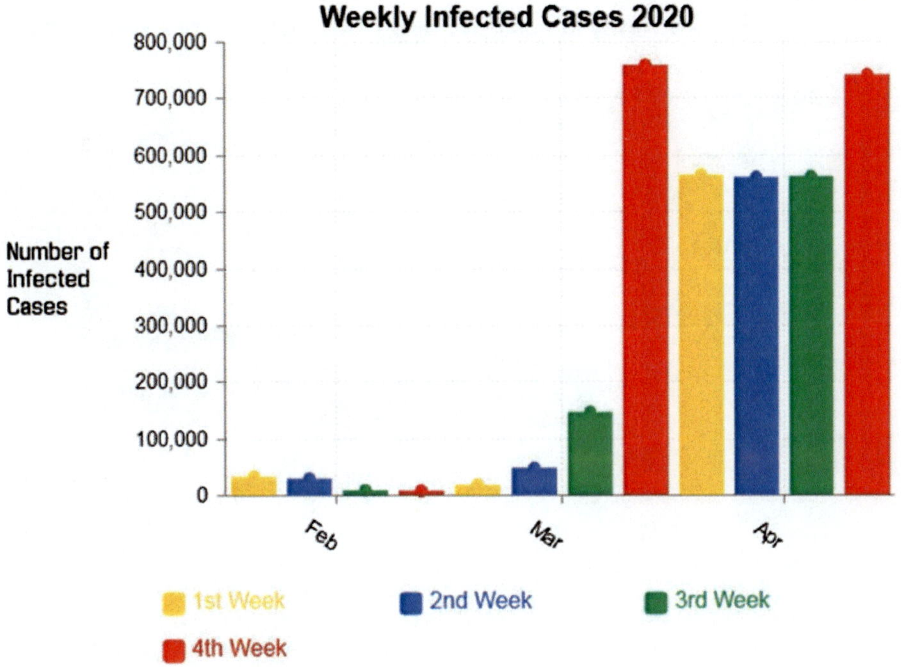

Fig. 1 Weekly new cases from February 2020 to April 2020

The fourth revolution technology provides a robust real-time daily detection for pandemic cases, moreover, tracking the sudden increase in one of the determined metrics in specific countries became accessible. Gathering data reports of the spreading of the pandemic daily and interpreting it as statistic graphs using technologies provide a robust analysis clear from human-error. Recently, public websites use this technology to publish these graphs globally using the daily reports that are announced by each country, compiling these reports using Cloud Computing and Big data technologies and then presenting them as visual graphs [17]. Some of these graphs, as shown in (Fig. 2) include new cases per day as well as accumulating the daily/weekly cases data to present the global daily/weekly cases which could probably help to monitor the domination of the virus.

Data manipulation of the daily updated virus reports illustrates the progress of the governments against the pandemic which might help to determine the most appropriate method to combat that disease. World Health Organization observes the virus data distribution on each country separately and its procedures as well as the proportional rate of newly infected versus newly recovered cases as shown in Fig. 3.

By considering the incoming data, the World Health Organization have noticed that the procedures which have been done by China to dominate the spreading of the virus have a significant impact and fast result. China has employed the fourth industrial revolution to be a way of eliminating the COVID-19 infection in the central root of spreading the virus, Wuhan.

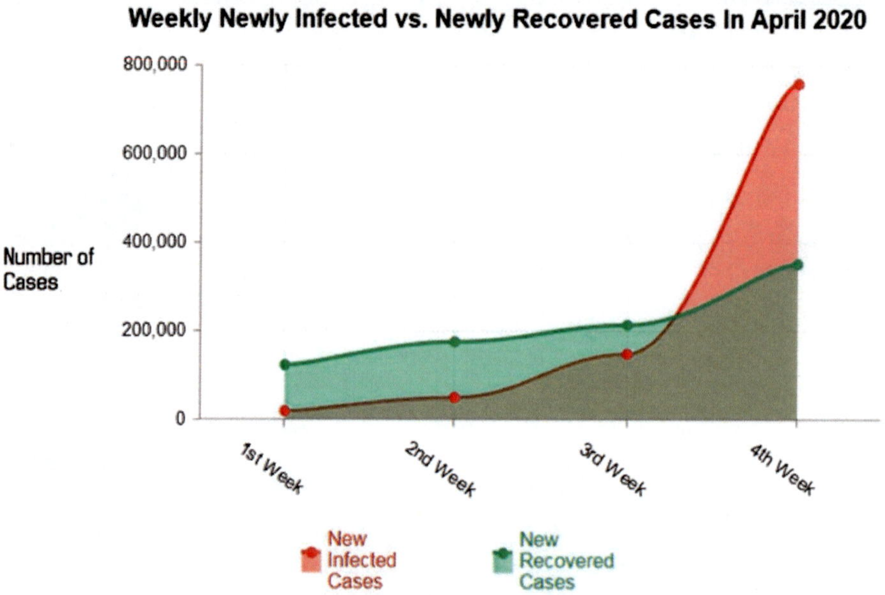

Fig. 2 Monitoring the newly recovered and infected cases over four weeks of April 2020

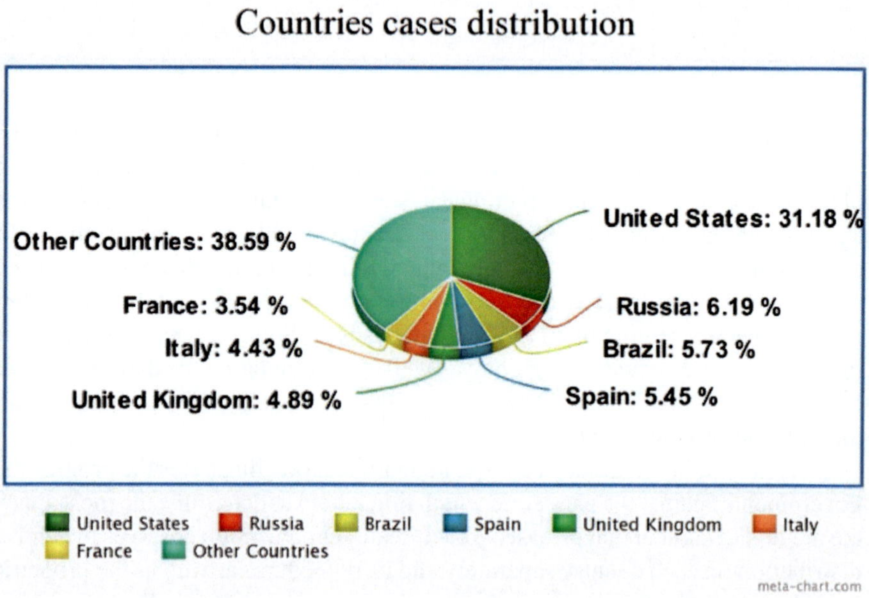

Fig. 3 Infected cases distribution since the pandemic outbreak has begun

3.1 Case Study: The Fourth Industrial Revolution Eradicates the Pandemic Outbreak in China

The fourth industrial revolution was applied starkly by China to contain the invading of the pandemic that led to the number shrinkage of cases dramatically in ten weeks along. A variety of technologies have contributed to support Chinese police to monitor the residents and track any suspected cases such as AI helmets, surveillance drones, QR drones, Robot Police machines, and electronic health code. These machines have been supplied by sensors and cameras provide Internet of Things technology. By using these smart robots and technology in addition to implementing lockdown as well as social distance, they became able to dominate the spreading of the virus in a short period. Figure 4 illustrates the number of infected cases in the lasted three months, starting from January to March, categorized by weeks. The virus invaded China in the first half of the period, but this domination was declined starkly in the last quarter of the period, in which China had a fully systematic and technological mechanism to prevent any potential situation that may cause infection spreading.

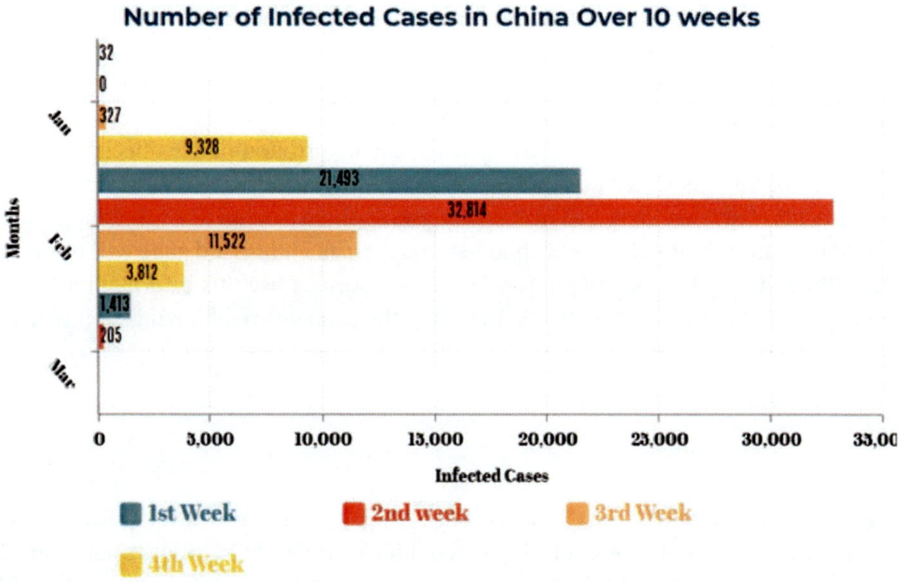

Fig. 4 Weekly new infected cases in China during the pandemic outbreak for ten weeks starting from the first week of January to the second week of March

3.1.1 Artificial Intelligence Helmets

Chinese inventors developed smart helmets to provide law enforcement officers real-time measurement of temperature for the population. These futuristic-looking helmets are not just used for protection, but infrared thermometers are attached to these helmets which have a capacity of thermal measurement five-meters-long. The smart helmets are equipped with Wi-Fi, Bluetooth, and 5G connectivity which allow data transmission and receiving information about any pedestrian instantly. As a result, Chinese police can identify suspected cases immediately and preventing the spreading of the virus.

3.1.2 Surveillance Drones

Chinese officers are patrolling streets without leaving their office station. Instead, they have used to send out artificial drones [18] which are automotive surveillance systems to measure pedestrians' temperature and ensure the commitment of residents to the government commands such as wearing masks, and keep the assigned safety social distance. Loudspeakers are attached to these surveillance systems to allow officers to speak to pedestrians as well as installing warning messages about the precautionary measures.

Besides, they also considered a manner to minimize the contact between officers and drivers at checkpoints using flying drones. To ensure the safety of officers, QR electronic code is attached by fly drones that surround police checkpoints. The essence of this electronic code is to collect personal information of truckers without interaction between them and officers by asking them to scan this QR code which is held by a drone using their smartphones, and then an authorized page will be opened to fill by the trucker personal information. Authorities use this information to track any potential coronavirus patients, therefore the probability of having an interaction is minimized.

3.1.3 Robot Police

The industrial revolution in China has been extended further to develop Robot Police machines to present physically in the country's streets to broadcast public health warnings to people outside and remind them to register their information online. The entry points of cities have been loaded by these bots in which infrared sensors are installed to get the temperature of drivers, and passengers as well.

3.1.4 Electronic Health Codes

To confirm that Chinese people follow the restrictions proposed by the government, Chinese cities have employed a distinct electronic health code that exposes the status of the constraints for individuals. This electronic health code is installed on two android applications that are developed for that purpose. The resident has to be assigned to one of three colored codes: Red, Yellow, or Green, these colored codes determine the situation of the resident. Those with green code have permission to go outdoors freely, whereas, the individuals who have a yellow code should stay home seven days, and red code holders have a 14-days quarantine period. This system renders a systematic traceable environment where the virus could be battled.

4 Problems and Challenges

1. Even though advanced technologies have a remarkable part in developing a pandemic vaccine [19], there is no reliable measurable mechanism to evaluate the effectiveness of the vaccine to resist the virus, which consequentially lengthens the period of admitting it. This enigma needs more research cooperation to maintain this gap and demonstrate the extension of those advanced technologies to expose and obtain resolutions.
2. The contributing of some of the fourth industrial revolution technologies such as Artificial Intelligence in genetic science is rarely used by researchers which consequently blocking many features could be used by geneticists to identify the genetic mutation of viruses in lesser time. Genetic mutation identification is a long-term process which helps to introduce virus symptoms as well as potential biological interactions that cause the infection, besides improving and accelerating the recognition of virus spreading behaviors.
3. Security is the most crucial concern in the case of applying the Internet of Things technology to dominate a specific region due to those numerous connections between sensors and hardware devices through online services [20]. Although, the benefits of applying this technology are abundant, the existence of security deficiencies and vulnerable systems may let to serious harms with critical losses. Accordingly, security assurance should be arranged adequately to keep the entire health system intact and wrap the data flow between systems.
4. Inconstant and variation of the incoming data for the pandemic virus make the prediction process tough to be investigated by the existing technology [21]. Even the powerful analysis mechanisms of those presented technologies such as AI technology could be fooled by the data variation, which means that the given data has no consistent shape or specific pattern to be identified by these technologies. This obstacle arises in the gathered data of the pandemic due to the disparity of dealing with the virus in each nation, some countries have an advanced and variety of the healthcare equipment others have don't apply the World Health Organization instructions as informed, and many similar reasons may cause lacking on the data standard. There is no concrete resolution for this problem, however, It is recommended to have a surveillance organization or intelligent systems with responsibilities of tracking the procedures of each country during the pandemic.

5. The immense data flow of the pandemic data over the globe needs to be allocated by a data center storage with high capacity with adequate computing power. This massive data stream and its complexity require a sophisticated and optimized computing version of the Cloud Computing technology to maintain the analysis process as well as the potential of involving a wide range of the existed data to the analysis process as much as possible. As a result, the results will be more robust and reliable rather than considering a small range of data due to the low computing performance. Concerning of allocating data throughout these procedures is vital using Big Data technology with a high storage scale with optimized storage mechanisms and algorithms, and eventually ensuring the stability of this process.
6. Although implementing the fourth industrial revolution technologies to overcome the pandemic outbreak has reasonable features, many people consider the collected data from these technologies invade their privacy. Extracting valuable data like individuals' location and personal details are breaking for their daily life privacy, and raise the feeling of surveillance their movements. Even though these sorts of information are required to install a systematic tracking environment for the pandemic, but It still influences the privacy, which makes it hard to find a concrete solution.

5 Conclusion

In general, the involvement of the fourth industrial revolution to combat the novel pandemic COVID-19 achieves remarkable outcomes of confinement the infection including utilizing the process of developing a vaccine, moreover, the reflection of using Big data, AI, and Cloud computing technologies aid the analysis procedures plus managing the global extensive pandemic data thoroughly. From conserving the economy by keeping businesses process to be done remotely with engaging in the projects cooperatively passing through online video stream platforms to support the remote teaching to saving the social relationships, daily habits, and regular services under the lockdown. Furthermore, the efficiency of this method has been approved by China where the COVID-19 pandemic is eradicated by a variety of advanced technologies such as surveillance drones, AI helmets, electronic healthcare code, and smart police robots, they have contributed to dominate the spreading of the pandemic in China. There is no doubt that the fourth industrial revolution should be considered as a short-term solution to control the COVID-19 outbreak, and accelerating its implementation globally will boost the prevention of spreading the virus.

References

1. Singhal, T.: A review of coronavirus disease-2019 (COVID-19). Indian J. Pediatr. (2020)
2. Wilder-Smith, A., Rocklöv, J., Liu, Y., Gayle, A.A.: The reproductive number of COVID-19 is higher compared to SARS coronavirus. J. Travel Med. **27** (2020)

3. Vaishya, R., Bahl, S., Suman, R., Vaish, A., Javaid, M., Haleem, A.: Industry 4.0 technologies and their applications in fighting COVID-19 pandemic. Diabetes Metab. Syndr.: Clin. Res. Rev. (2020)
4. McKay, R.A.: "Patient zero": the absence of a patient's view of the early North American AIDS epidemic. Bull. Hist. Med. (2014)
5. Vaishy, R., Bahl, S., Suman, R., Vaish, A., Javaid, M., Haleem, A.: Industry 4.0 technologies and their applications in fighting COVID-19 pandemic. Diabetes Metab. Syndr.: Clin. Res. Rev. (2020)
6. Javaid, M., Vaishya, R., Haleem, A., Vaish, A.: Emerging technologies to combat the COVID-19 pandemic. J. Clin. Exp. Hepatol. (2020)
7. Chen, Q., Xu, K., Chen, Y., Cheng, L., Liu, F., Wu, Z., Wu, H., Jin, C., Zheng, M., Wu, N., Jiang, C., Li, L., Yao, H., Lu, X.: Patient-Derived Mutations Impact Pathogenicity of SARS-CoV-2 (2020)
8. Bucher, K., Livingston, E.: Coronavirus disease 2019 (COVID-19) in Italy. JAMA Infogr. (2020)
9. https://www.researchgate.net/post/different_types_of_COVID-19_related_virus_genome_sequence_from_different_regions
10. Korber, B., Fischer, W.M.: Spike mutation pipeline reveals the emergence of a more transmissible form of SARS-CoV-2. bioRxiv (2020)
11. Haers, P.E., Swennen, G.R.J., Pottel, L.: Custom-made 3D-printed face masks in case of pandemic crisis situations with a lack of commercially available FFP2/3 masks. Int. J. Oral Maxillof. Surg. (2020)
12. Reynolds, K.J., Ruthenbeck, G.: Virtual reality for medical training: the state of the art. J. Simul. **9** (2020)
13. Alexandra, O., David, D.D., Matu, S.: Robot-based psychotherapy: concepts development, state of the art, and new directions. Int. J. Cogn. Ther. (2020)
14. Assi, E.B., Zhao, B., Sawan, M., Rong, G., Mendez, A.: Artificial intelligence in healthcare: review and prediction case studies. Engineering (2019)
15. Arun, T.G., Ozili, P.K.: Spillover of COVID-19: impact on the Global Economy. SSRN Electron. J. (2020)
16. Adhikari, R., Singh, R.: Age-structured impact of social distancing on the COVID-19 epidemic in India. Math. Epidemiol. (2020)
17. https://www.worldometers.info/coronavirus/worldwide-graphs/
18. Ruiz Estrada, M.A.: The uses of drones in case of massive epidemics contagious diseases relief humanitarian aid: Wuhan-COVID-19 crisis. SSRN Electron. J. (2020)
19. Saville, M., Hatchett, R., Lurie, N., Halton, J.: Developing Covid-19 vaccines at pandemic speed. New Engl. J. Med. (2020)
20. Hadjichristofi, G.C., Andrea, I., Chrysostomou, C.: Internet of Things: security vulnerabilities and challenges. In: 2015 IEEE Symposium on Computers and Communication (ISCC) (2017)
21. Huynh-The, T., Hwang, W.-J., Pathirana, P.N., Pham, Q.-V., Nguyen, D.C.: Artificial intelligence (AI) and big data for coronavirus (COVID-19) pandemic: a survey on the state-of-the-arts. IEEE Trans. Artif. Intell. (2020)

Social and Smart Networking and Cyber Security for Tracking COVID-19

The Future Scope of Internet of Things for Monitoring and Prediction of COVID-19 Patients

Ismail Elansary, Ashraf Darwish, and Aboul Ella Hassanien

Abstract The new outbreak of pneumonia triggered by a novel coronavirus (COVID-19) poses a major threat and has been declared a global public health emergency. This outbreak had first been discovered in December 2019 in Wuhan, China and until now has spread to the world. Emerging technology such as the Internet of Things (IoT) and sensor networks (SN) have been utilized widely in our everyday lives in a diversity of ways. IoT has also been an instrumental role in fighting against the COVID-19 pandemic currently out breaking across the globe, where it plays a significant role in tracking COVID-19 patients and infected people in hospitals and hotspots. This paper exhibited a survey of IoT technologies used in the fight against the deadly COVID-19 outbreak in different applications and discusses the key roles of IoT science in this unparalleled war. Research directions on discovering IoT's potentials, improving its capabilities and power in the battle, and IoT's issues and problems in healthcare systems are explored in detail. This study is intended to provide an overview of the current status of IoT applications to IoT researchers and the broader community and to inspire researchers to leverage IoT potentials in the battle against COVID-19.

Keywords COVID-19 · Coronavirus · Internet of things · IoT · Sensor networks · Healthcare monitoring

---

I. Elansary (✉)
Faculty of Science, Port Said University, Port Said, Egypt
e-mail: ielansary@sci.edu.eg
URL: http://www.egyptscience.net

A. Darwish
Faculty of Science, Helwan University, Cairo, Egypt

A. E. Hassanien
Faculty of Computers and Artificial Intelligence, Cairo University, Giza, Egypt

I. Elansary · A. Darwish · A. E. Hassanien
Scientific Research Group in Egypt (SRGE), Cairo, Egypt

A. E. Hassanien and A. Darwish (eds.), *Digital Transformation and Emerging Technologies for Fighting COVID-19 Pandemic: Innovative Approaches*, Studies in Systems, Decision and Control 322, https://doi.org/10.1007/978-3-030-63307-3_15

1 Introduction

The 2019 Corona Virus Disease (COVID-19) emerged from China, as the first case was registered in December 2019 in Wuhan, China. It eventually began to spread throughout China in various regions, and by the end of January 2020, the World Health Organization (WHO) announced the COVID-19 epidemic as a Public Health Emergency of International Concern (PHEIC) which was later a global pandemic [1]. By the beginning of May 2020, the WHO progress report demonstrates more than 4 million confirmed cases and more than 270,000 deaths worldwide [2]. The infection spreads every day and the health care system is struggling to take care of each infected person, particularly in extremely infected nations like the USA, Spain, Italy, UK, and the list goes on. Although the features of COVID-19 are not well defined, we can infer its behavior from statistics such as fast spread rate, the risk is higher for poor immune, diabetic and elderly persons, greatly varying recovery time trends and other knowledge we have is that it spreads to humans by droplets or direct touch. COVID-19 has dramatically changed the globe, not just in the healthcare sector but also in finance, schooling, travel, politics, etc.

Healthcare systems have put their tremendous efforts into treating the infected individuals and also testing the public for COVID-19 diagnosis to confrontation the situation. Countries are doing their part to extenuate the epidemic and even to meet the health care system requirements. Moreover, until now, there is unfortunately neither a cure medicine nor a prevention vaccine for this novel coronavirus. Most countries have attempted various forms of treatment and the use of specific effective drugs (without adequate confirmation). The protective measures determined by the WHO are to wash the hands frequently with soap and water or hand rubbing using alcohol following social distancing and exercising breathing hygiene. Masks will likewise play a part in protecting persons against infection, and learning when and how to use the mask is important. Besides this, the only way to avoid the spread of the infection is by remaining at home and avoiding social interactions, adopting social distancing, and isolating us from those who are infected or persons submitted to quarantine. It also takes a concerted effort from all social institutions, such as the healthcare system, government, and, most specifically, the public, to avoid the outbreak. And they can only be controlled to a restricted degree. Also, the exhaustive use of potential technologies along with effective health care medication and powerful management will be staggering to strengthen the line of defense to combat COVID-19.

IoT is an evolving technology that consists of a collection of connected items that connect anything, anybody, everywhere, any network, any device, and any time. IoT technologies can impact the overall spectrum of business as every system and entity can be recognized with massive advantages in the digital internet network. Such advantages usually consist of advanced connectivity of services, devices, and systems that go beyond machine to machine status. The IoT offers appropriate solutions for a range of applications and services, including waste management, smart cities, traffic problems, defense, smart infrastructure, transportation, emergency relief, healthcare, retail, and business management. Health care and medical constitutes one of the most interesting domains for IoT applications. IoT technology can develop therapeutic services like fitness activities, elderly treatment, health tracking remotely, and chronic disease management.

Compliance of prescription and diagnosis at home is another undoubtedly critical requirement. As a result, numerous medical and diagnostic sensors and instruments can be monitored by objects or mobile apps. Moreover, the IoT has the ability to improved consumer awareness and enhancement in the quality of human life at the least cost. Throughout the past few years, a large number of researchers have been investigating the potential of the IoT in health-care terms by addressing various real-world issues. Thus, the field is served by numerous applications and services.

The remaining of this chapter is organized as follows. Section 2 presents the related work which has been found in the literature review. Section 3 presents the IoT basics, history, and architecture and highlight the role of IoT in fighting COVID-19. Problems, challenges, and future perspectives are presented in Sect. 4. Section 5 concludes the outcomes of this chapter.

2 Literature Overview

Early identification of the signs of coronavirus would be one of the best ways to avoid coronavirus dissemination. Given that people's elevated body temperature is one of the most common symptoms, a real-time system for screening process supervision is required that automatically shows the thermal picture of people's temperature. Thus the screening phase would be less time-consumption and minimize the human interactions that may enable the COVID-19 to spread more quickly. In [2] real-time early detection and monitoring system was developed for COVID-19 utilizing an intelligent helmet equipped with a thermal imaging framework. The smart helmet can also detect high body temperature in the crowds and send the measured data to be displayed on a phone application. In [3] a smart system was designed that can detect and monitor the COVID-19 automatically from the thermal image utilizing smart glasses based on the IoT technology. Also, the suggested architecture has the potential to conduct face detection in crowds for alleged COVID-19 cases that having high body temperature. The architecture will provide information about the location visited by the suspected virus carriers via Google Location History (GLH) to provide accurate data about the process of detection. In [4] real-time early detection and tracking system was developed utilizing drone equipped with a thermal camera. The suggested application will measure temperatures at the ground surface from a height above ground. The drone could even transfer the calculated data to a server that will be viewed on a smartphone.

In [4] a system was designed for COVID-19 disclosure utilizing data from the embedded sensors in smartphones like microphones, cameras, temperature, and inertial sensors. Machine learning techniques are used for studying the symptoms of the illness and gaining information based on the data gathered. Compared with medical kits or CT scan procedures, this provides a low-cost and fast solution to coronavirus identification. This is arguably reasonable because the data gathered from the sensors of the smartphones have been used successfully in several individual systems, and the suggested solution incorporates these applications in a novel framework. For example, data gathered from the temperature-fingerprint sensor may be utilized for the estimation of the degree of fever [5].

In [6] wearable quarantine band which is based on IoT (IoT-Q-Band) was designed to find out the escaping people from quarantine. When planning it, they have taken into account the expense, damage to the global supply chain, and COVID-19 quarantine period recommended by the WHO. This wearable device monitors and records the quarantine people in real-time, with the bundled smartphone app. IoT-Q-Band is an economical solution that could benefit low-income regions to prevent the spread of COVID-19 due to it is a low-cost wearable band based on the IoT to find out and Track COVID-19 quarantine subjects absconding.

Photos and videos captured by a camera on smartphones or data obtained by inertial sensors onboard could be utilized to detect human fatigue [7, 8]. Also, authors in. [9] utilize videos from the smartphone's camera to predict nausea. Authors in. [10] used camera pictures and measurements of inertial sensors were utilized for tracking the position of the neck and predicting the level of human headache. Conversely, in [11, 12], audio data collected from the microphone of the smartphone is used to detect cough type. In another work, authors [13, 14] proposed the utilizing of standardization protocols of data exchange and AI to improve public awareness and community health management during the COVID-19 pandemic. For instance, additional benefits will be gained if AI is combined with thermal cameras which could have been deployed in several intelligent cities to detect the outbreak early. AI approaches will also show their tremendous efficacy in helping administrators to make smarter virus management decisions as tons of public health data are collected by data sharing between and across intelligent cities.

3 Basics and Background

3.1 Sensor Networks in Healthcare Applications

Wireless Sensor Networks (WSN) are a series of power-conscious, spatially dispersed wireless sensors that form an autonomous network independent of pre-existing infrastructures. A co-operative system is formed to record and monitor conditions at various locations. To this end, this system uses dedicated transducers with communication infrastructure. WSN's applications include data-intensive task performance such as seismic monitoring, terrain monitoring, habitat monitoring, and so on, as well as information gathering. Proactive computing largely relies on sensor networks (SN). In this technology, Systems could predict human needs such as healthcare and perform on their behalf if necessary. The integration of proactive computing and SN technology has the potential of life-changing and in providing a healthy lifestyle for improving the quality of life [15]. WSN technology has the potential of life-changing and could impact our lifestyle, industry, and work with a wide variety of applications in healthcare, transport, emergency, and disaster administration. WSN has various applications include agriculture (soil Irrigate and administration), ecological management, geophysical tracking (seismic activities), business processes (supply chain management), military networks, transport (traffic control and vehicle tracking), ecosystem surveillance (animal herd tracking). This involves the doctor's diagnosis and Prescription from inside or outside the hospital and performing video conferences with specialist surgeons/doctors from all around the world [15].

Body Sensor Network (BSN) is a WSN that could detect health parameters. Wireless Body Area Networks (WBAN), a special purpose WSN, combines wireless equipment and various networks to enable remote monitoring of patients in health-related situations. Moreover, the key aim of WBAN is to provide biofeedback information that continuously tracks health parameters such as body/intra-body temperature, pulse rate, arterial blood pressure, artificial pacemakers, and so on. Embedded SN contributed to the introduction of a leading method of public health surveillance. These networks can be programmed to enable accurate medication delivery, thus allowing accurate monitoring of vital information like Pulse Rate, Oxygen Level, periodic Blood Pressure (BP), etc. Figure 1 illustrates the architecture of the WBAN. As the figure indicates, WBANs have increasingly been used for health services and the monitoring of most areas of the human body. Moreover, viewed even from Fig. 1,

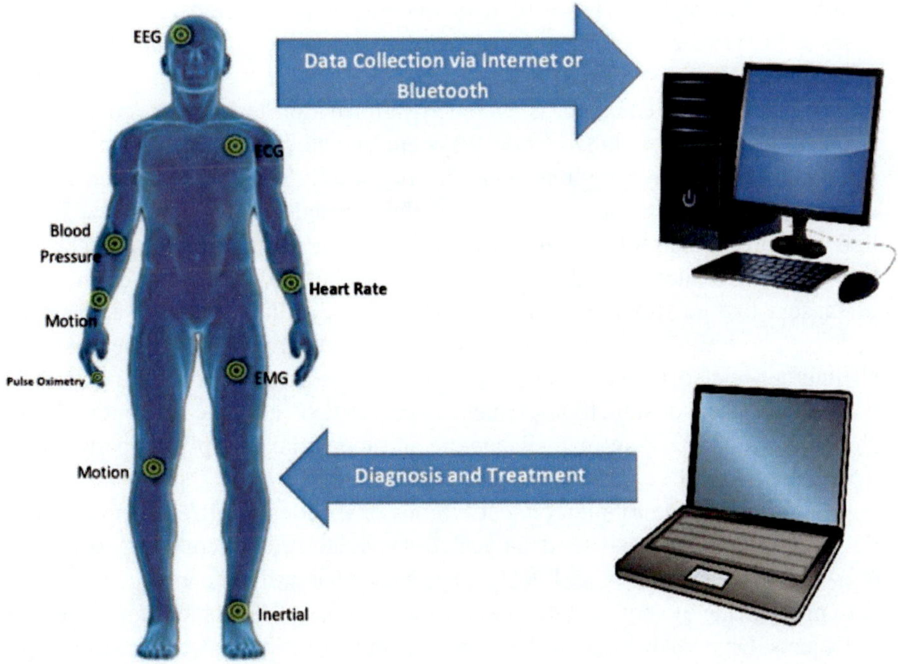

Fig. 1 Wireless body area network

the sensor nodes are available for measuring Blood Pressure (BP), Pulse Oximetry sensor to monitor the amount of oxygen in the human body continuously, Electroencephalography (EEG) sensor is an electrophysiological control tool for measuring brain electrical activity, Heart Rate (HR) sensor monitor the rate of the heart in Beats per Minute, Motion sensors sense movement alerting emergency personnel to the patient's movement which helps to prevent the patient from downfall, and the Electromyography (EMG) sensor measure the electric activity of muscle which may be used in prosthetic systems as a power signal.

3.2 Internet of Things Overview

Kevin Ashton in [16] is the first one to talk about IoT as an expression. Along with his counterpart Mr. Brock, Ashton launched the Massachusetts Automatic Identification Center with the aid of that system became monitoring moving objects were possible [17]. In 2002, the National Science Foundation of America (NSF) released a study on associated sciences focusing on the application of nanotechnology in information and communication technologies. Moreover, the 2005 International Telecommunications Union (ITU) report proposed the incorporation of the IoT with object recognition technology, cellular networks, sensors, embedded devices, and nanotechnologies. Over time, the move towards IoT technologies has grown tremendously. A national center for studying the IoT was founded in China in 2009, after which more than 90 cities in China based its strategic decisions and policies on improving the utilization of the IoT [17].

IoT has recently emerged as a modern information technology concept aimed at creating a complex global network system by gathering a range of physical and virtual 'objects' to—smartphone and sensors. Initially, IoT was suggested to refer to uniquely recognizable objects (things) and their virtual manifestations in an Internet-like structure, using radio-frequency identification (RFID) technologies. The definition of IoT was later expanded to include other kinds of 'things' including a range of sensors like positioning system (GPS) devices, actuators, global and handheld devices. The smooth incorporation and successful deployment of these sensors into an Internet-related network have posed many research issues, ranging from data collection, system design, to implementations [18]. Recently, IoT technologies have gained ground rapidly in many academic and industrial disciplines, particularly in healthcare [19, 20].

IoT has to be able to interconnect billions or trillions of different objects across the Internet, and a crucial need for scalable layered infrastructure is expected. The ever-increasing number of architectures proposed has still not converged into a standard model. One of some common IoT architectures is the 5-layer model consisting of Objects, Object Abstraction, Service Management, Application, Business layers. Objects Layer is the first layer that represents physical IoT sensors intended for collecting and processing data. Moreover, it involves actuators and sensors for conducting various functionalities like location queries, weight, motion, temperature, etc. Object Abstraction layer transfers information obtained from the Objects layer to the Service Management layer via secure channels. Information can be transmitted via different technologies like WiFi, RFID, Bluetooth, etc. Service Management Layer enables programmers of the IoT application to interact with heterogeneous objects without concern to a specified hardware platform. Also, this layer processes data obtained makes decisions, and provides the services by network wire protocols. Application Layer includes the services which the customers need. For the IoT, the value of this layer is that it has the potential to deliver high-quality smart services to satisfy the needs of consumers. Business Layer controls the overall activities and services of the IoT system. This layer's functions are to construct a business model, diagrams, flowcharts, etc. based on the data obtained from the Application layer [21].

IoT is implemented with several sensors, actuators, data analytics, and visualization software that allow IoT to be accessible anytime, anywhere, and on any entity. Furthermore, IoT has numerous applications such as healthcare, smart transportation, smart home environment, and smart environments like smart shopping, smart towns, and manufacturing enabled by IoT. We will briefly mention some of these applications.

Healthcare is one of the most required applications in the whole IoT model and a significant part of IoT's market share belongs to this sector. There are numerous start-ups and organizations set up to provide hospitals and homes with products and services for patient surveillance.

Various sensors are the basic components of industrial IoT to provide both production efficiency and protection. However, task automation and Robotization will result in job reductions. Besides that, new job opportunities also are generated in smart factory architecture, modeling, and analytics as well as everything relevant to those fields.

The deployment of smart cities became a new trend in constructions that achieve the expected results. However, these smart cities are expensive to build, and that is why not every country, whether developed or developing, can support them. In many places shared bicycles and shared power banks are deployed. IoT technologies also apply to public transport. Electronic tickets (or E-tickets) will save a large amount of time for passengers.

Environmental monitoring can be performed efficiently because of the wide variety of sensors available. Small smart sensors could track the weather forecast and other related data that could be used earlier to predict natural disasters.

3.3 *Internet of Things for Healthcare Applications*

Although healthcare accessible innovations introduced in IoT are still in its early stages, the future use in industry is increasingly emerging and evolving. Many IoT-enabled healthcare systems based research projects and industrial cases were developed and deployed. IoT-based health management solutions are applied in several fields, including caring for elderly and pediatric patients, monitoring of prolonged chronic disease, and private exercise and safety surveillance, among others.

(i) **IoT-based health care mobile system** combines communication technologies, mobile, and medical sensors for healthcare facilities. It can also be used by medical professionals or caregivers via P2P video and audio, which allows patient/caregiver contact anywhere at every time. This system helps patients to receive the diagnosis and medication more easily without a visit to the hospital or pharmacy, like refills for prescription drugs.
(ii) **IoT-based Ambient Assisted Living (AAA)** makes it easier for people with geriatric or disabilities to live longer, more comfortably. Higher age carries with it a higher chance of chronic illnesses like type II diabetes, stroke, cancer, etc. Integrating IoT technologies with AAL warrant real-time, location-specific tracking of patient living parameters (i.e., blood pressure) as well as environmental conditions.
(iii) **IoT-based Medication** can also be useful in stopping deadly adverse drug reactions (ADR) and in assessing prescription schedule compliance. Integrating wearable audio sensors, the technology of smart pill bottles, and classification ability will reliably warrant compliance with the prescription.
(iv) **Assist persons with Disabilities using IoT** through automated, reliable, resistive technologies IoT could significantly improve the quality of life for this population. For instance, smart gloves have been developed with cost-effective inertia sensors to allow those living with the loss of hearing to contact with those who are not well-spoken in sign language (SL). Moreover, IoT solutions will help people with mobility issues to use smart wheelchairs in smart communities, remotely gather information on specific special needs, or allow classrooms to make special needs schooling more available and effective for children with disabilities.
(v) **IoT Smart Medical Implants** are safe implantable therapeutic devices intended to enhance or repair functions of the human body. Examples of some of these devices include deep brain stimulation devices that utilize the pulses of regulated electrical to activate brain regions to reduce the repetitive changes induced by neurological conditions such as cochlear implants that utilize electrodes within the inner ear to improve hearing and pacemakers which is used to stabilize heart rate.

3.4 The Use of Internet of Things Against Fighting COVID-19

Previously discussed, the idea of the IoT makes use of the network for efficient data transfer and sharing. IoT uses a wide variety of connected devices to build a smart network for the management of the health system. It also allows patients, civilians, social workers, etc. to be in contact with the benefactors of the service to discuss any matter and cooperation. It warns and monitors all sorts of illnesses to enhance

Fig. 2 The IoT main applications for COVID-19 pandemic

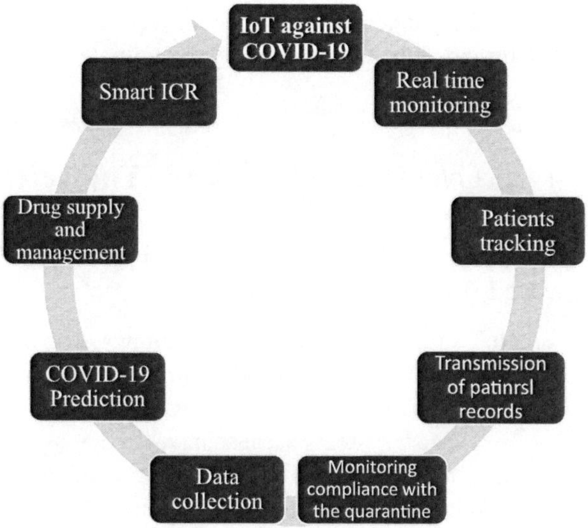

patient health. It gathers the patient's details and information remotely, without human contact. This data is also valuable for suitable decision-making processes. Therefore, by using IoT technologies in the COVID-19 pandemic, it is possible to completely guarantee the successful monitoring of patients as well as the suspected cases. Moreover, It can be used in COVID-19 diagnosis, treatment, provide health care for infected people, and pandemic management. Figure 2 represents the IoT major applications for the COVID-19 pandemic.

IoT is utilized for different applications to satisfy the significant requirements to mitigate the COVID-19 pandemic results. It has the power to forecast the upcoming status with the aid of correct data collected. IoT solutions are used to manage this pandemic properly. IoT services can be used to better monitor glucometer, blood pressure, heart rate, and other vital parameters especially monitor the decrease Oxygen level which is one of the common COVID-19 complications. Moreover, about 10–15% of COVID-19 patients need to put them on the Intensive Care Unit (ICU). IoT-based ICU equipped for immediate control of multiple health-related problems and alert generation procedures, including information distribution for specific vital circumstances. IoT helps track older people's health conditions; especially they are the most affected by COVID-19. The important applications of these technologies against COVID-19 are to track the real-time location of infected and suspected people for a seamless process of care without delay. Moreover, IoT improves the workflow of the treatment with effective results and also assists in the decision-making process during complicated cases. Another Significant IoT application against COVID-19 is a real-time screening of the suspected people. Moreover, During the COVID-19 epidemic, it has more relied on IoT-based platforms for drug supply and pharmaceutical management practices, covering four aspects: drug supply management, drug use management off-label, pharmaceutical care, and individuals. Moreover, during the pandemic, many people were forced to work through the home, so the IoT provided the tools necessary to facilitate the monitoring of COVID-19 patients.

4 Problems, Challenges, and Future Trends

This section presents the most related problems, challenges, and future trends for using IoT against COVID-19.

1. IoT efficiently links patients, clinics, doctors, and hospitals to organize and orchestrate diagnosis and treatment across a wide variety of locations. During the pandemic, data protection and confidentiality are the key concerns with healthcare systems because it passes across the unsecured system. The patient's care record which is made up of personal records, clinical conditions, diagnosis results, and associated medications, which considered sensitive data. Hackers may change the health-related details resulting in misdiagnosis, or incorrect disease assessment leading to unsuitable treatment, thus increasing the rate of mortality. Medical data transmission tracked through IoT devices is vulnerable to security concerns [22].
2. The privacy and security issues are found at the patient's records and hardware-level in terms of compatibility due to the needs of each component like storage space, operating system, and network architecture.
3. The integration problem at the devices level exists because of the design of the applications used because some of the devices support all programming languages and others support only particular languages. IoT tools and biosensors used in health-care systems such as with COVID-19 continuously track vital parameters. Thus, the uninterrupted wearing of sensors can cause skin rashes and also consume more power.
4. The researchers of cost analysis need to consider developing low-cost experiments of drug solutions allowed by IoT. During the epidemic, health-care facilities and clinics also contend with large numbers of patients, with numerous medical personnel undertaking various duties. Accurate identification of patients and personnel is key to ensuring adequate data protection in these cases. IoT on a smaller scale demands that handheld devices incorporate sensors for data collection and that centralized servers handle patient requests to ensure that all patients can access medical facilities using mobile devices such as smartphone.
5. A critical problem with using IoT in COVID-19 patient monitoring is related to the wide heterogeneity of systems, devices, services, and platforms operating available and possibly used for new or improved applications. Therefore, heterogeneity is also a non-negligible concern.

6. IoT is considered as one of the main sources to collect a huge amount of data. With the huge collected data of patient's records, researchers have to pay attention to storage, access, transportation, and processing of the huge amount of data they will produce.
7. Sensor networks are offering the ability to infer, measure, and collect the data of COVID-19 patients. The technological advances have made low-cost and low power miniaturized devices that can be used for use in large-scale, remote sensing applications. Smartphones are provided with a variety of sensors, GPS, accelerometer, digital compass, and cameras, enabling a wide range of mobile applications with COVID-19. In this context, the timely processing of huge and streaming sensor data, subject to energy and network constraints and uncertainties, has been identified as one of the main problems.
8. Research is still needed on the implementation of proper communication protocols such as IPv6 for individually addressing the things, the setting of various IoT standards for promoting interoperability and for scaling the cost of IoT facilities, and the assessment of risks and uncertainties. Therefore, there is a need for more new protocols in the future.
9. IoT applications involve several heterogeneous network technologies, where COVID-19 monitoring of patients requires continuity in the transmission of the collected data, and the overall consumption of bandwidth increases dramatically. However, the efficiency of the access management for enabling continuity and for optimizing the bandwidth usage is still an open problem.
10. The collected datasets from COVID-19 patients should be archived and stored in secure, easy to use to ensure that the integrity and security of the data and datasets can be kept at the optimum. Therefore, an intelligent system based on Artificial Intelligence techniques need to be investigated in the data warehouse to choose the right data and datasets for each service.
11. The lack of standards is defined as a big challenge towards applying IoT against COVID-19 by a large number of researchers. In this context, interoperability is still a challenge.
12. Software and systems architectures need to be defined, coded, and executed based on the standard such as map-reduce framework and any software frameworks that can reduce the processing time and complexity.
13. Visualization is needed to provide complex outputs so that the COVID-19 patients can easily use and learn. So that the analysis of patient's data can be computed as outputs or use outputs from statistical analysis and transform them into visualization.
14. IoT applications involve frequent COVID-19 patient's data transmission from them and other things to a cloud, such process quickly drains battery capacity on both the things and the gateway. Therefore, energy efficiency in both data processing and transmission is an important open problem.

5 Conclusion

COVID-19 outbreak has profoundly impacted people around the world and has fundamentally altered humanity's daily functions. IoT provides a comprehensive integrated healthcare network to combat a COVID-19 pandemic. The usage of IoT technology will reinforce the line of defense against the infection of COVID-19. Moreover, all electronic medical instruments are linked to the internet, and it automatically transmits a message to medical personnel during any critical situation. Infected cases could be treated remotely using devices with well-connected internet to monitor the COVID-19 patients and limit the outbreak of the disease. It treats all cases smartly to provide the patients with improved treatment in the end. Finally, this chapter presents the main problems, challenges and future trends that enable IoT fighting against COVID-19.

References

1. Elmousalami, H.H., Hassanien, A.E.: Day level forecasting for coronavirus disease (COVID-19) spread: analysis, modelling, and recommendations. arXiv preprint arXiv:2003.07778 (2020)
2. WHO: Coronavirus Disease 2019 (COVID-19) Situation Report—102 (2020). https://www.who.int/docs/default-source/coronaviruse/situation-reports/20200501-covid-19-sitrep.pdf?sfvrsn=742f4a18_4. Accessed 1 May 2020
3. Elavarasan, R.M., Pugazhendhi, R.: Restructured society and environment: A review on potential technological strategies to control the COVID-19 pandemic. Sci. Total Environ. **725**, 138858 (2020)
4. Abdulrazaq, M., Zuhriyah, H., Al-Zubaidi, S., Karim, S., Ramli, R., Yusuf, E.: Novel COVID-19 detection and diagnosis system using IoT based smart helmet. Int. J. Psychosoc. Rehabil. **24**, 2296–2303 (2020). https://doi.org/10.37200/ijpr/v24i7/pr270221
5. Abdulrazaq, M., Hazairin, N., Al-Zubaidi, S., Karim, S., Mustapha, S., Yusuf, E.: Toward a novel design for coronavirus detection and diagnosis system using IoT based drone technology. Int. J. Psychosoc. Rehabil. **24**, 2287–2295 (2020). https://doi.org/10.37200/ijpr/v24i7/pr270220
6. Maddah, E., Beigzadeh, B.: Use of a smartphone thermometer to monitor thermal conductivity changes in diabetic foot ulcers: a pilot study. J. Wound Care **29**(1), 61–66 (2020)
7. Maghdid, H.S., Ghafoor, K.Z., Sadiq, A.S., Curran, K., Rabie, K.: A novel AI-enabled framework to diagnose coronavirus COVID-19 using smartphone embedded sensors: design study. arXiv preprint arXiv:2003.07434 (2020)
8. Singh, V., Chandna, H., Kumar, A., Kumar, S., Upadhyay, N., Utkarsh, K.: IoT-Q-Band: a low cost internet of things based wearable band to detect and track absconding COVID-19 quarantine subjects. EAI Endors. Trans. Internet Things **20**, 163997 (2020). https://doi.org/10.4108/eai.13-7-2018.163997
9. Karvekar, S.B.: Smartphone-based human fatigue detection in an industrial environment using gait analysis. Available at: https://scholarworks.rit.edu/theses/10275/ (2019). Accessed 1 Feb 2020
10. Roldan Jimenez, C., Bennett, P., Ortiz Garcia, A., Cuesta Vargas, A.I.: Fatigue detection during sit-to-stand test based on surface electromyography and acceleration: a case study. Sensors **19**(19), 4202 (2019)

11. Story, A., Aldridge, R.W., Smith, C.M., Garber, E., Hall, J., Ferenando, G., Possas, L., Hemming, S., Wurie, F., Luchenski, S., Abubakar, I.: Smartphone-enabled video-observed versus directly observed treatment for tuberculosis: a multicentre, analyst-blinded, randomised, controlled superiority trial. Lancet **393**(10177), 1216–1224 (2019)
12. Lawanont, W., Inoue, M., Mongkolnam, P., Nukoolkit, C.: Neck posture monitoring system based on image detection and smartphone sensors using the prolonged usage classification concept. IEEJ Trans. Electr. Electron. Eng. **13**(10), 1501–1510 (2018)
13. Nemati, E., Rahman, M.M., Nathan, V., Vatanparvar, K., Kuang, J.: A comprehensive approach for cough type detection. In: 2019 IEEE/ACM International Conference on Connected Health: Applications, Systems and Engineering Technologies (CHASE), pp. 15–16. IEEE (2019)
14. Vhaduri, S., Van Kessel, T., Ko, B., Wood, D., Wang, S., Brunschwiler, T.: Nocturnal cough and snore detection in noisy environments using smartphone-microphones. In: 2019 IEEE International Conference on Healthcare Informatics (ICHI), pp. 1–7. IEEE (2019)
15. Rao, A.S.S., Vazquez, J.A.: Identification of COVID-19 can be quicker through artificial intelligence framework using a mobile phone-based survey in the populations when cities/towns are under quarantine. Infect. Control Hosp. Epidemiol. **41**, 826–830 (2020). https://doi.org/10.1017/ice.2020.61
16. Allam, Z., Jones, D.S.: On the coronavirus (COVID-19) outbreak and the smart city network: universal data sharing standards coupled with artificial intelligence (AI) to benefit urban health monitoring and management. In: Healthcare, vol. 8, no. 1, p. 46. MDPI (2020)
17. Tarannum, S., Farheen, S.: Wireless sensor networks for healthcare monitoring: a review. In: International Conference on Inventive Computation Technologies, pp. 669–676. Springer, Cham (2019)
18. Ashton, K.: That 'Internet of Things' thing. RFID J. **22**(7), 97–114 (2009)
19. Javdani, H., Kashanian, H.: Internet of things in medical applications with a service-oriented and security approach: a survey. Health Technol. **8**, 39–50 (2018). https://doi.org/10.1007/s12553-017-0180-8
20. Qi, J., Yang, P., Min, G., Amft, O., Dong, F., Xu, L.: Advanced internet of things for personalised healthcare systems: a survey. Pervasive Mob. Comput. **41**, 132–149 (2017). https://doi.org/10.1016/j.pmcj.2017.06.018
21. Al-Fuqaha, A., Guizani, M., Mohammadi, M., Aledhari, M., Ayyash, M.: Internet of Things: a survey on enabling technologies, protocols, and applications. IEEE Commun. Surv. Tutor. **17**(4), 2347–2376 (2015). https://doi.org/10.1109/comst.2015.2444095
22. Jagadeeswari, V., Subramaniyaswamy, V., Logesh, R., Vijayakumar, V.J.H.I.S.: A study on medical Internet of Things and Big Data in personalized healthcare system. Health Inf. Sci. Syst. **6**(1), 14 (2018)

The Truth About 5G and COVID-19: Basics, Analysis, and Opportunities

Haytham H. Elmousalami, Ashraf Darwish, and Aboul Ella Hassanien

Abstract 5G is a paradigm shift for data transfer and wireless communication technology where 5G involves massive bandwidths based on high carrier frequencies. Unlike 4G, 5G is highly integrative to produce a seamless user experience and universal high-rate coverage. The key role of 5G is increasing data capacity, improving data rate transfer, providing better service quality, and decreasing latency. Recently, COVID-19 is declared as an international epidemic. More than 4.5 million confirmed cases and +308,000 death cases have been recorded around more than 209 countries on 16th May 2020. There are several insane theories about 5G technology and human health. Therefore, people are burning valuable 5G infrastructure down out of fear for their health. People think that 5G towers are weakening the immune system and causing the global COVID-19 pandemic. This chapter reviews the data transmission revolution from 1G to 5G technology and discusses the impact of 5G technology on human health, pandemic, and business perspectives.

Keywords 5G · COVID-19 · Data transmission revolution · 5G coronavirus · Cell network generations · Emerging technologies

1 Introduction

In recent years, the growth of global mobile data is exponentially increasing during the digital transformation era. Artificial intelligence and machine learning (ML)

A. Darwish
Faculty of Science, Helwan University, Cairo, Egypt
e-mail: ashraf.darwish.eg@ieee.org

A. E. Hassanien (✉)
Faculty of Computers and Artificial Intelligence, Cairo University, Giza, Egypt
e-mail: aboitcairo@gmail.com
URL: http://www.egyptscience.net

H. H. Elmousalami · A. Darwish · A. E. Hassanien
Scientific Research Group in Egypt (SRGE), Giza, Egypt
e-mail: haythamelmousalami2014@gmail.com

A. E. Hassanien and A. Darwish (eds.), *Digital Transformation and Emerging Technologies for Fighting COVID-19 Pandemic: Innovative Approaches*, Studies in Systems, Decision and Control 322, https://doi.org/10.1007/978-3-030-63307-3_16

play critical roles in processing Big data and the internet of things (IoT) for different real-life applications [8]. It is expected growth by 23 times increase in 2021 in data volume against in 2005 [17]. The exponential growth will reach by 5 zettabytes (ZB) per month in 2030 as predicted by the International Telecommunication Union (ITU) [18]. The fifth-generation (5G) is the recent technology for mobile communications technology to speed the data transfer [3]. The objective of 5G is to meet the next 10 years' requirements technological applications such as machine-type communications (MTC) and Internet-of-Things (IoT). The core advantages of 5G technology are increasing data capacity, improving data rate transfer, providing better service quality, and decreasing latency.

Data transmission science and our ability to transfer data through the air has been greatly evolved over the past four decades. Different wavelengths of electromagnetic radiation have been used to transmit information for hundreds of years. In ancient Greece, they used signal fires to convey messages with visible light. Today, humans send messages with light through fiber optic cables [17]. These cables are capable of carrying insane quantities of information and these cables are the bedrock of the internet. However, all devices cannot be connected to the internet, because many devices need to be wireless [3].

The remainder of this chapter is organized as follows. Section 2 describes the basics, background of 5G against fighting COVID-19. Section 3 describes the role of 5G against fighting COVID-19. Section 4 concludes this chapter.

2 Basics and Background

2.1 Wireless Data Transmission Evolution

2.1.1 First Generation (1G)

To make wireless data transmission possible, cellular networks needed to be developed and cellular networks started in Japan in 1979, with the first-generation cellular network (1G). The G stands for generation and each subsequent generation refers to specific minimum speed connectivity and reliability necessary to classify that network as a particular generation. 1G began in Tokyo where high power radio towers would communicate directly with cell phones installed in cars and vehicles. The 1G towers used radio waves with frequencies sitting around on the electromagnetic spectrum and simply transferred the data in analog [1].

The analog data can be transmitted using a carrier frequency. For example, sending a simple sine wave of 100 Hz with significantly higher frequency. This can be applied with amplitude modulation (AM) or frequency modulation (FM). AM applies the data to the amplitude of the carrier frequency. Therefore, it will vary the amplitude of the carrier frequency, basically tracing the original wave with its peaks and troughs. FM applies the data to the frequency of the carrier wave varying the distance between peaks to trace the original wave. To transfer a call using this method, it is needed a dedicated frequency band, defined by the lowest and highest frequencies used [10].

2.1.2 Second Generation (2G and 2.5G)

Increasing the number of customers that can apply the same frequency band can be as simple as increasing the number of towers. Instead of using a single high-power tower to cover an entire city, multiple lower power towers could be used. Frequency bands could then be assigned to individual customers within each tower's range without interfering with the same band in neighboring cells [10]. This increased the number of users' networks could support; however, it cannot increase data transfer rates. 1G can is capable of about 2.4 kilobits per second. 2G ushered in a new era of mobile phones with the introduction of a fully digital system. Instead of encoding an analog signal into a frequency band, 2G encoded binary data. the digital data allowed for a new form of communication and data transfer. This was the era a new wireless communication language was born. In English, each character in that 160-character text was encoded with 7 bits. Therefore a 160-character text contained 1120 bits. When 2G was first introduced, it could achieve about 9.6 kbit/second [17]. It could handle that 1120-bit text message with ease. However, the 2G era lasted right up to the launch of the first iPhone, and speeds had increased to 200 kilobits/second. Thanks to improved internet protocols such as General Packet Radio Switching (GPRS), it referred to as 2.5G.

2.1.3 Third Generation (3G, H+ , and 3.5G)

3G made the change to a system that fully utilized the method of data packet switching, which General Packet Radio Switching (GPRS) utilized. Packet switching allowed thousands of customers to share many different frequency bands far more efficiently. The data was split into small data packets [1]. Each data packet contains a header, which contains the address of the destination and information on how to reassemble the data packets. Splitting the data up into smaller bite-sized chunks makes better use of the frequency bands. The data are divided into small chunks and sent it across many different frequency bands the moment a small availability appeared. This made the use of the frequency bands' capacity more efficient to carry more data. These protocols improved allowing more efficient use of the bandwidth. In 2005, High-Speed Packet Access of HSPA has been applied, which can be represented on the cell phone as an H+, which increased speeds up to 42 MBPS. This was labeled 3.5G.

2.1.4 Fourth Generation (4G)

4G introduced a new technology called Long Term Evolution (LTE), it introduced more frequency bands like the 700 MHz band that was previously used for analog TV broadcasts. Moreover, it introduced a new way of squeezing more data through the existing frequency bands with Orthogonal Frequency Division Multiplexing (OFDM). OFDM allowed us to spend far more data. The constructive and destructive interference where two waves meeting can combine and either enhance or cancel out the amplitude of each other. To prevent this signal traditionally had to be separated over time to prevent interference. However, OFDM allowed the signals to be squeezed together and overlapped, allowing the same amount of data to be sent over a shorter period [4].

2.1.5 Fifth Generation (5G)

Many base stations will help relieve congestion over a single frequency band. However, 5G will be using something called massive multiple input multiple outputs (massive MIMO). These are groups of antennas that are listening and broadcasting the same frequency bands. This would cause interference. However, 5G is looking to use beamforming which will allow the antenna to aim at the cell phones instead of broadcasting the signal in all directions. This coupled with the fact that the higher frequency waves can carry more data means 5G is reaching speeds of up 1800 Mbps in the US. The measure of frequency is hertz. One hertz (Hz) means is that one wave cycle is reaching per second, 190 Megahertz means 190 million wave cycles are arriving per second. Higher frequencies can carry more information because the information is encoded into the wave cycles [1].

Currently, the frequencies are between 700 MHz and about 2500 MHz. Therefore, that is means 700 million wave cycles per second to 2500 million wave cycles per second. However, 5G is looking to use frequencies as high as 90 Gigahertz. That is 90 billion wave cycles per second. 5G will allow higher download speeds and lower latency. This will be huge for time-critical technologies like self-driving cars that require rapid communication between vehicles in the network and allow more devices to join the network such as the internet of things (IoT). 5G has a lot of potential and applications related to this digital transformation era [3]. Accordingly, 5G has applied for Autonomous logistics, Edge computing, Autonomous machines (self-driving cars), Cloud gaming, Smart cities, virtual reality (VR), augmented reality (AR), drones' communications, smart homes and cities, artificial intelligence (AI) and IoTs as illustrated in Fig. 1.

Table 1 compares the data transfer, capabilities, and applications among different cellular network evolution ending up to 5G technology. The key capabilities of 1G is the analog voice and data transfer speed up to 2.4 kbps. Therefore, 1G has no impact on industrial applications. According to 2G, the key capabilities of 2G are digital voice, SMS, MMS, limited roaming, and encrypted communication with data transfer speed up to 40 kbps. Therefore, 2G applied for remote control and text

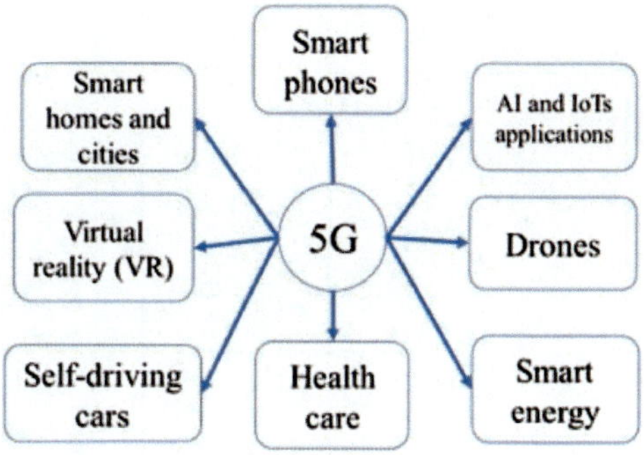

Fig. 1 5G applications

messages. In 1998, 3G technology has been evolved with mobile board band, locating services, multimedia streaming, and seamless global roaming using data transfer speed up to 21.6 Mbps. Accordingly, 3G has applied for video monitoring, remote condition monitoring, and emails. In 2008, 4G technology has been evolved with high-speed mobile internet, HD multimedia streaming, IP-based packet switching, and seamless global roaming using data transfer speed up to 1 Gbps. Accordingly, 4G has applied for mobile service technicians, service via smartphones, and wireless backhauls. Recently, late 2018, 5G technology has been developed private networks, Internet of Things (IoT), ultra-low latency, ultra-high reliability, millimeter-wave support, and massive machine-type communication. Accordingly, 5G has applied for autonomous logistics, autonomous machines (self-driving cars), cloud gaming, smart cities, virtual reality (VR), artificial intelligence applications.

2.2 Business and Cell Network Generations

Huge capital was invested with every new generation of the cellular network. However, huge investment was spent to cram more data onto a single frequency band over the years. When the customer is using that frequency band on the same tower, then it is needed to use a different frequency band. The more frequencies exist, the more calls the tower can handle based on the bandwidth. As the number of users grew, the system's capabilities were continually stretched. Adding more frequencies to grow bandwidth is an option. However, adding frequencies comes with its difficulties [3]. Frequencies need to be licensed and there is a lot of competition. Weather radar, military communication and security systems, GPS, television broadcasts, radio stations, radio astronomy, aviation systems, and air traffic control. They all need their frequency bands. To gain new frequency bands companies often had to go to auction and purchase the license with huge capital investment.

By the time the iPhone.2 launched and 3G was the new hot topic. 3G introduced additional frequency bands, it's estimated that in Europe alone companies paid over 100 billion dollars in auctions to gain new frequencies. Higher frequency waves have normally been shunned for these types of applications. High-frequency waves are

Table 1 Cellular networks evolution

Comparison criterion/cellular networks	1G	2G	3G	4G	5G
Data transfer	Up to 2.4 kbps	Up to 40 kbps	Up to 21.6 Mbps	Up to 1 Gbps	Up to 20 Gbps
Released	1979	1991	1998	2008	2018
Memory	–	–	<256 MB	<6 GB	>8 GB
Storage	No on-board storage	No on-board storage	<16 GB	<265 GB	>512 GB
Extensions	–	2.5G and 2.75G	3.5G	–	–
Capabilities	• Analog voice	• Digital voice • SMS and MMS • Limited roaming • Encrypted communication	• Mobile board band • Locating services • Multimedia streaming • Seamless global roaming	• High-speed mobile internet • HD multimedia streaming • IP-based packet switching • Seamless global roaming	• Private networks • Internet of Things (IoT) • Ultra-low latency • Ultra-High reliability • Millimeter-wave support • Massive machine-type communication
Applications	No impact on industrial applications	• Remote control • Text messages	• Video monitoring • Remote condition monitoring • Emails	• Mobile service technicians • Service via smartphones • Wireless backhauls	• Autonomous logistics • Edge computing • Autonomous machines (self-driving cars) • Cloud gaming • Smart cities • Virtual reality (VR) • IoT

not as good at traveling. They get blocked by practically everything including rain. To deal with this network, providers need to install huge numbers of transmitters. Studies have estimated that to bring 100 Mbps download speeds to 72% population coverage and 1 Gbps speeds to 55% of the US population, about 13 million utility pole-mounted 28 GHz base stations would be needed for 400 billion dollars.

3 The Role of 5G Against Fighting COVID-19

Every day new technologies emerge such as 5G and these technologies are initially confusing. People are burning valuable infrastructure down out of fear for their health. There are several insane theories such that 5G towers are weakening the immune system and causing the global COVID-19 pandemic. Moreover, people think that 5G is causing cancer. The electromagnetic spectrum starts on the far left with gamma radiation, which has very high frequency and short wavelengths. Higher frequencies and shorter wavelengths equate to higher energy, and indeed gamma radiation causes cancer [16]. The ionizing radiation is the main concern that's meaning it removes electrons and damages our DNA. However, 5G is operating over towards the lower energy frequencies such as visible light and microwaves. Visible light and microwaves are possible to cause heating with high enough powered beams of these wavelengths. This is the only non-crackpot theory where 5G sits in the realm of plausibility [4].

These frequencies are blocked by rain. Therefore, they certainly cannot penetrate the human skin and these transmitters simply do not have the power to cause heating that would be damaging. They are simply low power and there are plenty of studies that show that they are not harmful [1]. If humans are afraid of 5G in this way, humans might as well be afraid of the streetlights because they emit higher energy frequencies. It can seem like magic until we peel back the layers to their earliest iteration and see that they are the product of many years of problem-solving with each successive generation adding more complexity.

5G can revolutionize the healthcare sector based on adding better assistance to medical staff, patients' real-time monitoring, analysis of the recorded real-time data and produce corrective decisions and diagnosis. Moreover, the incorporating of 5G and IoT and AI improves the medical efficiency against COVID 19 epidemic. Figure 2 shows 5G applications for epidemic monitoring and health care against the COVID 19 epidemic where telemedicine, robots, and medical sensing can boost the efficiency of the medical service sector.

1. **5G Telemedicine**

Telemedicine is practicing of remotely curing and monitoring the patient's mobile applications, drones, and smart wearable's can be used for enhancing the telemedicine functionalities where 5G technology is the core of the communications in the telemedicine. On the other hand, 4G cannot support telemedicine, consultation teleconferencing, and real-time video conferencing because of its limited data transfer

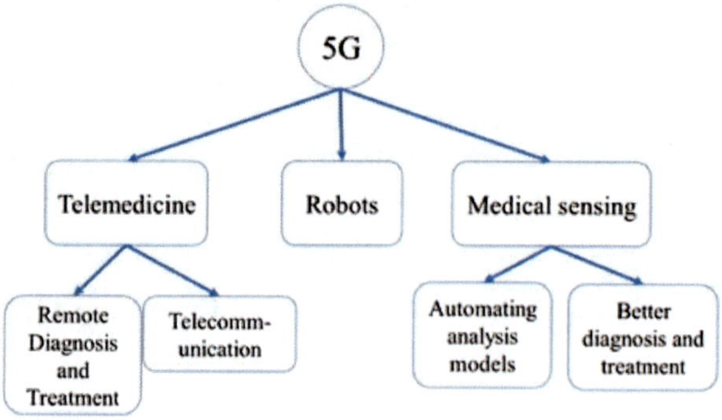

Fig. 2 5G applications for epidemic monitoring and health care

speed and limited bandwidth. Moreover, 5G technology can boost communication between cloud platforms and IoT devices due to its high-speed data transmission and ultra-low latency. For interactive applications, 5G supports augmented reality and virtual reality applications (AR/VR).

In November 2019, China launched 5G technology commercially and apply a 5G network for telemedicine against COVID 19 [6]. Several medical centers and hospitals have applied virtual platforms based on 5G telemedicine for COVID-19 patients. For Example, West China Hospital, Kunming Medical University, and an emergency facility in Wuhan have used 5G-based platforms, 5G remote consultation platforms, and teleconsultation for online free COVID-19 treatment and diagnosis [11, 21].

2. **5G Medical Imaging**

AI and big data analytics have a significant contribution to medical imaging processing with minimal human intervention. Leishenshan Hospital in Wuhan city has applied 5G-enabled platforms for medical imaging and COVID-19 real-time diagnosis [11]. Initially, thermal imaging techniques have applied for anti-aircraft defense and military applications. Accordingly, Thermal imaging technology has applied for various sectors such as healthcare sector. 5G-based thermal imaging can enable the real-time temperature of moving bodies. the data of moving body temperature have accumulated to the central monitoring devices based on 5G network and ultra-low latency. Moreover, drones, robots, and Unmanned Aerial Vehicles (UAVs) have been equipped with 5GC THERMAL IMAGING technology to monitor and record public temperature in public spaces in China during COVID 19 epidemic fighting [11].

3. **5G Robots**

5G-powered robots such as Drones, robots, and Unmanned Aerial Vehicles (UAVs) are promising attempts to fight the COVID-19 outbreak. The key objectives of the 5G-powered robots are to minimize humans interactions and automate the routine tasks in health care and logistic sectors. Advanced Info Services (AIS) In Thailand has applied COVID-19 at 20 hospitals based on 5G robots to enhance telemedicine protocols

[12]. In Shanghai, 5G-enabled intelligent robots deliver medicines to patients based on smart devices to combat the COVID-19 epidemic [15].

In Wuhan, Cloud Minds' 5G ROBOTS have been installed to disinfect and clean the premises, record the patient's temperature using 5G thermal imaging cameras. Cloud Minds' 5G ROBOTS used AI, 5G, and IoTs to track their patients COVID 19 vital signs such as heart rate, temperature, and blood oxygen percentages in their bodies on a real-time basis [5]. 5G police patrol robots have been designed in China based on cloud computing technologies, AI, IoT, 5G. The data have been collected using high-resolution cameras and infrared thermometers. Moreover, the robots can know if somebody is wearing a mask or not and sends an inform to the authorities to take law-actions.

4. 5G Coronavirus Conspiracy Theory

The COVID-19 has reported being a novel coronavirus of typical pneumonia since Dec 31, 2019. The COVID-19 started in Wuhan Chinese city. Simultaneously, Both the international world and other Chinese cities have received infected cases. On 11th March 2020, the world health organization (WHO) announced that the global COVID-19 outbreak is an international pandemic because of the speed and scale of transmission of the virus [9]). From 209 countries and territories, there are more than two million confirmed cases reported to WHO. Moreover, the number of cases announced outside China has almost doubled 13-folds in the past two weeks and the affected countries' number has almost tripled [13, 20]. More than four million confirmed cases and +290,000 death cases have been recorded around more than 209 countries on 10th May 2020 [7].

In the UK, there were attacks on mobile phone masts motivated by conspiracy theories linking coronavirus to 5G and it lowers human immunity. People think that the COVID-19 virus is created by 5G where it absorbs oxygen [19]. That's not dangerous and nonsense as well. The basics of a virus that it is a thing that has a physical existence that can be seen with an electron microscope and it cannot be transmitted through radio waves in general. The concern over 5G's risk of health comes from higher frequencies carrying greater radiation risks [16]. However, the electromagnetic spectrum is a simple way to separate safe radiation from unsafe radiation which is this higher wavelength and visible light ultraviolet or after x-rays. Gamma-rays are ionizing radiation and are dangerous directly for humans. The problem is that it damages DNA and causes cancers [14]. However, Gamma-rays do more direct damage to the more energetic and non-ionizing radiation. Another example is the microwave. The main effect is heating things that's why microwave oven works in part. However, it means that it is much easier to set limits if the energy is not enough to heat things to a dangerous level. Moreover, there is no relation between COVID 19 and 5G existence where the devastated countries such as Iran that does not have 5G. Therefore, the 5G coronavirus conspiracy theory is false [2]. Therefore, 5G is an important communication infrastructure in phone networks, emergency services, and our modern world with an acceptable impact on human health.

4 Concluding Remarks

Artificial intelligence plays a significant role in processing big data and the internet of things (IoT) for different real-life applications. This chapter reviews the data transmission revolution from 1G to 5G technology and discusses the impact of 5G technology on human health and pandemic. Therefore, 5G technology is a necessary infrastructure as a communication foundation for applying these applications. While 5G is being tried around the world and anticipated to be rolled out steadily in 2019, analysts around the world are starting to turn their consideration to what 6G may well be in +10 years, and there are activities in different countries focusing on the inquire about of conceivable 6G innovations. 6G will be the future with more speed and data transfer abilities. People are burning valuable infrastructure down out of fear for their health. There are several insane theories about 5G technology and human health. People think that 5G towers are weakening the immune system and causing the global pandemic. However, 5G waves and frequencies are with acceptable impact on human health. Moreover, this chapter shows that there is no relation between COVID 19 and 5G existence and the devastated countries such as Iran that does not have 5G. The intersection of telemedicine, robots, and UAVs need a strong communication channel that has ultra-low latency and high-speed data transfer to provide high-quality medical services against COVID 19. However, 5G faces several challenges as the followings:

- The lack of infrastructure 5G and the high costs of maintenance and installation.
- 5G needs another technology such as cloud computing technologies, AI, and IoT to revolutionize the healthcare services and other applications such as smart city management.
- No security guidelines or policies have been existed to use or install 5G technologies worldwide and use human data.

References

1. Elmousalami, H.H.: Comparison of artificial intelligence techniques for project conceptual cost prediction: a case study and comparative analysis. IEEE Trans. Eng. Manage. (2020a)
2. Tariq, F., Khandaker, M., Wong, K.K., Imran, M., Bennis, M., Debbah, M.: A Speculative Study on 6G. arXiv preprint arXiv:1902.06700 (2019)
3. Ting, D.S.W., Carin, L., Dzau, V., Wong, T.Y.: Digital technology and COVID-19. Nat. Med. **26**(4), 459–461 (2020)
4. Andrews, J.G., Buzzi, S., Choi, W., Hanly, S.V., Lozano, A., Soong, A.C., Zhang, J.C.: What will 5G be? IEEE J. Sel. Areas Commun. **32**(6), 1065–1082 (2014)
5. Agiwal, M., Roy, A., Saxena, N.: Next-generation 5G wireless networks: a comprehensive survey. IEEE Commun. Surv. Tutorials **18**(3), 1617–1655 (2016)
6. Deloitte, H.: Combating COVID-19 With 5G: Opportunities to Improve Public Health Systems [Online]. Available: https://www-_le.huawei.com/-/media/corporate/localsite/ua/pdf/deloitt%e-cn-consulting-5gassist-in-2019-ncov-en-200317.pdf?la=uk-ua (Mar 2020)

7. Boccardi, F., Heath, R.W., Lozano, A., Marzetta, T.L., Popovski, P.: Five disruptive technology directions for 5G. IEEE Commun. Mag. **52**(2), 74–80 (2014)
17 Roh, W., Seol, J.Y., Park, J., Lee, B., Lee, J., Kim, Y., …, Aryanfar, F.: Millimeter-wave beamforming as an enabling technology for 5G cellular communications: theoretical feasibility and prototype results. IEEE Commun. Mag. **52**(2), 106–113 (2014)
9. Li, D.: 5G and Intelligence Medicine_How the Next Generation of Wireless
10. Tortermvasana, K.: AIS Uses 5G, Robots in Pandemic War [Online]. Available: https://www.bangkokpost.com/%20business/1895175/ais-uses-5g-robots-in-pand%emic-war and https://www.bangkokpost.com, Apr 2020
11. Chunming, Z., He, G.: "5G Applications Help China _ght against COVID-19", CAICT, 2020 [Online]. Available: http://www.caict.ac.cn/english/yjcg/qwsj/202003/P020200312474854300829.%pdf
12. Li, Q., Guan, X., Wu, P., Wang, X., Zhou, L., Tong, Y., … Xing, X.: Early transmission dynamics in Wuhan, China, of novel coronavirus–infected pneumonia. New England J. Med. (2020)
13. Nordrum, A.: 5 Myths About 5G. IEEE Spectrum, May (2016)
14. Arthur, C., Shuhui, R., R..: China, Robot Delivery Vehicles Deployed to Help With COVID-19 Emergency. UNIDO, Apr 2020 [Online]. Available: https://www.unido.org/stories/china-robot-delivery-vehiclesdeployed-%20he%lpcovid-19-emergency
15. Gupta, A., Jha, R.K.: A survey of 5G network: architecture and emerging technologies. IEEE Access **3**, 1206–1232 (2015)
16. Liu, S., Yang, L., Zhang, C., Xiang, Y.T., Liu, Z., Hu, S., Zhang, B.: Online mental health services in China during the COVID-19 outbreak. Lancet Psychiatry **7**(4), e17–e18 (2020)
17. Wu, J.T., Leung, K., Leung, G.M.: Nowcasting and forecasting the potential domestic and international spread of the 2019-nCoV outbreak originating in Wuhan, China: a modeling study. Lancet **395**(10225), 689–697 (2020)
18. Dong, E., Du, H., Gardner, L.: An interactive web-based dashboard to track COVID-19 in real-time. Lancet Infect. Dis. (2020)
19. Xuequan, M.: "Shanghai Uses 5G Robots on the Frontline to Combat Covid-19. Xinhua Net, Mar. 2020 [Online]. Available: http://www.xinhuanet.com/english/2020-03/01/c_138833060.htm
20. Ahmed, W., Vidal-Alaball, J., Downing, J., Seguí, F. L.: Dangerous messages or satire? Analyzing the conspiracy theory linking 5G to COVID-19 through social network analysis. J. Med. Internet Res. (2020)
21. Elmousalami, H.H., Hassanien, A.E.: Day level forecasting for Coronavirus Disease (COVID-19) spread: analysis, modeling, and recommendations (2020c). arXiv preprint arXiv:2003.07778

Blockchain Use Cases for COVID-19: Management, Surveillance, Tracking and Security

Mohamed Torky, Ashraf Darwish, and Aboul Ella Hassanien

Abstract Blockchain has become a key technology in building and managing healthcare systems. the distinguished attributes of the blockchain (e.g. security, decentralization, time stamping, and transparency) make it the best technology for managing COVID-19 pandemic in real-time. This chapter investigates five blockchain use cases for fighting against the COVID-19 virus spread. Finally, this chapter is ended with discussing the recent blockchain platforms that can be utilized for Managing epidemic diseases, HashLog, and XMED Chain.

Keywords Blockchain · COVID-19 · Coronavirus · Security · Cryptography · Data management

1 Introduction

The world is witnessing now the invasion of the COVID-19 Virus epidemic, which represents a real threat to the lives of peoples and the destruction of the global economy at all levels. COVID-19 virus is identified as the most serious new disease within Coronaviruses (CoV) family. This family is a large family of viruses that cause illness ranging from the common cold to more severe diseases such as Middle East Respiratory Syndrome (MERS-CoV) and Severe Acute Respiratory Syndrome

---

M. Torky (✉)
Higher Institute of Computer Science and Information Systems, Culture & Science City, Giza, Egypt
e-mail: mtorky86@gmail.com
URL: http://www.egyptscience.net

A. Darwish
Faculty of Science, Helwan University, Cairo, Egypt

A. E. Hassanien
Faculty of Computers and Artificial Intelligence, Cairo University, Giza, Egypt

M. Torky · A. Darwish · A. E. Hassanien
Scientific Research Group in Egypt (SRGE), Giza, Egypt

A. E. Hassanien and A. Darwish (eds.), *Digital Transformation and Emerging Technologies for Fighting COVID-19 Pandemic: Innovative Approaches*, Studies in Systems, Decision and Control 322, https://doi.org/10.1007/978-3-030-63307-3_17

(SARS-CoV). The coronaviruses family are divided into four clusters, alpha (α), beta (β), gamma (γ) and delta (δ) coronavirus [1]. The variety of coronavirus patterns is a result of the genomes of coronaviruses, the behavior of RNA-dependent RNA polymerase, and the high frequency of homologous RNA recombination [2]. Based on the genome and RNA combinations, the COVID-19 virus is reported to be a new member of the β cluster of coronaviruses and named as SARS-CoV-2 and the disease as COVID-19 according to the International Committee on Taxonomy of Viruses (ICTV) [3–5]. According to the World Health Organization (WHO), The novel coronavirus (COVID-19) infected more than two million individuals with a mortality rate of about 6.9% across 213 countries till 25 April 2020 [6].

The infection dynamics of the novel coronavirus (SARS-COV2) or COVID-19 reflects higher spread rates than the previous SRAS-CoV. The rapid growth of the infection dynamics of the COVID-19 virus could be due to the distinguished DNA sequence of SARS-COV2 compared to SARS-COV which makes the COVID-19 virus has a faster spreadability.

The rapid growth of infection kinetics of the COVID-19 virus revealed a big weakness in the healthcare systems in most countries around the world. These weaknesses can be summarized as follows [7]:

1. ***Outdated health Surveillance system***: Many surveillance systems used to monitor and control potential new epidemics and control existing diseases are not updated for along time. For example, China's current disease surveillance system was not updated five decades-old ago. Moreover, these surveillance systems have inaccurate results, security, privacy, and data access problems.
2. ***Non-optimal data management***: Digitally processing the Covid-19 virus raises important challenges regarding data control, data storing, gathering data, verifying that data, and then cleaning up that data. Mitigating all these challenges using traditional ways is so far from the optimal techniques that should be used to digitally treat with the epidemics. Moreover, epidemiologists need high-quality data to model viruses accurately and transparently to enable decision-makers and health authorities to execute the valid treatment with the epidemic dynamics.
3. ***Lack of data transparency***: The rapid growth of infection kinetics of COVID-19 confirms that the real infection and death statistics are guessed to be much higher than reported. This is due to the many disruptions of the digital systems while processing virus data. This reflects a clear limitation in data transparency and validity. Moreover, unfortunately, it is impossible to know just how serious this outbreak is without access to a secure, decentralized surveillance system.
4. ***Lack of dependence on advanced technology***: In many countries, healthcare surveillance systems don't depend enough on the upcoming new technologies such as Artificial intelligence (AI), big data management, and Blockchain technology. Hence, the low investments in this important technology leading to less effective healthcare systems.
5. ***Locality and Centrality of Healthcare Systems***: Epidemic diseases can spread quickly across political borders. Hence, the locality of traditional health care systems run by governments can miss the control of outbreaks because epidemic

diseases still have infection ability across physical borders as COVID-19 disease does now. So, A decentralized system is the fastest way to digitally process and report outbreaks data.

Although some countries such as the United States and Italy have good healthcare systems, they were the most affected countries by the COVID-19 epidemic. On the other hand, although China has a less efficient healthcare system compared to the USA, it could suppress the infection spread efficiently and control and smooth the infection curve. The main reason is China was the best county that exploited the huge strengths of the upcoming technology especially the blockchain for treating with COVID-19 epidemic smartly. For example, In January 2020, the Shandong University of Finance, Alibaba, and other Chinese companies released about 29 blockchain-based mobile applications for tracking and controlling the infection data of COVID-19 [8]. Also, Life Technology launched the nation's first blockchain epidemic monitoring platform, which can track the progress of COVID-19 in all Chinese provinces in real-time, and register the relevant epidemic data on the chain so that the data can be traced and cannot be tampered with, this is in addition to using AI, Big data, Cloud computing, and 5G-based applications [9].

Blockchain technology proves day after day that it can play a pivotal role in ensuring the time stamping, de-centrality, security, and transparency of the epidemic information and the traceability of the epidemic dynamics and can help in preventing future pandemic [10]. Storing epidemic data into sequences of hashed blocks and processing it through P2P networks make blockchain technology could be used to mitigate many critical challenges in healthcare-related processes such as patient record management, healthcare surveillance, monitoring disease epidemics, controlling crises and many more.

In this chapter, Blockchain use cases for fighting against the COVID-19 epidemic are discussed, and some emerged blockchain platforms that can be used for managing healthcare systems against pandemic disease are reviewed and discussed.

This chapter is organized as: Sect. 2 gives an overview of the blockchain technology. Section 3 discusses a set of blockchain use cases for fighting against the COVID-19 virus. Section 4 discusses the recent blockchain platforms that are used for controlling and managing healthcare systems against epidemic disease. Finally, a general conclusion is formulated in Sect. 5 to summarize the key contributions of blockchain against the COVID-19 outbreak.

2 Blockchain Technology: An Overview

Blockchain has appeared as a powerful technology in processing digital coins through P2P-online financial transactions since emerging Bitcoin in 2008. The theory of blockchain refers to processing an infinite number of transactions or digital assets through a P2P network. Those transactions/digital assets then stored in a cryptographically secured data unit called "*Blocks*" after verifying them using a blockchain

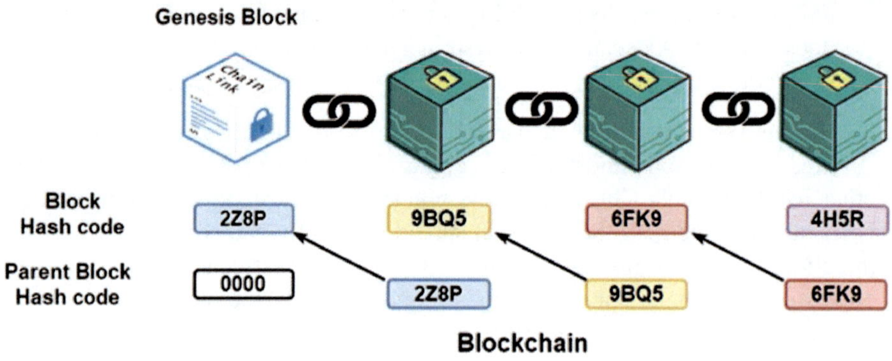

Fig. 1 Blockchain as a sequence of hashed blocks: a proposed model

algorithm [11]. Implementing blockchain as Distributed Ledger makes the history of any transaction or digital asset untampered, secured, and transparent through the decentralized design and cryptographic hashing of blocks. Figure 1 depicts a simple design of blockchain as a sequence of hashed blocks. Each block is identified by two hash codes, block hash, and parent hash.

Blockchain consists of five major features: Nodes, Transaction/digital asset, block, algorithm, and miners.

1. **Nodes**: Decentralization is one of the most important attributes of blockchain. With blockchain, there is no unique owner to the blockchain. Instead, it is a distributed ledger across a set of many nodes (or peers) connected to the chain. Nodes can be represented in the blockchain system as an electronic device that maintains copies of the statues of the blockchain and maintains the network working. Each node is represented by a unique alphanumeric identification number that identifies transactions executed by that node.
2. **Transactions/digital assets**: Sometimes is called *crypto-asset* [12]. Crypto asset refers to a transaction or digital asset that can be implemented in the form of cryptographic tokens. These tokens can be transmitted between nodes' accounts through the blockchain network.
3. **Blocks**: The Blockchain consists of sequences of many blocks and each block has two basic parts:

   - *Block header* involves six components:
     1. Block version, which identifies the consensus protocol to follow.
     2. Markle Tree Root Hash, which used to verify the hash code that identifies all block transactions.
     3. Timestamp, which is used to immutably track the creation and update time of the block for block integrity guarantee.
     4. N-Bits: identifies the hash code-threshold that specifies the block validity.
     5. Nonce: It is an arbitrary number that can be used just once in the cryptographic communication. It is randomly generated when a block is created, which then generates a block header hash.

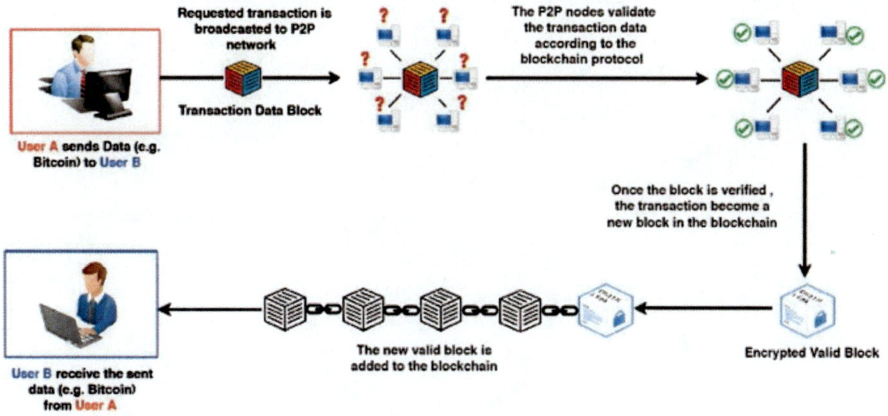

Fig. 2 Blockchain methodology: a proposed model

6. Parent Block Hash is a 256-bit hash code that refers to the previous block. It used to connect the blocks in a sequence order.

- *Block body*: contains all number of transactions that are confirmed and validated within the block.

4. **Block Algorithm:** is a set of rules and procedures to carry out blockchain operations.
5. **Miners**: are specific nodes which perform the block verification process before adding anything to the blockchain structure.

Figure 2 explains how the blockchain system is working when a specific transaction occurs between two users in the blockchain system.

User 'A' transfers a transactional data (e.g. bitcoin) to a user 'B' through the P2P network. The requested transactions are broadcasted to all nodes in the form of a transaction data block. Then, the P2P validates the transaction data according to the used blockchain protocol. Once the block is validated by all peers the candidate transactional data block becomes a new valid block and added to the blockchain and User 'B' receive the number of bitcoins.

Blockchain technology is distinguished by some attributes such as security, transparency, trusted transaction, cost-effectiveness, decentralization, and real-time tracking as depicted in Fig. 3. These main attributes make it applicable not only in financial transactions and digital currency but in many fields. One of these fields is managing healthcare systems.

In the coming section, we will explain how this important technology can be used for mitigating some vulnerabilities of healthcare systems against COVID-19 Virus

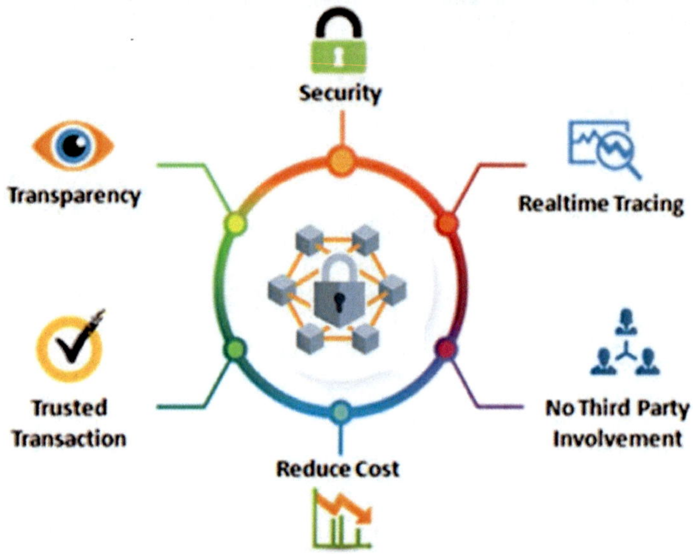

Fig. 3 Blockchain attributes

3 Blockchain Use Cases for COVID-19 Virus

Blockchain could provide means to enhance many public health activities associated with controlling and preventing epidemic diseases such as COVID-19 Virus. Based on powered solutions of the blockchain, various aspects and challenges of manipulating the COVID-19 virus can be addressed and mitigated. Blockchain technology has many potential advantages for improving healthcare systems such as, access to information, healthcare surveillance, record management, monitoring and tracking disease outbreaks, management crises, and many more. In this section, four blockchain use cases for fighting against the COVID-19 virus are explained and discussed in some details.

3.1 Recoding Management: Blockchain is a Single Source of Truth

Containing the COVID-19 virus should be manipulated as a data recording and management issue. The most important advantage for blockchain in managing healthcare systems is that it can be utilized as a single public source of truth for the data provenance. For example, it could be used for recording confirmed COVID-19 cases, processing real-time data, and ensuring its integrity while recognizing and removing misinformation about the COVID-19 coronavirus.

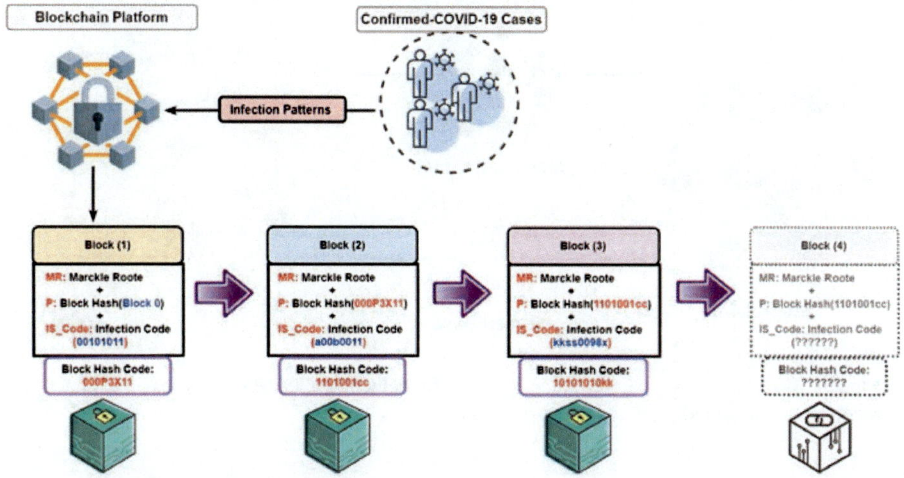

Fig. 4 Recording confirmed COVID-19 cases in the blockchain as a sequence of hashed blocks: a proposed model

In pandemics like COVID-19 coronavirus, there are plenty of gigabytes of incoming data, that needs to untraditional techniques to manage and store it efficiently. Based on blockchain technology, data gathering and recording will become automated and impossible to tamper data of any block in the blockchain system. Figure 4 depicts how the detected confirmed cases of COVID-19 can be digitized as *infection patterns* and stored as a sequence of hashed blocks.

By using blockchain, it is become possible to ensure transparency without compromising patient security and privacy. Block data could include patients 'data, treatments given, and any case progress and update. Blockchain will also make sure that data are archived and protected against any unauthorized access, but still keeping it available for all stakeholders of the whole healthcare system. Blockchain will enable all users to be aware of all the real data of infection spread in real-time such as confirmed infected cases, death numbers, recoveries, etc. The connectivity using blockchain networks and the access to the wealth of data it provides would allow decision-makers and health authorities to quickly track the spread of epidemic diseases and help researchers to develop the appropriate vaccine and drugs.

3.2 Blockchain Healthcare Surveillance System

Blockchain can also be used for surveilling epidemic diseases. A blockchain healthcare surveillance system can offer various tools to control and prevent infection spread of future outbreaks. A blockchain healthcare surveillance system would allow local and national health authorities to access the surveillance data for monitoring infection dynamics of epidemic diseases as with the COVID-19 virus. A global blockchain

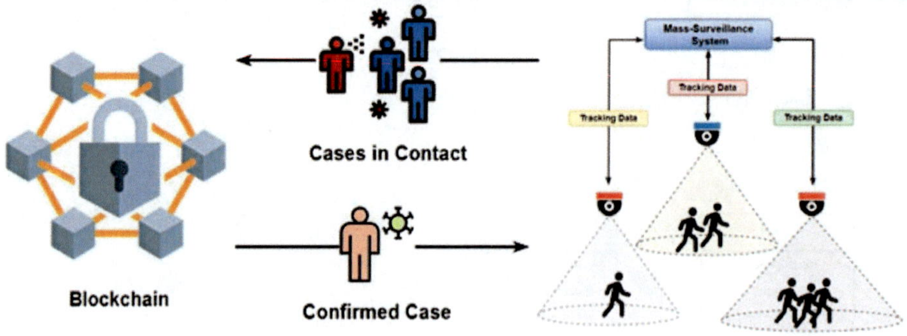

Fig. 5 Blockchain as a surveillance system for detecting COVID-19 cases in contact with a recognized confirmed case: a proposed model

healthcare surveillance system could easily allow governments and health authorities to monitor and track epidemic diseases in poor areas where connectivity is weak, and costs must be kept low. Doctors and local practitioners can retrieve real-time information about the pandemic on surrounding areas, regardless of governmental or political barriers. Moreover, global organizations like the World Health Organization (WHO) could access the data with ensuring security and integrity of data, Moreover, it can reports real and transparent information that helps lab researchers to reach the appropriate vaccine. Because the system is decentralized and secured through blockchain, pandemic data remains immutable and multiple organizations can report the data transparently. Figure 5 explains how the communication between blockchain and the Mass-Surveillance system networks [13] can help in detecting probably infected persons who are cases in contact with a recognized confirmed case.

3.3 *Tracking COVID-19- Infection Spread Using Blockchain*

Tracking global health information surveillance, especially for the infectious spread of Covid-19 can be also executed using blockchain technology. The transparency of storing and retrieving infection spread data based on blockchain will result in more accurate reporting and more efficient responses and treatments. So, tracking infection kinetics of the COVID-19 coronavirus by using blockchain would allow for the fast processing of infection data, allowing early detection of infections before they spread to the level of an epidemic outbreak.

Blockchain could track down where the virus originated, probably It could enable doctors to review patients' symptoms and monitor diagnostic data in real-time, as well as integrating patient history information, and allocate it in a distributed way and have that information available to different parties, including authorities such as the WHO. Also, blockchain can help in developing treatments, diagnostics, and tests quickly. This could enable governments and health authorities to keep track of the virus kinetics and infection spread. For instance, A team of researchers at the

University of Cape Town in South Africa has designed a blockchain-based application that will allow citizens to verify their COVID-19 infection status. The platform is called *Covi-ID*. The system is still under development and aims to manipulate several challenges around the global COVID-19 coronavirus pandemic [14]. The Covid-ID application intends to enhance contact tracing of infected people, while users will be able to provide a verified COVID-19 status. Moreover, the app will reward persons for precaution behaviors against infection spread, like remaining at home during lockdown periods. COVID-ID is being built on a decentralized network called the self-sovereign identity blockchain network or *Sovrin* [15].

COVID-ID App [16] is implemented through two design views as follows:

1. *Creating a COVID-ID account and receiving a COVID-19 status*. Through this design view, the users create a Covi-ID account with one of the organizations that offer Custodial digital wallets that called *self-sovereign identity wallet*. The organizations that grant *self-sovereign identity wallets* may be issued by government institutions, Universities, Posts, Banks, etc. the user can create his health identity account by using Covi-ID app and all health identity information is recorded in his self-sovereign identity wallet and can be managed by his created account on Covi-ID. Once the user created his healthcare identity account, he can be tracked by the communication between the blockchain and Covi-ID App. For example, a specific symptomatic person can be checked by doctors and testing kits. His/her health identity can be verified by scanning his/her QR code associated with testing kit results (i.e. Positive/Negative COVID-19 case). After that, the new update in the user's health identity is recorded in the blockchain and also send to Covi-ID App, which updates his/her *self-sovereign identity wallets* created with the corresponding custodian organization. Figure 6 explains how users create their health identity accounts and managing their self-sovereign identity wallets using the COVID-ID application.
2. *Verifying a user's Covid-19 status*: through this design view of the COVID-ID App, the person's statues of the COVID-19 virus are checked and verified by a security man who uses a *verifier device* which connected with Covi-ID App. When a specific person wants to enter a public place (e.g. supermarket, university, Gym, hospital, transport stations, etc.), the person is asked to show his QR code by the security man who then scans the QR code using his verifier device. By showing the QR code, this means that the person agrees to grant access to his health information recorded in his *self-sovereign identity wallet*.

The verifier device then calls Covi-ID App to perform a query on the person's self-sovereign identity wallet created by the corresponding custodian. The query results present the person's COVID-19 statue which can be visualized in three color indexes, Red, Green, and Yellow. The red color means this person is a confirmed infected COVID-19 case and he will not be allowed to enter the place and will be asked to return home and continue self-isolation unless in medical emergencies. The green color means that the person is sound and free of COVID-19, hence he allowed to enter the place. The yellow color means that this user is suspicious COVID-19 case, hence implementing additional social distancing measures is required for allowing

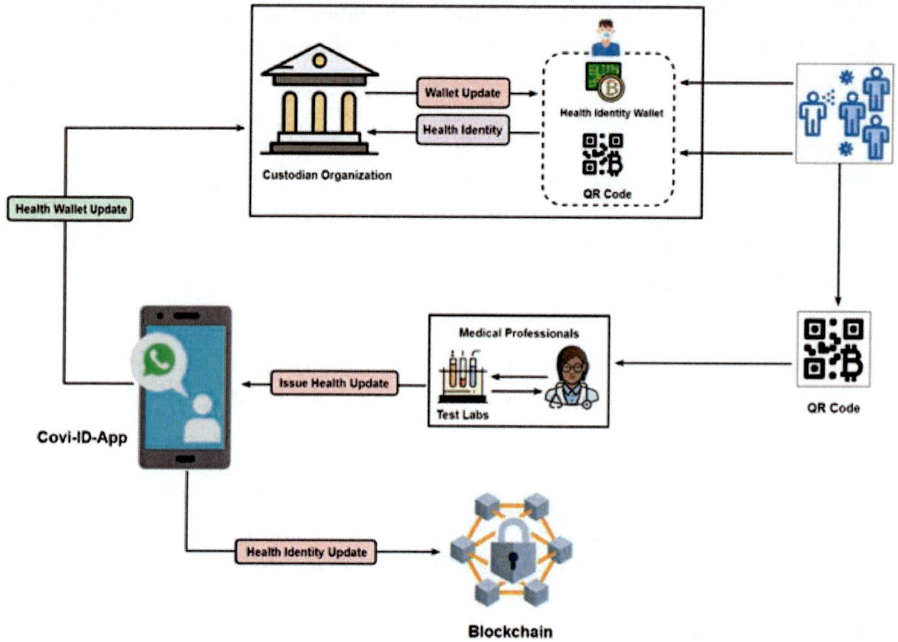

Fig. 6 Creating a Covi-ID account and receiving a COVID-19 status: design view 1: a proposed model

this person to enter the place. Within this framework, Fig. 7 explains how the user's Covid-19 status is verified using the Covi-ID app.

3.4 *Securing Medical Supply-Cain Using Blockchain*

Tracking and tracing medical supply chains can be managed using blockchain effectively and transparently [17]. Blockchain-based supply chain systems could be used to record, review, and tracking of demands, supplies, and logistics of epidemic drugs and health tools. It is known that medical supply chains depend on multiple parties from drug and vaccine manufacturers and distributors to warehousing and delivery logistics. The whole medical supply chain process of recording and verifying medications and health tools is tamper-proof, while also allowing the stakeholder to monitor and track the new updates of the medical supply chain process. The blockchain could ensure that testing devices, vaccines, drugs, and other relief efforts are reaching the critical places at the right times and in the quantities needed. Securing medical supply chain stages and verifying valuable medical assets using blockchain will help in recognizing fake drugs and detecting suspicious medication, this could have life-saving effects. Figure 8 explains how blockchain can manage medical supply-chain phases and help in recognizing fake drugs.

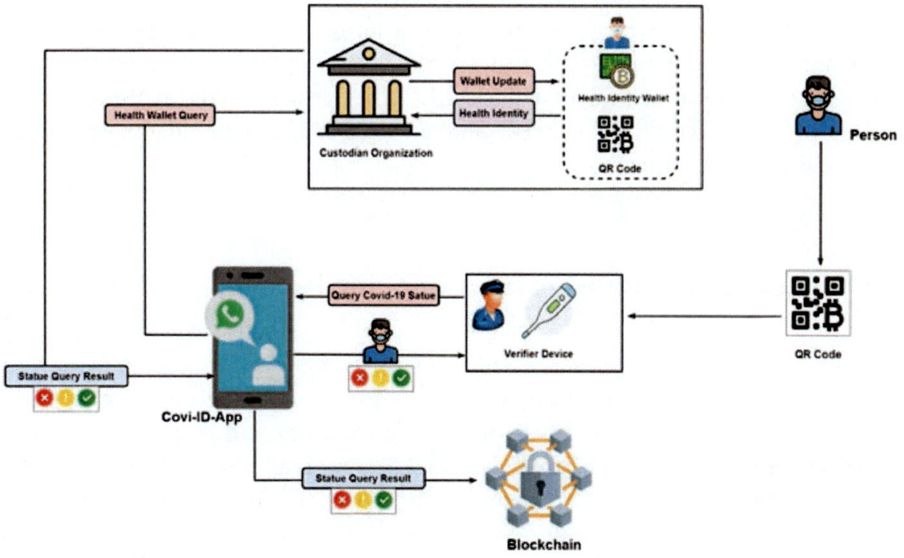

Fig. 7 Verifying a user's Covid-19 status: design view 2: a proposed model

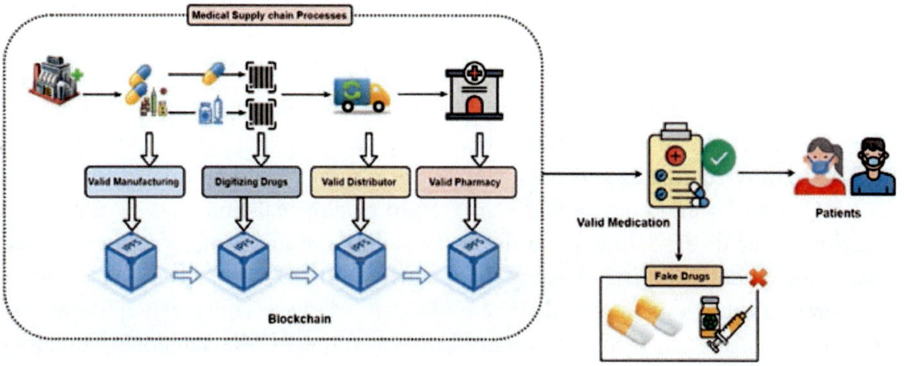

Fig. 8 Securing medical supply-cain processes using blockchain: a proposed model

Integrating a surveillance system with blockchain technology will change the global medical supply chain system processes and could change the classical techniques the world uses to respond to epidemics diseases.

3.5 *Tracking Zoonotic Diseases Using Blockchain*

Zoonotic diseases like the Covid-19 virus (which believed that transferred from bats) could be detected in animals before they leap to humans. In response to this challenge, a blockchain monitoring system can be used for recording and tracking veterinary field records of animals. Since a lot of animals are migratory and not remaining in the same area, - a P2P global blockchain system would allow for tracking zoonotic disease growth across the world in a transparent and real-time manner. Animal diseases could be "flagged" and removed in animal populations before they make the jump to humans.

4 Blockchain Platforms for Managing Epidemic Diseases

Blockchain is rapidly been addressed as a base technology for developing new platforms able to manage epidemic diseases and containing the outbreak. Using the blockchain-based platforms in the health care industry will help to prevent future pandemics and provide untraditional solutions for mitigating big challenges in healthcare systems against recent pandemics such as COVID-19 and the future epidemic diseases. In this section, we discuss two important blockchain-based platforms for managing epidemic diseases.

4.1 HashLog

HashLog is a platform launched by Acoer, the developer of blockchain-enabled health care applications for global health organizations to fight against the coronavirus. The visualization processor of HashLog able to communicate in real-time with Hedera Hashgraph's blockchain [18] to ensure real-time access, data visualization of the disease dissemination, and tracking the infection dynamics. This can assist the epidemiologists to verify the integrity of disease records. Each transaction is stored through a verified hash code on Hedera's ledger. This mean epidemiologist can authenticate data to be legitimate. This technique allows scientists, researchers to understand the spread of the coronavirus and its kinetics over time through a visualized maps on visualization HashLog dashboard.

4.2 XMED Chain

XMED Chain [19] is the first global medical blockchain platform that depends on artificial intelligence, big data, and blockchain technology for developing healthcare mobile applications. The patient can create an XMED account through his smartphone then he can upload his medical records through his XMED account. These medical records are encrypted and recorded on the global blockchain which also can be shared globally without breaking the patients' privacy. XMED Chian's objective is promoting the standardization of medical data manipulation and implement systems where information can be interchanged transparently in a secure way.

The XMED chain recently attempts to mitigate several of the essential challenges in the Healthcare system that include the following:

1. XMED chain provides the best medical solution due to the use of blockchain, big data technologies, and AI that can achieve excellent international medical data management.
2. Allows for secure and safe deliberation of medical data, through the privacy-preserving and encryption techniques of blockchain technology.
3. The XMED Chain platform permits for efficient data sharing among medical Stakeholders such as Medical Institutions, Healthcare Suppliers, Insurance Companies, etc.
4. XMED Chain users can have access to multidimensional global medical services just by using his own XMED Chain—application.

5 Conclusion

This chapter is designed to determine the effects of blockchain opportunities for controlling and managing the infection kinetics of the COVID-19 coronavirus. The investigation of blockchain contributions in controlling the COVID-19 pandemic has shown important opportunities. firstly, recording pandemic data as a single source of truth. Secondly, utilizing blockchain as a healthcare surveillance system. Thirdly, tracking COVID-19- Infection spread. Fourth, securing medical supply-chain systems. Fifth, Tracking zoonotic diseases before the jump to the human.

Finally, the chapter is ended with discussing the recent two blockchain-based platforms, HashLog, and XMED Chain that can be used for managing healthcare systems against pandemics.

References

1. Shereen, M.A., Khan, S., Kazmi, A., Bashir, N., Siddique, R.: COVID-19 infection: origin, transmission, and characteristics of human coronaviruses. J. Adv. Res. **16**(24), 91–98 (2016). https://doi.org/10.1016/j.jare.2020.03.005
2. Sino Biological, Coronavirus. https://www.sinobiological.com/research/virus/coronavirus-overview. Accessed 25 April 2020
3. Cui, J., Li, F., Shin, Z.-L.: Origin and evolution of pathogenic coronaviruses. Nat. Rev. Microbiol. **17**(3), 181–192 (2019)
4. Lai, C.-C., Shih, T.-P., Ko, W.-C., Tang, H.-J., Hsueh, P.-R.: Severe acute respiratory syndrome coronavirus 2 (SARS-CoV-2) and coronavirus disease-2019 (COVID-19): the epidemic and the challenges Int. J. Antimicrob. Agents, 105924 (2020)
5. World Health Organization. Laboratory testing for coronavirus disease 2019 (COVID-19) in suspected human cases: interim guidance, 2 March 2020. World Health Organization; 2020
6. World Health Organization (WHO), Coronavirus disease (COVID-19) Pandemic, https://www.who.int/emergencies/diseases/novel-coronavirus-2019. Accessed 25 April 2020

7. De Meijer, C.R.W.: Blockchain and Coronavirus: Could It Prevent Future Pandemics, https://www.finextra.com/blogposting/18570/blockchain-and-corona-virus-could-it-prevent-future-pandemics. Accessed 17 April 2020
8. Alper, T.: Here is How China Uses Blockchain Technology to Fight the Coronavirus, https://cryptonews.com/news/here-is-how-china-uses-blockchain-technology-to-fight-the-co-5905.htm. Accessed 18 April 2020
9. World Economic Forum. How next-generation information technologies tackled COVID-19 in China, https://www.weforum.org/agenda/2020/04/how-next-generation-information-technologies-tackled-covid-19-in-china/. Accessed 14 April 2020
10. Torky, M., Hassanien, A.E.: COVID-19 Blockchain Framework: Innovative Approach. arXiv preprint arXiv:2004.06081, 5 April 2020
11. Zheng, Z., Xie, S., Dai, H., Chen, X., Wang, H.: An overview of blockchain technology: Architecture, consensus, and future trends. In: 2017 IEEE International Congress on Big Data (BigData Congress), 25 June 2017, pp. 557–564. IEEE
12. Hasan, H.R., Salah, K.: Proof of delivery of digital assets using blockchain and smart contracts. IEEE Access. **22**(6), 65439–65448 (2018)
13. Bigo, D., Carrera, S., Hernanz, N., Jeandesboz, J., Parkin, J., Ragazzi, F., Scherrer, A.: Mass surveillance of personal data by EU member states and its compatibility with EU law. Liberty and Security in Europe Papers, vol. 61, 6 Nov 2013
14. Jenkinson, G.: Coronavirus-Tracking Apps Take Different Approaches in Keeping People Safe, [online], https://cointelegraph.com/news/coronavirus-tracking-apps-take-different-approaches-in-keeping-people-safe. Accessed 13 May 2020
15. Sovrin Foundation. Sovrin: A Protocol and Token for SelfSovereign Identity and Decentralized Trust, 2018 [online], https://sovrin.org/wp-content/uploads/2018/03/Sovrin-Protocol-and-Token-White-Paper.pdf. Accessed 13 May 2020
16. Writer, S.: How Covi-ID Tracks Your COVID-19 Health Status? 2020, [online], https://techfinancials.co.za/2020/04/21/how-covi-id-tracks-your-covid-19-health-status/. Accessed 17 May 2020
17. Jayaraman, R., Saleh, K., King, N.: Improving opportunities in healthcare supply chain processes via the internet of things and blockchain technology. Int. J. Healthcare Inf. Syst. Inform. (IJHISI). **14**(2), 49–65 (2019)
18. Hedera Hashgraph. [online], https://www.hedera.com/. Accessed 25 May 2020
19. Steemit, XMED Chain: The World 1 st Global Medical Blockchain, [online], https://steemit.com/healthcare/@foxicoreviews/xmed-chain-the-world-1-st-global-medical-blockchain. Accessed 25 May 2020

Cyber Security in the Age of COVID-19

Arome J. Gabriel, Ashraf Darwsih, and Aboul Ella Hassanien

Abstract As a containment strategy for the dreaded Corona Virus Disease 19 (COVID 19) which is spreading rapidly and causing severe damage to life and economy of nations, places of public gathering like schools, places of religious worship, open physical markets, offices as well as venues for social meetings (such as clubs) are been closed down, to promote social distancing in most nations across the globe. Therefore, most public/private organizations, and even individuals have resorted to the use of diverse Information Technologies (IT) for connecting themselves and other life essentials. Educational, agricultural, religious and even health institutions now deliver their services to users/clients and receive payments via online platforms, students study from home, even employees of most organizations now work remotely (maybe from their homes). Moreover, there is a sharp growth in demand for food deliveries and online grocery. The massive adoption of IT by almost all aspects of human life especially during this epidemic has also led to increased cyber security concerns. Cybercriminals and other individuals with malicious intent now take COVID-19 as an opportunity to perpetrate cybercrimes, especially for monetary gains. Domestic violence seems to be on the rise perhaps due to the lock-down, contact tracing approaches are massively been developed and used, healthcare systems are being attacked with ransom ware and resources such as patient records confidentiality, and integrity are being compromised. Individuals are falling victim

A. J. Gabriel (✉)
School of Computing, Federal University of Technology, Akure, Nigeria
e-mail: ajgabriel@futa.edu.ng

A. Darwsih
Faculty of Science, Scientific Research Group in Egypt, Helwan University, Cairo, Egypt
e-mail: ashraf.darwish.eg@ieee.org
URL: http://www.egyptscience.net

A. E. Hassanien
Faculty of Computers and Artificial Intelligence, Cairo University, Cairo, Egypt
e-mail: Aboitcairo@gmail.com

A. Darwsih · A. E. Hassanien
Scientific Research Group in Egypt (SRGE), Cairo, Egypt

A. E. Hassanien and A. Darwish (eds.), *Digital Transformation and Emerging Technologies for Fighting COVID-19 Pandemic: Innovative Approaches*, Studies in Systems, Decision and Control 322, https://doi.org/10.1007/978-3-030-63307-3_18

to phishing attacks through COVID-19 related content. This paper presents an extensive study of major cyber security concerns that are and could take place during the COVID 19 pandemic as well as strategies for mitigating them.

Keywords COVID 19 · Cyber security · Data Privacy · Cybercrime · Information Security

1 Introduction

The dreaded Corona Virus Disease 19 (COVID 19) was declared a pandemic by the World Health Organisation (WHO) on the 11th of March, 2020, roughly five months after it broke out in Wuhan, China. It has since spread explosively across other nations and continents of the world causing tremendous damage to life and the world economy at large. In most nations across the globe, spots of open get-together like schools, churches, mosques, markets, other workplaces, and even social hubs or clubs have all been shut down, to promote social distancing and ultimately curb the rapidly spreading and devastating COVID 19 pandemic. Various public enlightenments or adverts are relayed daily on mass media (radio, television, newspapers) to sensitize people on the dangers of, and strategies for preventing or reducing the transmission of the dreaded Covid-19 epidemic. Indeed, even hospitals and other healthcare facilities, observe a lot of precautionary measures (hand-washing, temperature measurement, and even recording of travel histories), before admitting or isolating/quarantining visitors/patients as the case may be.

COVID-19 is a member of the family of coronavirus that can cause severe illness to humans and eventually lead to the death of the infected person. The first recorded incidence of viral attack from the family of coronavirus is the 2003 Severe Acute Respiratory Syndrome (SARS) [1–3]. Then came the second incidence in 2012, which was the Middle East Respiratory Syndrome (MERS) in Saudi Arabia [4, 5]. COVID-19 is the third and current viral attack from the coronavirus family. Currently, researchers are making concerted efforts to understand the spread pattern and predict the growth rate in different communities and countries. The statistics reported by WHO on the 22nd May 2020 reveals that there are four million, nine hundred and ninety-three thousand, four hundred and seventy (4,993,470) confirmed cases of infected people and three hundred and twenty-seven thousand, seven hundred and thirty-eight (327, 738) death cases worldwide. In Africa, the cumulative number of infection cases was seventy-one thousand, seven hundred and fifty-two (71, 752) confirmed cases of infected people, and one thousand, nine hundred and eighty-one (1981) death cases.

Several strategies have been proposed and adopted in the various life domains, towards curbing and possibly eliminating the COVID-19 epidemic. These strategies include, total lockdown that implies, restriction in movement of individuals, closure of public places of gatherings like the (markets, places of worships, offices, banking halls, major grocery stores, and even schools), contact tracing of infected persons and those who came near them. As the world grapple with how to answer questions that pertains to the spread pattern of the epidemic, how long it will take before an infected person recovers or die, when the epidemic will subside or disappear completely, or even, places of the highest concentration of infected persons, other strategies like, real-time data collation, modelling, and predictions became very relevant, a very significant concern that needs serious and urgent attention is the cyber security implications/consequences of COVID-19. One issue is however common and requires much attention than it is currently getting. That is the cyber security and privacy issues that are creeping up.

Cybercrime seems to be the foremost threat to businesses or organizations on the planet, and probably the most serious issue with humanity. The effect on society is shown in the Annual Official Cybercrime Report, of Cyber-security Ventures. The report opined that cybercrime will cost the world $6 trillion every year by 2021, a huge rise from the $3 trillion it was in 2015. This places cybercrime as more lucrative (hence, attractive) than the trade of illegal drugs speaks to the best exchange of financial riches ever, dangers the motivations for advancement and speculation, and will be more beneficial than the worldwide exchange of all major unlawful medications joined. Cybercrime costs consist of so much including, financial frauds, theft/loss of data, Intellectual property theft, reduced productivity, disruption of business, loss of time, loss of money, loss of convenience, reduced availability of systems, and even misappropriation of funds. Furthermore, several cyber threats have even become prevalent in this COVID 19 era. This includes, but is not limited to, phishing, spoofing, sniffing, Denial of Service (DoS), worms, malware, computer viruses, flooding, Man-in-the-Middle, and even False Data Injection (FDI) attacks. Although diverse security implications of the adoption and use of Information Technologies exist and have been studied by several researchers and other stakeholders. Understanding the seriousness of these security and privacy issues during a pandemic of the kind we are witnessing through COVID-19 is not as easy as it may seem. This is largely because cybercrimes exhibit diverse characteristics which makes it easy for unsuspecting persons and systems to fall victim.

The major objective of this book chapter therefore is, to highlight the major cyber security implications of the COVID-19 pandemic on businesses and humanity at large. The chapter provides insight into the security and privacy aspects of most information technologies been adopted and used by businesses and organizations especially due to the COVID-19 outbreak. It also produces at-a-glance report for future researchers and a motivation for further research in pursuing the provision of cyber security and privacy measures suitable for times of epidemic outbreaks or even wars.

This book chapter consists of seven (7) sections. Section 2 presents the impacts of COVID-19 related to fake news and misinformation. Section three discusses the remote working from home and cyber security matters involved. Section 4 highlights the significance of contact tracing to the COVID-19 pandemic. It also discusses the information security concerns that come with computer-based or mobile contact tracing. In Sect. 5, the role of medical imaging in the diagnosis of COVID-19 is highlighted. The issue of ensuring the security of medical images in-store or while been transferred is also discussed. Section 6 presents other general cyber security issues that came with the epidemic. Finally, the conclusion is presented in Sect. 7.

2 Impact of Covid-19 Related Fake News or Misinformation

The dread that accompanied the sudden outbreak and rapid spread of the newest member of the Severe Acute Respiratory Syndrome Coronavirus 2 (SARS-CoV-2) infection, which is now known as Corona Virus Disease 2019 (COVID-19), has resulted in an explosive demand and desire for information amongst the people. As authorized agencies are working hard to make desired accurate information available, there is an ugly trend of growth in generation and spread of rumors, misinformation hoaxes, and/or fake news as regards to the cause, spread pattern, prevention and even remedy of the epidemic. These have all become a serious threat or clog in the wheel of efforts at combating the COVID 19 epidemic [6]. More dangerously is the fact that these rumors or fake news spread more speedily on the internet than the true news from trustworthy sources. This indeed sabotages efforts at mitigating the consequences of the epidemic. The menace of fake news was highlighted in the study reported in [7], and it must be noted here that there is now a long list of consequences of misinformation or fake news on the world's effort at promptly curbing the COVID-19 pandemic. Firstly, the issue of media fatigue. Fake news leads to relaxation of healthy conduct/etiquettes (social distancing, hand washing, and even use of hand sanitizer), as well as stimulating wrong behaviors that ultimately result in an escalation of the spread of the virus among individuals. The case of an overdose of Chloroquine in Nigeria as reported in [8], as a result of the dissemination of unfounded rumors in the media on the effectiveness of the drug in curing COVID-19. Secondly, misinformation or fake news results in heightened fear often resulting in death or suicide. The case of an Indian father who reportedly committed suicide upon discovering that he tested positive to COVID-19 is an example [9]. Thirdly, misinformation, or fake news often results in panic buying of food items and stationeries, which ultimately disrupts the supply chain [10]. Fourthly, COVID-19 related rumors or fake news often leads to confusion for the health authorities and the general public. Fifthly, misinformation or rumors also reduces the legitimacy of new scientific findings of the cure or vaccine for this disease. Sixthly, fake news and rumors also create a social stigma around COVID-19 [11]. Several efforts are directed at identifying and eliminating fake news and their sources. For instance, the World Health Organization (WHO), in conjunction with major tech companies such as Twitter, Youtube, Facebook, LinkedIn, Google, Microsoft, and Reddit, have agreed on doing everything possible to eliminate all forms of COVID-19 related fake news [12]. The research community is not left out in the fight against fake or unfounded news. Once any piece of news is discovered to be unfounded, such news is taken down from the Internet, leaving behind only news that is scientifically proven. Indeed, innovative tools like natural language processing or text-mining may be helpful in the detection of online content that has no scientific basis [7: Shu et al. 2020]. Verified sources of fake news and misinformation could also be prosecuted by relevant regulatory establishments and law enforcement organizations [13].

The authors of the work reported in [14] carried out an experimental study on misinformation in social media. They opined that misinformation can increase humankind's major challenges. The authors pointed out that an example of insight is that of the dreaded COVID-19 pandemic. The major goal of this study was to probe why people trust and spread false (and true) news content about COVID-19. Results obtained from two (2) studies with over 1600 participants, showed support for the idea that people share false claims about COVID-19 in part because they simply fail to think sufficiently about whether or not content is accurate when deciding what to share. Results also showed that nudging people to think about accuracy before sharing anything online, is a simple way to improve choices about what to share on social media. Accuracy reminders are easy to implement on social media platforms and could have an instantaneous constructive effect on curtailing the menace of misinformation about the COVID-19 outbreak. One limitation to their study is, however, the fact that only the United States of America was considered. There is, therefore, a need to conduct experiments on other parts of the world.

3 Remote Working from Home and Cyber security Issues

Since governments of nations across the world have resorted to enforcing lockdown measures as a means to curbing the COVID-19 pandemic, organizations, businesses, and even individuals have resorted to the use of the various computer or mobile applications and computer networks (such as the Internet) for running their day The rapid upsurge in the adoption of computer networks has resulted in a corresponding increase in cyber-attack surfaces, resulting in increased opportunities for cybercriminals to perpetuate and even get away with various heinous cybercrimes. This is causing varying degrees of damages to the individual, national, and even global economies. Some of the consequences of their crimes include; *Business disruption, Fraud*, (that may involve impersonation, man-in-the-middle attacks), *Critical data breach*, (which could result in loss of sensitive business and personal data, or even reputational damages), monumental loss of resources (lives, properties, time, money), as well as *Third-party failures*.

The Cyber-security Ventures in their annual official report have described Cybercrime as the foremost threat to businesses or organizations on the planet, and probably the most serious issue with humanity. They further predicted that cybercrime will cost the world $6 trillion every year by 2021, a huge rise from the $3 trillion it was in 2015 [15]. The COVID-19 pandemic could result in the doubling of this figure if appropriate precautionary or preventive measures are not put in place.

Employees, students, businesses, and other individuals who will work remotely probably from home ought to know the most proficient method to identifying and responding to the diverse cyber-attacks (such as SPAM, phishing, etc.). Employers could use suitable information technologies to monitor the activities of their employees even as they work from home [16, 17]. Cyber-safe remote working from home would require that certain steps be diligently and intentionally taken by both employees, their organizations, and even customers.

These steps for ensuring cyber-safe remote working from home are hereby further emphasized;

(a) In the context of COVID-19, the WHO warns that people must avoid touching objects and surfaces that are not clean. To avoid the havoc of Computer Viruses as well, people should avoid opening emails from unknown parties or visiting untrusted websites. Individuals must carefully determine who to trust in enterprise networks. They must only click on links whose source they can trust. This is because some cybercriminals now set up keyloggers disguised as COVID-19 related titles. Once clicked and unpacked can result in the personal credentials of that individual been gathered without his notice. Furthermore, individuals must carefully verify the authenticity of websites (or links) before visiting them. This is because several websites exist that are malicious and engaged in Phishing attacks.
(b) The current increasing demand for virtual conferencing and other collaboration, tools (e.g., Zoom app) might uncover more vulnerabilities or loopholes for hackers to take advantage of. Businesses and organizations hastily adopting consumer-grade video conferencing could make it stress-free for attackers to masquerade (pretend to) as members of their staff to carry out unauthorized access or perpetrate crime.
(c) Insecure remote access to corporate networks is potentially another dangerous problem. Employees must make sure that they access IT resources within the organization and elsewhere on the internet. In a safe manner. This can be achieved via the use of a Virtual Private Network (VPN) solution with an encrypted network connection.
(d) Businesses and Organisations must ensure that they bring their cyber-security strategies or approaches up-to-date by incorporating support for remote working from home. This arrangement could include location-based, time-based, and other access control strategies, the utilization of personal or mobile gadgets. Utilization of employees' gadget could leave the entire organization vulnerable to attacks. As such, careful efforts are made at ensuring loopholes are blocked.
(e) Employees should endeavor to connect with colleagues for official matters only via IT tools provided by employers. There is often a range of software installed in the background of company IT that keeps people secure. Full protection cannot be guaranteed if personal gadgets are used.
(f) Workers working remotely from home must not reveal login IDs or passwords to family members or anyone for that matter.
(g) Organizations must raise awareness amongst their employees warning them of the heightened risk of COVID-19 themes phishing attacks. Even customers should be given safety tips.

(h) The use of strong passwords, and preferably multi-factor authentication, by customers and employees alike, for all remote access accounts must be encouraged.
(i) Organizations should make available straightforward directives on safety practices and how to use remote working.
(j) Organizations and their employees should endeavor to use up-to-date antivirus and network gateway applications (like firewalls).
(k) Organizations and businesses should provide call-in helplines or online chat provisions which employees and customers can seek help or clarification from concerning security.
(l) Organizations should encourage their employees to imbibe the habit of disabling USB drives to eliminate the risk of malware, proposing the transfer of data via collaboration tools for instance.
(m) Organizations, businesses, and individuals should ensure that several levels of verification or confirmation be adopted especially when dealing with transactions that involve a huge amount of money. This is necessary as several variations of financial fraud are possible especially during this COVID-19 pandemic.
(n) Organizations and individuals should ensure that they apply critical security patches and update their network gateways as well as anti-virus software across their entire IT architecture.
(o) Organizations and businesses should ensure they back up all critical systems and authenticate the integrity of backups, preferably organizing for offline storage of backups frequently.

Organizations and businesses should vigilantly expect the greater risk of malware, and even ransomware during the COVID-19 pandemic.

4 Contact Tracing and Privacy Concerns

It is now well over four (4) months since the dreaded COVID-19 pandemic took the world by surprise. We are currently at a stage where a lot of efforts are being directed at control of the epidemic. Achieving this control of COVID-19 involves identification and quarantining of infected persons as well as individuals that have in any way been in contact with those infected persons. It is to this effect that much effort has been directed at the development of intelligent applications (usually smartphone-based) for contact tracing. This allows for the automatic detection of individuals that a person with COVID-19 has probably infected. Although these intelligent applications useful methods to combating COVID-19, they raise several concerns that include security and privacy.

Contact Tracing represents the first and most effective move towards combating an epidemic. This is especially true since, at that instance, there is a lack or scarcity of resources for mass testing and vaccines for instantaneous use. Mobile contact tracing is quite efficient and quick, and hence, represents a huge improvement on the traditional manual interview-based contact tracing procedure, which is flawed and merely relies on symptoms [18].

The World Health Organisation (WHO) as reported in [19] opined that contact-tracing follows three (3) major stages: The first stage has to do with identifying the contacts. That is, finding those individuals that the COVID-19 infected person/patient had contact with. Then, the second stage involves creating a list or record of those individuals who might have had contact with the patient around that period. These individuals are then duly informed. Finally, the third stage involves follow-up on these individuals on the list created at the second stage.

One fundamental issue to consider when developing solutions/applications for contact tracing is that of security and/or privacy. Successful implementation and adoption of IT applications such as tracing solutions could largely be dependent on how well the protection of privacy can be guaranteed [20]. While some existing contact tracing solutions considered some form of privacy, others completely ignored considerations for the privacy of individuals involved. Some countries have even adopted the notion of mass surveillance to track people in the name of contact tracing [21]. The following are some contact tracing proposals and/or software applications.

4.1 Epic

EPIC is an efficient privacy-preserving contact tracing application developed and reported in [22] for identifying infection. The authors reported that their contact tracing solution allows users to upload their data to the server in a secure manner. Also, where, an individual becomes infected, other persons can check whether they ever came in contact with him. In their system, no superfluous information is disclosed to the server, and the result of the contact tracing procedure is denoted by a matching score. Smartphones, Bluetooth devices, short-range wireless devices, and a server are the components involved in the EPIC model. An individual who was to confirm if he has been in contact with an infected person is required to send a request which includes his public key. The authors reportedly used homomorphic encryption to guarantee the information (queries) exchanged between individuals and the server. However, several privacy challenges still exist in this model: Firstly, while the queries are encrypted, the timestamps are stored in plain-text on the server. The implication, therefore, is that at a particular timestamp the location of the user is available to the server (or the authorities). The server can get compromised by individuals with malicious intent, thus leading to leaking or exposure of the timestamps from which other sensitive private information such as location, age, name, place of work, and so on can be deduced or inferred. Furthermore, there is no evidence to show that the EPIC application supports scalability (can withstand the increasing population of testing). It must be noted that scalability is a requirement for full-scale adoption of contact tracing systems.

4.2 TraceTogether

A mobile application for contact tracing was developed by the Government Technology Agency and Health Ministry of Singapore and reported in [23]. The functionality of this application involves the interchange of time-varying tokens thru Bluetooth connections among phones nearby. The components that make up the application include; citizens of Singapore as well as their Ministry of Health. This application falls into the category of centralized solutions. Users are expected to trust the Singaporean Health Ministry for the protection of their personal information. Encryption scheme(s) determined by the central health ministry is/are adopted for protecting users' information. Indeed, the government could force users to release their data to the contact tracing application as necessary. The TraceTogether application adopts encryption schemes for ensuring the protection of users' personal information. However, security/privacy issues exist in this application. The fact that this application uses time-varying tokens implies that snoopers also enjoy some form of privacy, and their efforts at tracking down innocent users get aided. Also, the centralized control vested in the hands of the health ministry implies that this software application suffers other security/privacy issues that include; linkage attack [24]. Furthermore, the large amount of location data made available to the government via this application creates the probability of a lot more security/privacy consequences; for instance, a malicious user can manipulate (i.e. add or delete) the data aggregated by this tracing application. Relay attacks could also be another probable attack. The trade-off between privacy and functionality is high.

4.3 Reichert's MPC Based Solution

The authors of the work in [25] proposed a privacy-preserving contact tracing application. This application adopts the principle of Multi-Party Computation (MPC), where two parties under semi-honest security settings are involved. The scenario in this application is such that, location-based-services used stores their history locally. The data points of infected persons can be used by the Health Authorities to initiate an MPC session with any individual who intends to trace themselves. The entities involved here are the Users and the Health Authority (HA). This application here is geo-location based and has a central control manned by the Health Authority. To prevent all personal information from been revealed or exposed, the Garbled circuit (GC) scheme is adopted. This contact tracing solution is a theoretical cryptography application. It suffers from several attacks that include; Denial of Service (DoS) or the Distributed Denial of Service (DDoS) attack with an increase in complexity and/or computational overhead.

4.4 CAUDHT

A Decentralized proximity-based Contact Tracing solution that works using a Distributed Hash Table (DHT) was presented in [26]. The major components of this solution are, the users, the DHT as well as the Health Authority. The use of the DHT implies that the Health Authority can only carry out confirmation of the results of established positive persons. This thus, limits how much data is available to the central authority via the protocol. This solution is resistant to linkage attacks and offers a lot of personal data privacy guarantees. This is largely due to the use of the DHT and Blind Signatures.

4.5 Berke et al.'s Location-Based System

Cryptography and GPS based Contact Tracing Solution was reportedly developed by the authors of [27]. The functionality of this system is founded on detailed GPS locations as well as a private set intersection. This enables their proposed system notification when a user came in close contact with an infected person(s). The major entities involved in this location-based (approach) contact tracing application are smartphone-carrying users as well as a server for storing encoded or encrypted personal/private data. This is a lifesaving application. However, since this application continuously collects data about users' locations, attackers might want to leverage on that to reconstruct location histories of users diagnosed as carriers and possible re-identification of those from their shared anonymized data. To cater to the general privacy preservation concerns, this application adopted the Diffie-Hellman protocol [28].

4.6 DP-3T

The DP-3T is reportedly the best-known contact tracing system currently. Reports have it that, a group of twenty-five (25) scholars from eight (8) European institutes teamed up to develop this contact tracing application that supports privacy preservation. Unlike other contact tracing applications, the DP-3T application works transparently and it is thus a more trustworthy system. Besides, it is a decentralized and even open-source system. The major actors in DP-3T are, the users, the health authority, as well as a server, used majorly for matching activities among users. Vaudenay [29] did point out about thirteen (13) attacks possible with DP-3T, especially the fact that the decentralized nature of the DP-3T could be more of a disadvantage than an advantage. However, their claim was refuted by the DP-3T community who claimed that the only probable attack against their system could be the behavior of their solution when "Several contact events" are involved.

5 Covid-19 Medical Imaging: Security Issues

Medical imaging plays a significant part in human health management globally. Infectious diseases such as tuberculosis, asthma, influenza, and other health issues that have to do with the lungs, chest, or entire human respiratory system are best examined using medical imaging. As the COVID-19 epidemic is rapidly spreading across the nations of the world, epidemiologists, other medical personnel, and even researchers have shifted attention to medical imaging such as X-ray and computed tomography (CT) as a viable tool in the fight against COVID-19. The efficacy of imaging tools in the fight against COVID-19 is further enhanced by emerging Artificial Intelligence (AI) technologies. The medical imaging procedure involves several stages that include, *image acquisition*, *segmentation*, as well as *diagnosis*, and maybe *follow-up*. Although acquisition, usage, and storage of medical images offer a lot of benefits in health management, it also comes with several concerns (challenges) chief among which is, the concern for security and/or privacy of these medical images. A review of existing literature shows that although a lot of COVID-19 related medical images based research works have been conducted in the past four to five months alone, yet almost none of these works considered the security and/or privacy of these images.

The research work reported in [30] underlined the significance of medical images and why they need to be secured. Medical images can be used by radiologists, virologists, epidemiologists, modelers, and other stakeholders for making clinical decisions, such as for disease diagnosis, tracing, and predictions. The authors opined that medical images are very personal and sensitive data that requires strict security measures either while in-store or while being transferred between medical laboratories, professionals, and specialists during referral cases or consultation. To cater for medical image security, their work adopted quantum cryptography. The paper however failed to present an evaluation report of their proposed approach. Neither did the paper present a comparative assessment of their method with medical image security methods.

It must be noted that a lot of personal private information about patients can be inferred or deduced from these images, hence the need to secure or protect these images from unwanted unauthorized malicious entities. This section of our chapter presents a summary of some imaging COVID-19 related works and further underlines the need for privacy or security considerations concerning these medical images.

The fact that there are so many of these COVID-19 related medical imaging research works underlines the significance of medical imaging to the prediction, diagnosis, and ultimately, combating of the pandemic. Indeed only some of these COVID-19 and other infectious disease-related medical imaging research works are presented in Table 1. Despite the many medical imaging research works, consideration was not given to the security/privacy concerns that come with medical image storages and/or exchange.

Table 1 Some COVID-19 and other infectious disease-related medical imaging research works

Research work	Type of image used	Application in COVID-19
Zheng et al. [31]	CT lung images	Diagnosis
Cao et al. [32]	CT lung images	Quantification
Huang et al. [33]	CT lung images	Quantification
Qi et al. [34]	CT lung images	Quantification
Gozes et al. [35]	CT lung images	Diagnosis
Li et al. [36]	CT lungs/lesion	Diagnosis
Chen et al. [37]	CT lungs/lesion	Diagnosis
Jin et al. [38]	CT Lung images	Diagnosis
Shan et al. [39]	CT lung images	Quantification
Tang et al. [40]	CT lung images	Quantification
Shen et al. [41]	CT lung images	Quantification
Ghoshal and Tucker [42]	X-ray images	Classification: COVID-19/others
Narin et al. [43]	X-ray images	Classification: COVID-19/normal
Zhang et al. [44]	X-ray: chest images	Classification: COVID-19/others
Wang and Wong [45]	X-ray images	Classification: COVID-19/bacterial pneumonia/viral Pneu./normal
Jin et al. [46]	CT	Classification: COVID-19/others
Wang et al. [47]	CT	Classification: COVID-19/viral pneumonia
Song et al. [48]	CT	Classification: COVID-19/bacterial pneumonia/normal
Xu et al. [49]	CT	Classification: COVID-19/influenza/normal
Shi et al. [50]	CT	Classification: COVID-19/CAP
Tang et al. [51]	CT	Severity assessment

6 General Covid-19 Related Security and Privacy Issues

Quite a lot of cyber security and privacy issues abound and are escalating at an alarming rate at this crucial period in the world. Indeed, removing identifying information, such as social security number, name, and address, does not guarantee privacy, since exploiting the remaining information, e.g., zip code, gender, etc., can still identify a person uniquely [52, 53]. In this section, we present some common probable attacks and likely solutions or mitigation strategies (Table 2).

Table 2 Some common cyber-attacks, their targets, and mitigation strategies

Attack name (or class)	Targets	Solution
Robust statistical de-anonymization attacks	High-dimensional micro-data, such as; • Individual preferences, recommendations, transaction records, and so on • Demographic information (e.g., age, gender, occupation, etc.) • User historical information (e.g., purchasing, browsing, location, etc.)	Cryptography based solutions k-anonymization [53, 54] t-closeness [55]
Inference attacks	Allows for more information, inference not only from original data available in store but by exploiting prediction or classification outputs	Differential privacy [56] Trusted processors based privacy-preservation solutions (e.g., Intel SGX-processors) [57–59]
Data injection attacks	Malicious entities can contaminate collated data by injecting them with false data and profiles. This can result in very serious negative consequences, as, wrong data becomes available to be fed into statistical/predictive models. This, in turn, leads to wrong predictions	Data injection attack detection frameworks Attack detection models Anomaly detection models

6.1 Sensitive Information Exposure

During contact tracing, most persons trust that the information submitted or registered about them will be used solely for COVID-19 mitigation or other purposes beneficial to them, however, their information can be used for purposes different than that. This is considered by [60] as *exposure risk*. According to [61], keeping patients personal information on a centralized entity can lead to exposure risks that fall into any of the following five (5) configurations:

- *Deception by the recipient (misleading service):* The system can lie about its privacy policies and trick users to reveal personal information, using it later for a different purpose from profiling for advertisement display. For example, selling the information or sharing it with other organizations.
- *Mission creep:* Initially, the policy of usage of personal information is defined clearly by the system, but later the systems expand its goals in a previously unforeseen manner, changing the use of personal information for other purposes related to the new goals of the organization.

- *Accidental disclosure:* Information about users can be made available accidentally, for example leaving private information on a server that can be accessed by a search engine over the Internet.
- *Disclosure by malicious intent:* Storage servers' security can be breached and users' personal information can be stolen.
- *Forced disclosure:* Systems must disclose the information for legal reasons.

Users can be targets of unwanted commercial solicitations (spam). Most of these unsolicited messages carry COVID-19 as a keyword. This is in a bid to prey on people's fear of the COVID-19 pandemic, which has heightened the hunger for related news. Unsuspecting individuals become victims of diverse cybercrimes just when they click to read such messages or news. Indeed, malware, worms, and Trojan horses can be deposited on their electronic gadgets. Even servers used by health employees at health care facilities can get infected or compromised leading to sensitive information leakages, data loss (which ultimately may lead to wrong or incomplete data been fed to statistical models, and hence wrong predictions). Theft of users' identity as well as a fraud of various kinds are a huge possibility.

6.2 Data Injection Attacks

Another set of attacks that could sabotage efforts at curbing COVID 19 and other health epidemics is the group referred to as data injection attacks. Some common probable data injection attacks include the following;

- False Data Injection Attacks
- Profile Injection Attacks
- Push Attacks
    - Random Attack
    - Average Attack
    - Bandwagon Attack
    - Segment Attack
- Nuke Attacks
- Obfuscated Attacks (Attacks with Unknown Models)
    - Noise Injection
    - User Shifting
    - Target Shifting.

Detailed explanations of these attacks can be found in [62].

6.3 Solutions and Mitigation Strategies

Several approaches exist for mitigating common cybercrimes in these trying times. Experts and even end-users must consider and equip themselves and their systems with this knowledge. Cryptographic, Steganographic, or even a combination of both, could be effective tools for information security challenges [63, 64]. The use of malware and virus detection models could also suffice [65]. In addition to these, some authors opined that effective modeling and configuration of risk management policies and even Internet security gateway rules/policies must be combined with cryptographic schemes to ensure robust anti-cybercrime structures [66, 67].

Sensitive information exposure or leakage detection/prevention strategies could include the following;

(a) Policy-based strategies

  (i) Privacy through law and regulation
  (ii) Privacy through awareness and user control

(b) Architectural Strategies

  (i) Pseudonymous personalization
  (ii) Anonymization (can be achieved via perturbation/obfuscation)

    - Heuristic transformations [68]
    - k-anonymity [53, 54]
    - Random perturbation techniques [69], and
    - differential privacy [56]:

  (iii) Client-side personalization (which can reduce private data exposure due to centralization)

    - Use of aggregation servers with anonymization and/or cryptography
      – Client-side profiling with anonymization (perturbation, randomization, the substitution of profile information)
    - P2P Architectures
      – pseudonymous and anonymization techniques
      – cryptography.

The use of security primitives and anonymity techniques is fundamental to security and privacy in information technology-based systems. Some of these security primitives and anonymization techniques include but are not limited to;

(i) homomorphic encryption [70],
(ii) secure multiparty computation,
(iii) garble circuits [71],
(iv) Differential privacy [56],
(v) k-anonymization [53, 54] and,

(vi) t-closeness [55].

To cater for data injection attacks, the following approaches suffices;

(a) Development and adoption of Data Injection Attack Detection frameworks towards;
    (i) Minimizing the impact of an attack,
    (ii) Reducing the likelihood of a successful attack, and
    (iii) Minimizing any negative impact resulting from the addition of the detection scheme
(b) Development and adoption of Attack Model Detection
(c) Development and adoption of Anomaly Detection strategies that could include;
    (i) Statistical Process Control
    (ii) Time Interval Detection Schemes.

6.4 Trust and Reputation Models for Enhancing the Adoption of IT in the Battle Against COVID-19

Adoption of trust as well as reputation models in some of the systems in existence and those been developed could also go a long way in enhancing the acceptance of contact tracing and other applications geared towards mitigating COVID-19 and other infectious epidemics. Some trust and reputations models as highlighted by Braga et al. in [72], includes the following;

(a) *Trust Models*
    (i) StereoTrust Model.
    (ii) Qualitative Assessment Dynamics.
    (iii) Sinalpha + Social Tuner + Contextual Fitness.
    (iv) User Trust Model.
    (v) Computational Trust Based on Quantum Decision Theory.
    (vi) Hidden Markov based Trust Model.
    (vii) BlackBox Trust Model.
    (viii) Hypertrust Model.
    (ix) Topological Potential Weighted Community-Based Recommendation Trust Model.
    (x) Qualitative Trust Model.
    (xi) Trust Assessment of Security for e-Health Systems.
(b) *Reputation Models*
    (i) Accumulated Reputation Model.
    (ii) NDR and NDRU.
    (iii) Negative Multinomial Reputation.

(iv) Vertical Handover Decision Making.
(v) Anti-Distributed Voice Spam.
(vi) ANN-based Reputation Bootstrapping.
(vii) Flow-Based Reputation with Uncertainty.
(viii) @Eigentrust Dynamic Evolutionary Model in P2P.
(ix) A Topic-Biased User Reputation Model in Rating Systems.
(x) Utility-Based Reputation Model for Grid Resource Management.

7 Conclusion

The enforcement of lockdown (restriction in movement) by the government of most nations across the globe in a bid to combat the dreaded COVID-19 pandemic has led many businesses, organizations, workers, and customers alike to adopt IT applications and computer networks for their everyday work and transactions. This explosive increase in the use of IT has led to so many cyber, and information security as well as privacy concerns. The attack surfaces and loopholes available to cybercriminals to perpetuate various cybercrimes have greatly increased, and the consequences to lives and economy can be gruesome. This book chapter highlighted these cyber security matters arising, especially as it affects individuals and their economies. Cyber security issues that come into play during COVID-19 induced digital or mobile contact tracing, COVID-19 induced remote work from home, COVID-19 medical imaging, and even the cyber security implications of the spread of COVID-19 related fake news (misinformation) were discussed. Some of the efforts currently been made to combat the excesses of cybercriminals and fraudsters and other cyber security issues as well as some recommendations and/or tips to cybercrime-safe cyberspace were presented.

Organizations and businesses must endeavor to carry out a thorough assessment of their cyber posture especially as remote access to organizational servers and resources are fast becoming a norm. They should endeavour to create massive all-encompassing cyber security awareness for their workers, customers, and other third-parties. Organizations must ensure that every single point of failure in their security architecture is carefully managed. Finally, healthcare service providers must carefully plan robust cyber security schemes that will guarantee the safety and privacy of patients' personal information.

References

1. Tong, T.R.: Severe Acute Respiratory Syndrome Coronavirus (SARS-CoV). Perspect. Med. Virol. **16**, 43–95 (2006). https://doi.org/10.1016/S0168-7069(06)16004-8
2. Tsang, K.W.: Severe Acute Respiratory Syndrome (SARS). Int. Encycl. Public Health 691–695 (2008). https://doi.org/10.1016/B978-012373960-5.00219-7

3. Cheng, V.C., Lau, S.K., Woo, P.C., Yuen, K.Y.: Severe acute respiratory syndrome coronavirus as an agent of emerging and reemerging infection. Clin. Microbiol. Rev. **20**(4), 660–694 (2007). https://doi.org/10.1128/CMR.00023-07
4. Oh, M.D., Park, W.B., Park, S.W., Choe, P.G., Bang, J.H., Song, K.H., Kim, E.S., Kim, H.B., Kim, N.J.: Middle East respiratory syndrome: what we learned from the 2015 outbreak in the Republic of Korea. Korean J. Intern. Med. **33**(2), 233–246 (2008). https://doi.org/10.3904/kjim.2018.031
5. Aleanizy, F.S., Mohmed, N., Alqahtani, F.Y., El Hadi Mohamed, R.A.: Outbreak of Middle East respiratory syndrome coronavirus in Saudi Arabia: a retrospective study. BMC Infect. Dis. **17**(1), 23 (2007). https://doi.org/10.1186/s12879-016-2137-3
6. Sandhya, R.: Public health experts worry about the spread of COVID-19 misinformation. https://www.rollcall.com/2020/03/18/public-health-experts-worry-aboutspread-of-COVID-19-misinformation/. Published in 2020
7. Shu, K., Sliva, A., Wang, S., Tang, J., Liu, H.: Fake news detection on social media. ACM SIGKDD Explor. Newsl. (2017). https://doi.org/10.1145/3137597.3137600
8. Busari, S., Adebayo, B.: Nigeria records chloroquine poisoning after Trump endorses it for coronavirus treatment. CNN. https://www.cnn.com/2020/03/23/africa/chloroquine-trump-nigeria-intl/index.html. Published 2020
9. Joe, W.: Coronavirus: Indian man "died by suicide" after becoming convinced he was infected. The Telegraph. https://www.telegraph.co.uk/global-health/science-and-disease/coronavirus-indian-man-died-suicide-becoming-convinced-infected/. Published 2020
10. Spencer, S.H.: False claims of nationwide lockdown for COVID-19.https://www.factcheck.org/2020/03/false-claims-of-nationwide-lockdown-for-covid-19/
11. Kasulis, K.: Patient 31 and South Korea's sudden spike in coronavirus cases. Aljazeera. https://www.aljazeera.com/news/2020/03/31-south-korea-sudden-spike-coronavirus-cases-200303065953841.html
12. The Verge. Major tech platforms say they're 'jointly combating fraud and misinformation' about COVID-19. https://www.theverge.com/2020/3/16/21182726/coronavirus-covid-19-facebook-google-twitter-youtube-joint-effort-misinformation-fraud
13. Collinson, S., Khan, K., Heffernan, J.M.: The effects of media reports on disease spread and important public health measurements. PLoS ONE (2015). https://doi.org/10.1371/journal.pone.0141423
14. Pennycook, G., McPhetres, J., Zhang, Y., Rand, D.: Fighting COVID-19 misinformation on social media: experimental evidence for a scalable accuracy nudge intervention. PsyArXiv [Working Paper], 1–24 (2020). https://doi.org/10.31234/OSF.IO/UHBK9
15. Ahmad, T.: Corona virus (COVID-19) pandemic and work from home: challenges of cybercrimes and cyber-security. Available at SSRN 3568830, 2020—papers.ssrn.com
16. Gabriel, J.A., Egwuche, O.S.: Modeling employees' activities of public service sector using production rules. In: Annals Computer Science Series 13th Tome 2nd Fasc, Romania, 2015
17. Iwasokun, G.B., Egwuche, O.S., Gabriel, A.J.: Neural network-based health personnel monitoring system. Afr. J. Comput. ICT IEEE **8**(1), 79–87 (2015)
18. Roomp, K., Oliver, N.: ACDC-tracing: towards anonymous citizen-driven contact tracing. Arxiv 2020 preprint, available at https://arxiv.org/ftp/arxiv/papers/2004/2004.07463
19. Raskar, R., Schunemann, I., Barbar, R., Vilcans, K., Gray, J., Vepakomma, P., Kapa, S., Nuzzo, A., Gupta, R., Berke, A.: Apps gone rogue: maintaining personal privacy in an epidemic. arXiv preprint arXiv:2003.08567 (2020)
20. Gabriel, A.J., Alese, B.K., Adetunmbi, A.O., Adewale, O.S., Sarumi, O.A.: Post-quantum crystography system for secure electronic voting. Open Comput. Sci **9**, 292–298 (2019). https://doi.org/10.1515/comp-2019-0018
21. Asher, A.I.: 11 countries are now using people's phones to track the coronavirus pandemic, and it heralds a massive increase in surveillance. Accessed 15 May 2020. Online. Available: www.businessinsider.com/countriestracking-citizensphones-coronavirus-2020-3?r=DEIR=
22. Altuwaiyan, T., Hadian, M., Liang, X.: Epic: efficient privacy-preserving contact tracing for infection detection. In: 2018 IEEE International Conference on Communications (ICC), IEEE, pp. 1–6 (2018)

23. Tracetogether Accessed 23 April 2020. [Online]. Available https://www.tracetogether.gov.sg/
24. Dwork, C., Roth, A.: The algorithmic foundations of differential privacy. Found. Trends Theor. Comput. Sci. **9**(3–4), 211–407 (2014)
25. Reichert, L., Brack, S., Scheuermann, B.: Privacy-preserving contact tracing of covid-19 patients. In: 2020. Cryptology eprint archive. Available at https://eprint.iacr.org/2020/375.pdf
26. Brack, S., Reichert, L., Scheuermann, B.: Decentralized contact tracing using a dht and blind signatures. Cryptology eprint archive, 2020. Available at https://eprint.iacr.org/2020/398.pdf
27. Berke, A., Bakker, M., Vepakomma, P., Raskar, R., Larson, K., Pentland, A.: Assessing disease exposure risk with location histories and protecting privacy: a cryptographic approach in response to a global pandemic. arXiv preprint arXiv:2003.14412 (2020)
28. Diffie, W., Hellman, M.: New directions in cryptography. IEEE Trans. Inf. Theory **22**(6), 644–654 (1976)
29. Vaudenay, S.: Analysis of DP-3T. Accessed: 23 April 2020. [Online]. Available: https://eprint.iacr.org/2020/399 (2020)
30. Alowolodu, O.D., Adelaja, G.K., Alese, B.K., Olayemi, O.C.: Medical image security using quantum cryptography. Issues Informing Sci. Inf. Technol. **15**, 57–67 (2018)
31. Zheng, C., Deng, X., Fu, Q., Zhou, Q., Feng, J., Ma, H.: Deep learning-based detection for COVID-19 from chest CT using weak label. MedRxiv (2020)
32. Cao, Y., Xu, Z., Feng, J., Jin, C., Han, X., Wu, H.: Longitudinal assessment of COVID-19 using a deep learning–based quantitative CT pipeline: illustration of two cases. Radiol. Cardiothorac. Imaging **2**, e200082 (2020)
33. Huang, L., Han, R., Ai, T., Yu, P., Kang, H., Tao, Q.: Serial quantitative chest CT assessment of COVID-19: deep-learning approach. Radiol. Cardiothorac. Imaging **2**, e200075 (2020)
34. Qi, X., Jiang, Z., Yu, Q., Shao, C., Zhang, H., Yue, H.: Machine learning-based CT radiomics model for predicting hospital stay in patients with pneumonia associated with SARS-CoV-2 infection: a multicenter study. MedRxiv (2020)
35. Gozes, O., Frid-Adar, M., Greenspan, H., Browning, Y., Zhang, H., Ji, W.: Rapid AI development cycle for the coronavirus (covid-19) pandemic: initial results for automated detection and patient monitoring using deep learning ct image analysis. arXiv:2003.05037 (2020)
36. Li, L., Qin, L., Xu, Z., Yin, Y., Wang, X., Kong, B.: Artificial intelligence distinguishes COVID-19 from community acquired pneumonia on chest CT. Radiology 200905 (2020)
37. Chen, J., Wu, L., Zhang, J., Zhang, L., Gong, D., Zhao, Y.: Deep learning-based model for detecting 2019 novel coronavirus pneumonia on high-resolution computed tomography: a prospective study. MedRxiv (2020)
38. Jin, S. Wang, B. Xu, H. Luo, C. Wei, Y. Zhao, W., "AI-assisted CT imaging analysis for COVID-19 screening: Building and deploying a medical AI system in four weeks," *MedRxiv*, 2020
39. Shan, F., Gao, Y., Wang, J., Shi, W., Shi, N., Han, M.: Lung infection quantification of COVID-19 in CT images with deep learning. arXiv:2003.04655 (2020)
40. Tang, L., Zhang, X., Wang, Y., Zeng, X.: Severe COVID-19 Pneumonia: assessing inflammation burden with volume-rendered chest CT. Radiol. Cardiothorac. Imaging **2**, e200044 (2020)
41. Shen, C., Yu, N., Cai, S., Zhou, J., Sheng, J., Liu, K.: Quantitative computed tomography analysis for stratifying the severity of coronavirus disease 2019. J. Pharm. Anal. (2020)
42. Ghoshal, B., Tucker, A.: Estimating uncertainty and interpretability in deep learning for coronavirus (COVID-19) detection. arXiv:2003.10769 (2020)
43. Narin, A., Kaya, C., Pamuk, Z.: Automatic detection of coronavirus disease (COVID-19) using X-ray images and deep convolutional neural networks. arXiv:2003.10849 (2020)
44. Zhang, J., Xie, Y., Li, Y., Shen, C., Xia, Y.: COVID-19 screening on chest X-ray images using deep learning based anomaly detection. arXiv:2003.12338 (2020)
45. Wang, L., Wong, A.: COVID-net: a tailored deep convolutional neural network design for detection of COVID-19 cases from chest radiography images. arXiv:2003.09871 (2020)
46. Jin, C., Cheny, W., Cao, Y., Xu, Z., Zhang, X., Deng, L.: Development and evaluation of an AI system for COVID-19 diagnosis. MedRxiv (2020)

47. Wang, S., Kang, B., Ma, J., Zeng, X., Xiao, M., Guo, J.: A deep learning algorithm using CT images to screen for Corona Virus Disease (COVID-19). MedRxiv (2020)
48. Song, Y., Zheng, S., Li, L., Zhang, X., Zhang, X., Huang, Z.: Deep learning enables accurate diagnosis of novel Coronavirus (COVID-19) with CT images. MedRxiv (2020)
49. Xu, X., Jiang, X., Ma, C., Du, P., Li, X., Lv, S.: Deep learning system to screen Coronavirus disease 2019 peumonia. arXiv:2002.09334 (2020)
50. Shi, F., Xia, L., Shan, F., Wu, D., Wei, Y., Yuan, H.: Large-scale screening of COVID-19 from community acquired pneumonia using infection size-aware classification. arXiv:2003.09860 (2020)
51. Tang, Z., Zhao, W., Xie, X., Zhong, Z., Shi, F., Liu, J.: Severity assessment of coronavirus disease 2019 (COVID-19) using quantitative features from chest CT images. arXiv:2003.11988 (2020)
52. Wang, J.: Privacy-preserving recommender systems facilitated by the machine learning approach. PhD Thesis at the Université du Luxembourg (2018)
53. Sweeney, L.: k-anonymity: a model for protecting privacy. Int. J. Uncertainty Fuzziness Knowl. Based Syst. **10**(05), 557–570 (2002)
54. Bayardo, R.J., Agrawal, R.: Data privacy through optimal k- anonymization. In: ICDE 2005 Proceedings 21st International Conference on Data Engineering, 217–228. IEEE, 2005
55. Li, N., Li, T., Venkatasubramanian, S.: t-closeness: privacy beyond k-anonymity and l-diversity. In: ICDE 2007 IEEE 23rd International Conference on Data Engineering, **128**, 106–115. IEEE (2007)
56. Dwork, C., Roth, A.: The algorithmic foundations of differential privacy. Found. Trends® Theor. Comput. Sci. **9**(3–4), 211–407 (2014)
57. Hunt, T., Zhu, Z., Xu, Y., Peter, S., Ryoan, E.W.: A distributed and box for untrusted computation on secret data. In OSDI, 533–549 (2016)
58. Hynes, N., Cheng, R., Song, D.: Efficient deep learning on multi-source private data. arXiv preprint arXiv:1807.06689 (2018)
59. McKeen, F., Alexandrovich, I., Berenzon, A., Rozas, C.V., Shafi, H., Shanbhogue, V., Savagaonkar, U.R.: Innovative instructions and software model for isolated execution. HASP@ISCA **10** (2013)
60. Lam, S., Frankowski, D., Riedl, J.: Do you trust your recommendations? An exploration of security and privacy issues in recommender systems. In: Muller, G. (ed.) Emerging trends in information and communication security, volume 3995 of Lecture Notes in Computer Science, Ch. 2, p. 14, vol 29. Springer Berlin/Heidelberg, Berlin, Heidelberg (2006 Cited on page 36)
61. Foner, L.N.: Political artifacts and personal privacy: the yenta multi-agent distributed matchmaking system. PhD thesis, Program in Media Arts and Sciences, School of Architecture and Planning, Massachusetts Institute of Technology (June 1999, Cited on pages 4 and 36)
62. Williams, C.: Profile injection attack detection for securing collaborative recommender systems center for web intelligence. DePaul University School of Computer Science, Telecommunication, and Information Systems Chicago, Illinois, USA
63. Gabriel, A.J., Alese, B.K., Adetunmbi, A.O., Adewale, O.S.: Post-quantum crystography based security framework for cloud computing. J. Int. Technol. Secured Trans. **3**(4), 344–350 (2014)
64. Gabriel, A.J., Alese, B.K., Adetunmbi, A.O., Adewale, O.S.: Post-quantum crystography: a combination of post-quantum cryptography and steganography. In: The 8th International Conference for Internet Technology and Secured Transactions (ICITST-2013), Technically Co-sponsored by IEEE UK/RI Computer Chapter, 9th-12th Dec 2013, London, UK, pp. 454–457
65. Adebayo, O.T., Alese, B.K., Gabriel, A.J.: A model for computer worm detection in a computer network. Int. J. Comput. Appl. (0975–8887). **66**(2), 22–28 (2013)
66. Alese, B.K., Gabriel, A.J., Olukayode, O., Daramola, O.A.: Modelling of risk management procedures for cybercrime control systems. In: The 2014 International Conference of Information Security and Internet Engineering, World Congress on Engineering, ISBN 978-988-19252-7-7; pp. 505–509. (United Kingdom, 2014)

67. Alese, B.K., Gabriel, A.J., Adetunmbi, A.O.: Design and implementation of internet protocol security filtering rules in a network environment. Int. J. Comput. Sci. Inf. Secur. USA **9**(7), 134–143. Available at www.academia.edu (2011)
68. Bakken, D.E., Rarameswaran, R., Blough, D.M., Franz, A.A., Palmer, T.J.: Data obfuscation: anonymity and desensitization of usable data sets. IEEE Secur. Priv. **2**(6), 34 (41, November 2004, Cited on page 39)
69. Agrawal, R., Srikant, R.: Privacy-preserving data mining. In: International Conference on Management of Proceedings of the 2000 ACM SIGMOD
70. Gentry, C.: A fully homomorphic encryption scheme. PhD thesis, Stanford, CA, USA (2009). AAI3382729
71. Yao, A.C.-C.: How to generate and exchange secrets. In: 27th Annual Symposium on Foundations of Computer Science, 1986, pp. 162–167. IEEE, 1986
72. Braga, D.D., Niemann, M., Hellingrath, B., Neto, F.B.: Survey on computational trust and reputation models. ACM Comput. Surv. **51**, 5, 101 (Nov. 2018, 40 pages). https://doi.org/10.1145/3236008

Tracking of COVID-19 Geographical Infections on Real-Time Tweets

Khaled Ahmed, Sara Abdelghafar, Aya Salama, Nour Eldeen M. Khalifa, Ashraf Darwish, and Aboul Ella Hassanien

Abstract Coronavirus COVID-19 is a global pandemic stated by the World Health Organization (WHO) in 2020. The COVID-19 devastating impact was not only affect human life but also many aspects of it such as social interaction, transportation options, personal saving and expenses, and more. The power of social media data in such world pandemic outbreaks provides an efficient source of tracking, raising awareness, and alerts with potentials infection location. Social networks can fight the pandemic by sharing helpful content and statistics based on demographics features of users around the world. There is an urgent need for such frameworks for tracking helpful content, detecting misleading content, ranking the trusted user content, presenting accurate demographics statistics of the outbreak. In this paper, the real-time tweets of Coronavirus pandemic (COVID-19) analysis will be presented. The proposed framework will be used to track the geographical infections, trends of the content, and the user's categorization. The framework will include analysis, demographics features, statistical charts, classifying the content of tweets related to its usefulness. The performance of the proposed framework is evaluated based on different measures such as classification accuracy, sensitivity, and specificity. Finally, a set of recommendations will be presented to benefit from the proposed framework with its full potentials as a tool to stand against the COVID-19 spreading.

Keywords COVID-19 disease · COVID-19 · Twitter analysis · Data science · Misleading detection · Tweets ranking

K. Ahmed (✉) · A. Salama · N. E. M. Khalifa · A. E. Hassanien
Faculty of Computers and Artificial Intelligence, Cairo University, Giza, Egypt
e-mail: khaled.elahmed@gmail.com
URL: http://www.egyptscience.net

S. Abdelghafar
Computer Science Department, Faculty of Science, Al-Azhar University, Cairo, Egypt

A. Darwish
Faculty of Science, Helwan University, Cairo, Egypt

S. Abdelghafar · A. Salama · N. E. M. Khalifa · A. Darwish · A. E. Hassanien
Scientific Research Group in Egypt (SRGE), Giza, Egypt

A. E. Hassanien and A. Darwish (eds.), *Digital Transformation and Emerging Technologies for Fighting COVID-19 Pandemic: Innovative Approaches*, Studies in Systems, Decision and Control 322, https://doi.org/10.1007/978-3-030-63307-3_19

1 Introduction

The recently discovered Coronavirus pandemic (COVID-19) has caused a global infection alarm. More than 1 million, 800 thousand confirmed cases, and more than 117,000 COVID-19 deaths have been registered in 213 countries [1] according to the World Health Organization (WHO). In such pandemics, public awareness should be elevated. Social media analysis is considered one of the tracking tool as stated in scientific researches, including geographical identification to raise awareness of infected place to alarm user to avoid those places [2–4], while some studies focus on the estimation for the content trends and users categorization [5–9].

This chapter introduces a data analysis approach and tweet statistics on COVID19. A misleading detection model based on the meaning of tweets, classification, and profile ratings will be presented. The COVID-19 twitter data set [10], which comprises 22,985 tweets, was obtained between January 01 and 24 March 2020 will be used in this research. Recall, precision, f-measure and Root Mean Square Error (RMSE) will be used to evaluate the performance of the proposed framework based on the conducted experiments.

The rest of this chapter is structured in the following Sections. The suggested solution is laid out in detail in Sect. 2. The dataset description, evaluation metrics, and experimental results are presented in Sect. 3. Section 4 concludes this work and discusses future work.

2 The Proposed Tracking of COVID-19 Geographical Infections System

The proposed system is divided into three components, the first component is the data analysis based on the tweets data set such as the number of CVOID-19 tweets per day, the highest tweets area over a geographical map, top-ranked user value, lowest-ranked user value, and tweet value per date. Figure 1 illustrates the different phases of the framework. The second component is the detection of misleading tweets based on the tweet value [11, 12].

The tweets were preprocessed by removing extra spaces, new lines, and converting the text to Unicode format, then the text is mapped with it a label. The label has two values (useful or misleading). Afterward, the word bagging process is initiated, along with the normalization process. The binary classier based on the average perceptron algorithm is trained with 65% of the dataset size, while the rest of the data is used for testing. The third component is dividing and ranking the users into three classes (ranked the helpful content to three classes A, B, and C).

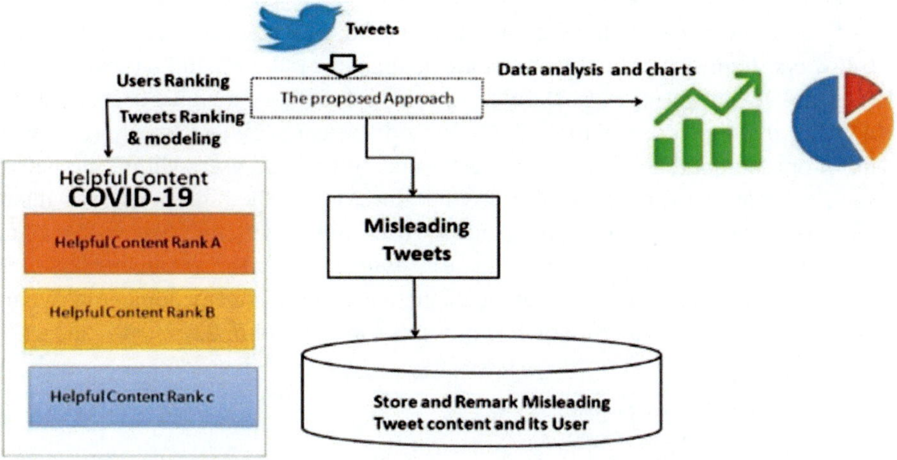

Fig. 1 The architecture of the proposed approach

2.1 Tweets (Modeling, Ranking) and Users Ranking

The proposed approach is based on three algorithms, The first algorithm is used to model the tweets into a unified pattern then rank the tweet based on its tweet value. Algorithm 1 starts by taking an input set of tweets, applying preprocessing word bagging, stop words checking, normalization to the data into suitable pattern to be used for the classifier for the training phase. Tweet value indicates the importance of the tweets within the data sets, it is calculated for each tweet in the data set based on number of likes, tweet text, tweet permalink, re-tweet count, and likes count, so the higher tweet value the higher rank and importance tweet, tweets are divided into three classes based on tweet value (A, B and C) which is presented in Algorithm 1.

Algorithm 1: tweets modeling and ranking

```
input Tweets data set
data Pre -processing ()
word bagging process
normalization process
calculate TweetValue()
OrderUsefultweetsDescending( tweetValue)
Int i=0;
while i ≤ UsefulTweetsList.Length() do
     if i≤ 25 then
         usefulRankA.Add(tweets[i])
         else if i > 25 AND i ≤ 50 then
           usefulRankB.Add(tweets[i])
          else
         usefulRankC.Add(tweets[i])
Endif
Endwhile
```

User value is calculated for each user in the data set based on the number of followers, total tweets, re-tweets number, images (yes, no), links (yes, no), chats (yes, no), text tweets (yes, no), and followers' number so the higher user value in the network the higher user rank in the proposed approach, by this feature (user value) the proposed approach can track, cluster and control the users within the network such as create a list of the lowest users' values, list of the highest users' values or clustering the users into three main classes user rank A with the highest 25% of user value, Class B the less ration of the top users with 25% and the rest are in class C. The proposed approach can as well divide the network users into any number of dynamic clusters based "user value" which is presented in Algorithm 2.

Algorithm 2: User ranking

```
input Tweets data set
data Pre -processing ()
word bagging process
normalization process
calculate UserValue()
OrderUsers Descending( UserValue)
Int i=0;
while i ≤ ListList.Length() do
    if i≤ 25 then
        usefulRankA.Add( user [i])
        else if i > 25 AND i ≤ 50 then
          usefulRankB.Add( user [i])
         else
        usefulRankC.Add( user [i])
Endif
Endwhile
```

2.2 *Misleading Tweets*

Input data set is labeled based on tweets value into two classes (useful tweet, misleading tweet) to be used for training the binary classier with tweets text and class label which is presented in Algorithm 3.

Algorithm 3 : Twitter misleading detection and tweets ranking

```
Apply algorith1 from step 1 till step 5
binary classier Training
if PredictLabelClass== Useful then
    add to useful content
    store tweet value  , username
else
    add to misleading content  List
    store tweet value  , user name
end if
```

3 Experimental Results and Discussion

3.1 *Data Characteristics*

The COVID-19 twitter data set used in the research was launched in early 2020 published by tweet binder. Tweet binder is a social media software company specialized in social media analytics. The dataset consists of 22,985 tweets that were collected for three months. Those tweets had more than 1000 re-tweets on the twitter platform. It contains, twitter username, tweet date, tweet text, tweet permalink, re-tweet count, and likes count. Table 1 presents an example of data from the COVID-19 twitter dataset used in this research. Also, the data set consists of users data that have more than 1000 re-tweets. The data includes username, name, location, total tweets, re-tweets number, images (yes, no), links (yes, no), chats (yes, no), text tweets (yes, no), and followers' number.

3.2 *Data Cleaning Process*

A clean process on the location column in the dataset is performed before the analysis of the location of the tweets. There are a few noises in the location column such as "everywhere," "world" and "home." Figure 2 displays an ambiguous, noisy data sample. In some locations text, the IP addresses, emails, telephone numbers, or cities such as "Africa, Iran and Germany" are also included. Besides, some locations have also been given in Arabic, French and Italian languages other than English.

Consequently, we have used some cleaning processes to define where most of the data have been tweeted. For this cause, as previously stated, we have excluded records without location or records with noisy data. We have also removed all special characters, street names, and unnecessary additional details. Also, some tweets have relative names such as 'England' and 'UK,' or 'USA' and 'America.' It will affect the number of tweets on the charts. As a result, the locations in related places had to be merged in one location. The number of tweets with its location was used to illustrate the traffic of tweets on the map as it will be shown in the experimental results and Fig. 3.

3.3 *Performance Evaluation Measurements*

Precision and recall are the evaluation measures frequently used as they represent the efficiency of any proposed framework. Precision evaluates how good the classifier in defining correctly the positives [13] is. While, recall represents the ability of the classifier to claim all the positives [14]. Therefore, it depends on the machine problem

Table 1 Data examples from the COVID-19 twitter dataset

Twitter username	Tweet date	Tweet text	Tweet permalink	Retweet count	Likes count
The Spectator Index	23/3/2020 22:11:21	United States: Man, in his sixties dies after trying to self-medicate against coronavirus by taking chloroquine phosphate in the form of fish tank cleaner	https://www.twitter.com/user/status/1242212365426253824	1158	2607
The White House	23/3/2020 22:08:41	Live: Press briefing with coronavirus task force https://t.co/Ciwe6T4Dr2	https://www.twitter.com/user/status/1242211691649052672	2411	5061

@georgien	Georgie Ndirangu	Africa ✈ World ✈ Africa
@AyoBanl	Ayò Bánkólé	African Man | Global Thoughts
@afrkml	Ns. Rizal do 런쥔	afriezalkamilgo@gmail.com
@Proserpi	Lupesb EU 🇪🇺❤	All around the world
@LewisHa	Lewis Hamilton	All around the World!!
@screenra	Screen Rant	All over
@FrankLu	Frank Luntz	All over
@naoseja\	Não sei, tô louca	ativem as notificações 🔔
@c_droste	Christian Drosten	France Germany
@fepache(	Felipe Pacheco	France, Germay
@oriolserr;	Oriol Serra	Berlín, REPÚBLICA Fed. d'Alemany;

Fig. 2 Noisy sample of tweets' location

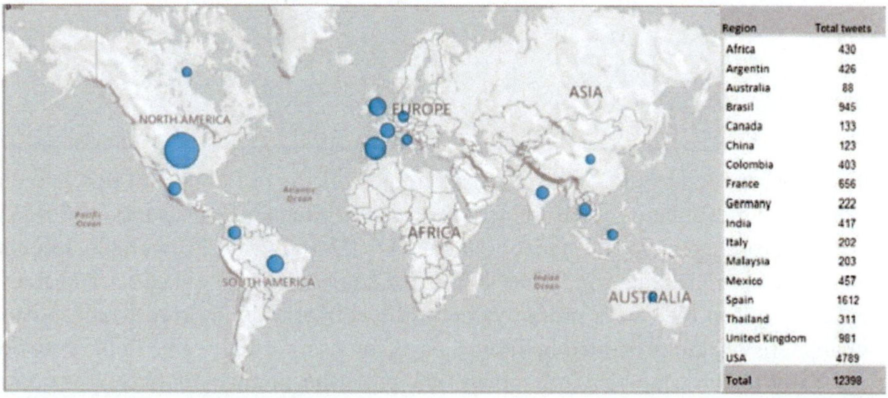

Fig. 3 Number of tweets per country

whether to favor recall over precision or vice versa. Equation 1 and Eq. 2 present Precision and Recall respectively as the following:

$$\text{Percision} = \frac{TP}{TP + FP} \tag{1}$$

$$\text{Recall} = \frac{TP}{TP + FN} \tag{2}$$

where TP is the true positive that is correctly classified as positive, TN is the true negative that is correctly classified as negative. However, FP is the false positive that is classified as positives and they don't belong to the class. And the same for FN, they are the false negatives which are classified as negative and they belong to the given class. Also, another important evaluation measure is F-measure that can be described as the mean of both recall and precision [15] and calculated as presented in Eq. 3.

$$\text{F-measure} = \frac{2 * recall * percision}{percision + recall} \quad (3)$$

Besides the precision, recall, and F-measure, RMSE is used to evaluate the accuracy of the proposed classifier model that must be measured first as it has a direct effect on the validity of the predicted values. RMSE is measured to determine the accuracy as shown in Eq. 4 [16].

$$RMSE = \sqrt{\frac{1}{\mathrm{n}} \sum_{i=1}^{n} (\mathrm{Y_{pi}} - \mathrm{Y_{Ai}})2} \quad (4)$$

where $Y_P = [y_{P1}, y_{P2}, \ldots, y_{Pn}]$ is the predicted value of n samples and $Y_A = [y_{A1}, y_{A2}, \ldots, y_{An}]$ is the actual value.

3.4 Results

The experimental results of the proposed framework are presented in Table 2. Table 2 presents the evaluation measurements (Precision, Recall, F-measure), and RMSE. Table 3 presents the WHO statistics about COVID-19 tweets. Top-ranked user characteristics are presented in Table 4. Table 5 shows the top-ranked user characteristics based on the user's profile value in the network, while lowest-ranked users' characteristics are presented in Table 6 based on users' tweet value in the network.

WHO tweets statistics are shown in Fig. 4, the number of tweets per day is presented in Fig. 5. The lowest count of re-tweets for users of twitter is presented in Fig. 6, the summation of total COVID-19 tweets impact on the network per month

Table 2 Classification performance evaluation results

Evaluation measures	Results
RMSE	28.9
Recall	0.83
Precision	0.82
F-Measure	0.82

Table 3 WHO tweets statistics for the COVID-19

	Retweet count	Likes count	Tweet value
January	152,472	198,323	441
February	141,556	226,033	862
March	400,119	622,823	464
Total	694,147	1,047,179	1767

Table 4 Top ranked users re-tweets count

Most retweeted users	Retweet count
@realDonaldTrump	965,938
@OH mes2	666,441
@spectatorindex	619,282
@KKMPutrajaya	545,316
@tedlieu	486,116
@elonmusk	350,555
@WHO	305,479
@SethAbramson	298,224
@funder	297,570
@briantylercohen	265,325

Table 5 Top ranked users based on users value

User	User value	Tweets value	Impacts	Followers
@BarackObama	149,555.48	937,730.30	1,029,551,294	114,525,349
@Cristiano	111,363.10	146,999.29	83,133,379	83,133,379
@realDonaldTrump	101,549.23	4,655,495.21	3,416,987,529	74,921,695
@cnnbrk	79,763.11	2,926,715.96	1,592,790,699	56,965,914
@narendramodi	74,891.55	1,480,665.47	1,021,192,777	53,890,511
@BillGates	69,580.83	168,244.42	98,312,130	49,206,297
@CNN	65,902.53	6,233,159.84	3,867,253,802	46,209,546
@nytimes	65,698.71	5,259,362.62	3,561,917,622	45,744,661
@BBCBreaking	60,700.61	2,155,949.50	1,344,547,747	42,176,801
@SrBachchan	57,985.11	69,582.13	40,613,193	40,613,193
@NiallOfficial	57,036.70	57,036.70	39,469,616	39,469,616
@SportsCenter	52,593.58	1,130,020.55	654,471,734	36,385,762
@NASA	52,094.19	57,303.61	35,758,385	35,758,385
@espn	51,047.00	183,769.14	105,390,928	35,130,322
@imVkohli	49,498.41	59,176.76	68,258,110	34,213,243
@elonmusk	47,845.39	110,541.26	97,827,343	32,725,028
@NBA	44,545.43	58,799.97	30,242,570	30,242,570
@ConanOBrien	42,572.87	85,145.74	57,203,251	28,601,626

is shown in Fig. 7 while top-ranked user based on several re-tweets is presented in Fig. 8 and finally COVID-19 tweets impact per date is presented in Fig. 9.

The highest number of COVID-19 tweets was 20,000,000 tweets in a period of (12 and 13 March), the lowest number of tweets was at 4 January, the sum of tweets value was the highest at March and lowest at January WHO authority shared the top

Table 6 Lowest misleading users based on users value

User	User value	Tweets value	Impacts	Followers
@pawat58720145	0.00	0.00	1	1
@sadbadme	0.00	0.00	1	1
@archawi51088663	0.00	0.00	0	0
@gegyrigle	0.00	0.00	1	1
@DelsVeja	0.00	0.00	0	0
@PVxNASgVi9jVIDs	0.01	0.01	1	1
@itwasmymess	0.01	0.01	3	3
@tsah85801624	0.01	0.01	3	3
@chmpooth	0.01	0.01	2	2
@arveecasil09	0.01	0.01	3	3
@jibbie Fp	0.01	0.01	3	3
@vickyc828	0.01	0.01	2	2
@Piyamas76998789	0.02	0.02	5	5
@JohnJaide	0.02	0.02	4	4
@kukulcanondero	0.02	0.02	6	6
@haruhiandshiro	0.02	0.02	5	5
@knxxminnie	0.02	0.02	5	5
@fafa82134746	0.02	0.02	5	5

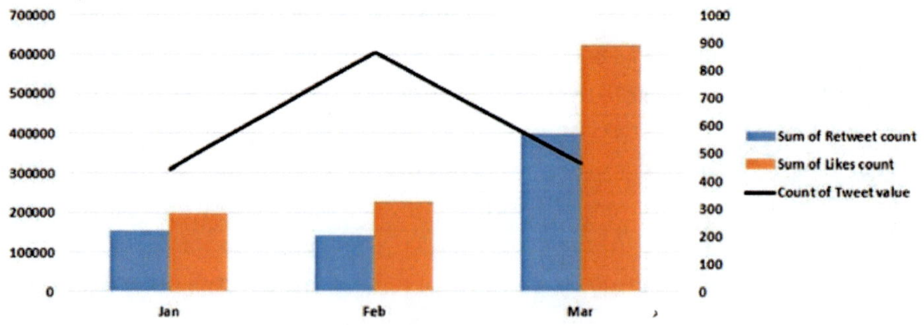

Fig. 4 Count of WHO tweets over the months of the COVID-19 dataset

liked tweets, the highest number of re-tweeting in March while the value of the top-ranked tweets was at February. Based on the presented map the top-ranked countries that re-tweets about COVID-19 are the USA, Span, England, and Brazil while the lowest was Australia.

Figure 4, Who authority Tweets about Covid-19 in January 2020 to March 2020 are analyzed using the proposed approach in three main features such as re-tweets count, number of likes for tweets, and tweets value. In January 2020 the minimum number of the three features is achieved. In February 2020 tweets had achieved the

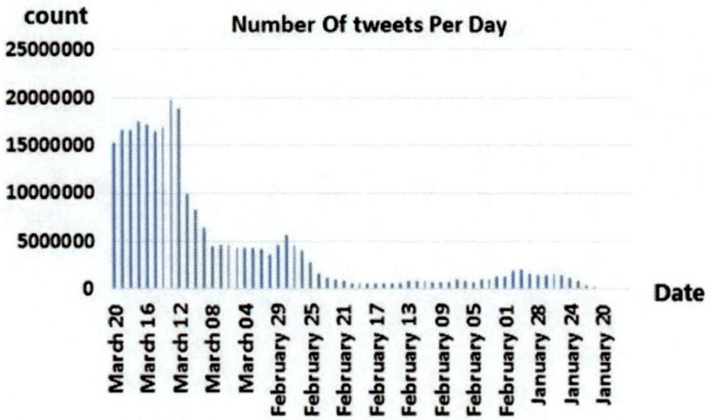

Fig. 5 Number of tweets per day

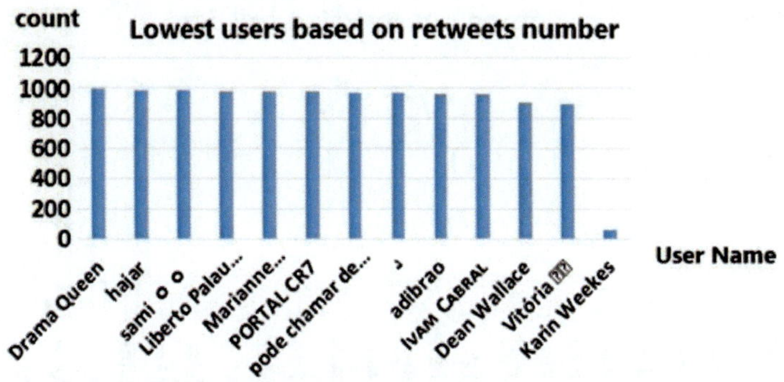

Fig. 6 Lowest ranked users based on the re-tweets count

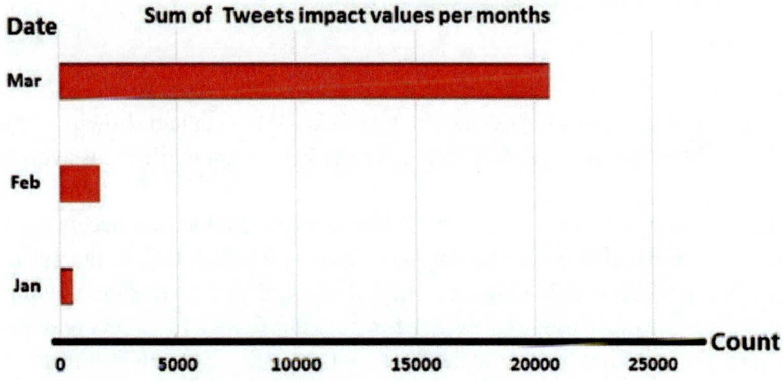

Fig. 7 Total COVID-19 tweets value per month

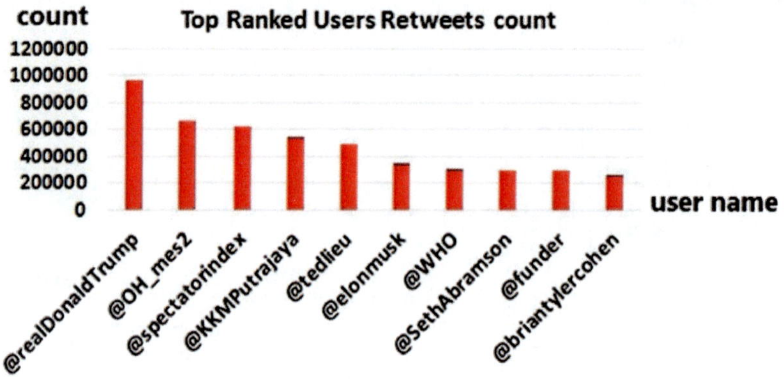

Fig. 8 Top-ranked user based on several re-tweets

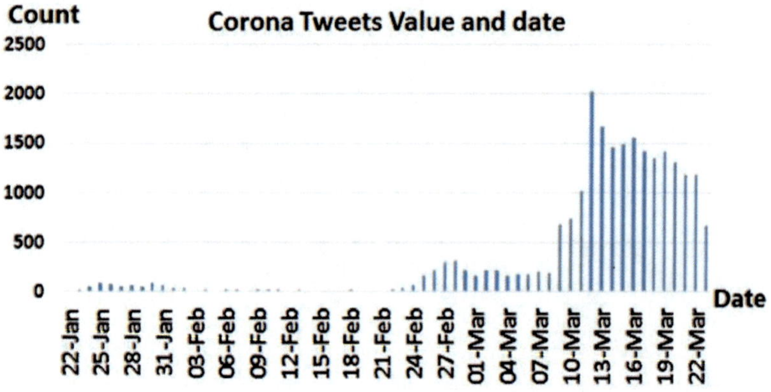

Fig. 9 COVID-19 tweets value per date

value of the maximum tweet, and the minimum number of re-tweets, while it has an average number of likes for tweets. In March 2020 tweets had shown the maximum number of likes for tweets and the highest re-tweets count while it has average tweets values.

Figure 5 shows the biggest number of tweets was 20,000,000 tweets in the period of (12,13 march) 2020 while the minimum tweets for COVID-19 start from January till 20, January 2020 till 25 February 2020. The number of tweets per day presents an indicator for the daily usage of twitter for tracking Covid-19 tweets between users.

Figure 6 shows ranking for twitter users with several re-tweets feature for COVID-19 tweets, the lowest user was Karin.

Figure 7 presents that the summation of tweets value for COVID-19 achieved maximum value in March 2020, it achieved the minimum value in January 2020, while it achieved an average rank in January 2020.

Figure 8, shows that the proposed approach delivers ranking for users with re-tweets count that they achieved, the top-ranked user was Donlad Trump while the lowest user in twitter network was Briantylercohen. Re-tweets count feature shows the importance of tweets in the network and shows the number of users that ranked the tweets as well as an important tweet.

Figure 9, presents a diagram of tweets value and its date, the useful tweets date about Covid-19 that were gained top tweets value, was in (12,13, 16,19,22) march 2020. While, the lowest tweets value about Covid-19 was in the period of (22 January till 24 February) 2020.

3.5 Recommendations

The proposed framework may be implemented as a Twitter or Facebook intelligent layer in real-life applications by scanning the content of the tweet or post, providing a true or false value banner, to let social media users share or not share the tweets or posts. The proposed model will be improved in detecting the miss-leads based on the training of larger data sets. Blocking prohibited misleading contents and users based on lower-level clusters with misleading contents and lowest network user value will be an added value for any social media platform.

4 Conclusion and Future Work

Social media have tremendous power in shaping our lives and especially in global crises such as the COVID-19 pandemic. Therefore, there is a need for an extra layer of intelligence to analyze social media content in real-time. In this paper, a framework has been introduced to analyze the tweets to determine the influential bloggers which have a major effect on individuals and their decisions. Tweets are unstructured text without a pattern. Therefore, to convert tweets to be structured text, a lot of processes need to conduct first such as cleaning and analyzing the text. Then the model will generate a set of meaningful statistical figures, maps, and tables. Also, the detection of misleading tweets using a binary classifier based on tweets values is introduced. The achieved results based on the different performance evaluation measures prove the feasibility of the proposed framework. Tweets are ranked into three classes based on their content to help other users after avoiding misleading tweets. Finally, a set of recommendations are presented for real-life usage. For future work, sentiment analysis would be applied to the tweets so it would also help in ranking bloggers. Besides, more data cleaning will be applied to the data to include Franco Arabic and other languages to improve the classification accuracy of the proposed framework.

References

1. world health organization. online Available: https://www.who.int/emergencies/diseases/novelcoronavirus-2019. Last Accessed on 30 March 2020
2. Yufang, W., Kuai, X., Yun, K., Haiyan, W., Feng, W., Adrian, A.: Int. J. Environ. Res. Public Health **17**, 678. https://doi.org/10.3390/ijerph17030678 (2020)
3. Broniatowski, D.A., Paul, M.J., Dredze, M.: National and local influenza surveillance through Twitter: an analysis of the 2012–2013 influenza epidemic. PLoS ONE **8**(12), e83672 (2013). https://doi.org/10.1371/journal.pone.0083672
4. Smith, M., Broniatowski, D.A., Paul, M.J., Dredze, M. Towards a real-time measurement of public epidemic awareness: monitoring influenza awareness through twitter. In: AAAI Spring Symposium on Observational Studies through Social Media and Other Human Generated Con-Tent, George Washington University, Washington, DC, USA (2016)
5. John, S.P., Roberto, V., Christleya, R.M., Alan, D.R.: Mapping tweets to known disease epidemiology; a case study of Lyme disease in the United Kingdom and the Republic of Ireland. J. Biomed. Inform. X, 4. https://doi.org/10.1016/j.yjbinx.2019.100060 (2019)
6. Chen, L., Hossain, K.T., Butler, P., Ramakrishnan, N., Prakash, B.A.: Syndromic surveillance of flu on twitter using weakly supervised temporal topic models. Data Min. Knowl. Discov. **30**, 681710 (2016)
7. Hayate, I., Wakamiya, S., Aramaki, E.: Forecasting word model: Twitter-based influenza surveillance and prediction. In: Proceedings of the COLING 2016, the 26th International Conference on Computational Linguistics: Technical Papers, Nara Institute of Science and Technology, Nara, Japan, 7686 (2016)
8. Lee, K., Agrawal, A., Choudhary, A.: Forecasting influenza levels using real-time social media streams. In: Proceedings of the 2017 IEEE International Conference on Healthcare Informatics (ICHI), Park City, UT, USA, 2326, pp. 409414 August 2017
9. Wang, B., Yin, P.H., Bertozzi, A.L., Brantingham, P.J., Osher, S.J., Xin, J.: Deep learning for real-time crime forecasting and its internalization. Chin. Ann. Math. Ser. B **40**, 949966 (2019)
10. Covid-19 Twitter Evolution of Coronavirus: online Available: https://www.tweetbinder.com/blog/covid-19-coronavirus-twitter. Last Accessed on 30 March 2020
11. Boididou, C., Papadopoulos, S., Zampoglou, M., Apostolidis, L., Papadopoulou, O., Kompatsiaris, Y.: Detection and visualization of misleading content on Twitter. Int. J. Multimedia Inf. Retrieval **7**(1), 71–86 (2018)
12. Buntain, C., Golbeck, J.: Automatically identifying fake news in popular Twitter threads. In: 2017 IEEE International Conference on Smart Cloud (SmartCloud), pp. 208–215. IEEE (2017, November)
13. Goadrich, M., Oliphant, L., Shavlik, J.: Learning ensembles of first-order clauses for recall-precision curves: a case study in biomedical information extraction. In: Proceedings of the 14th International Conference on Inductive Logic Programming (ILP). Porto, Portugal (2004)
14. Raghavan, V., Bollmann, P., Jung, G.S.: A critical investigation of recall and precision as measures of retrieval system performance. ACM Trans. Inf. Syst. **7**, 205229 (1989)
15. Tharwat, A.: Classification assessment methods. Appl. Comput. Inform. 1–13 (2018)
16. Markus, B.H., Lancianese, S.L., Nagarajan, M.B., Ikpot, I.Z., Lerner, A.L., Wism, A.: Prediction of biomechanical properties of trabecular bone in MR images with geometric features and support vector regression. IEEE Trans. Biomed. Eng. **58**(6), 1820–1826 (2011)